ADVANCES IN
CYCLOADDITION

Volume 6 • 1999

ADVANCES IN CYCLOADDITION

Editor: MICHAEL HARMATA
Department of Chemistry
University of Missouri-Columbia

VOLUME 6 • 1999

JAI PRESS INC.
Stamford, Connecticut

100 Prospect Street
Stamford, Connecticut 06904-0811

ISBN: 0-7623-0531-2

ISSN: 1052-2077

Transferred to digital printing 2006

CONTENTS

LIST OF CONTRIBUTORS

Kay M. Brummond — Department of Chemistry
West Virginia University
Morgantown, West Virginia

Ian R. Hunt — Department of Chemistry
University of Calgary
Calgary, Alberta, Canada

Brian A. Keay — Department of Chemistry
University of Calgary
Calgary, Alberta, Canada

Lily Lee — Department of Chemistry
Boston University
Boston, Massachusetts

José L. Mascareñas — Departamento de Química Orgánica
Universidad de Santiago de Compostela
Santiago de Compostela, Spain

Albert Padwa — Department of Chemistry
Emory University
Atlanta, Georgia

James H. Rigby — Department of Chemistry
Wayne State University
Detroit, Michigan

John K. Snyder — Department of Chemistry
Boston University
Boston, Massachusetts

Christopher S. Straub — Department of Chemistry
Emory University
Atlanta, Georgia

PREFACE

Who would have thought that there would be two volumes of this series published in the same year? It's quite exciting!

This volume presents work from six different groups working on various aspects of cycloaddition chemistry. José Mascareñas gives us a very interesting account of the chemistry of β-alkoxy-γ-pyrones and related species. Al Padwa and Chris Staub discuss further advances in rhodium carbenoid chemistry and the unusual cycloaddition processes possible with these intermediates. Higher order cycloadditions mediated by transition metals highlight Jim Rigby's update on his group's efforts in this area. Lily Lee and John Snyder present us with a detailed account of the indole ring as a dienophile, challenging us to consider the untapped potential in this area. Brian Keay and Ian Hunt discuss the intramolecular Diels–Alder reactions of furan; a report that is both top-notch science, and what could be a great learning tool for students who need to see how fundamental chemical principles can and should be applied to synthetic problems. Finally, Kay Brummond introduces us to a new version of the Pauson–Khand reactions, one that will no doubt be further exploited in productive ways by her group well into the future.

All of the initial editing for this volume was done while I was on research leave at the Georg August Universität in Göttingen. I need to thank the Alexander Humboldt Foundation for a fellowship and Professor Reinhard Brückner (then Göttingen, now Freiburg) and Professor Lutz F. Tietze (Göttingen) for their hospitality.

Finally, this work is a continuing series. Though submissions are by invitation only, I would be happy to accept suggestions or nomination for contributors to future volumes.

Michael Harmata
Editor

THE [5+2] CYCLOADDITION CHEMISTRY OF β-ALKOXY-γ-PYRONES

José L. Mascareñas

Advances in Cycloaddition
Volume 6, pages 1–54.

ISBN: 0-7623-0531-2

ABSTRACT

Among the different cycloaddition methods for assembling seven-membered carbocycles, those that lead to 1,4-oxygen-bridged systems are particularly attractive since the resulting bicycles offer numerous possibilities for a stereoselective, divergent conversion into a variety of skeletons. Most of the methods to assemble these oxabicycles rely on the classical [4C+3C] cycloaddition of furans to oxoallyl cations, although other [4+3] annulation routes have recently been developed. In contrast, strategies based on [5+2] annulations have been pursued to a lesser extent, a fact that could be partly due to the scarcity of readily available molecules that can participate as five-carbon partners in the cycloaddition. Despite being relatively stable, β-hydroxy-γ-pyrones can play such a role in [5+2] cycloadditions with alkenes or alkynes, owing to their formal intrinsic performance as an α,β-unsaturated ketone and an enol. The annulation can be accomplished by simple thermolysis, which works satisfactorily in the case of intramolecular reactions, or by prior conversion of the pyrone into more reactive intermediates that can undergo the cycloaddition under milder conditions. The rich functionalization of the resulting oxabicyclo[3.2.1]octane adducts offers unique opportunities for their divergent transformation into a variety of valuable skeletons, including highly substituted seven-membered carbocycles, stereochemically rich tetrahydrofurans, and O-bridged medium-sized carbocycles.

I. INTRODUCTION

The blossoming in the range of synthetic methods and strategies that has taken place during recent decades has culminated in the development of impressive total syntheses of extremely complex molecules such as halichondrine,[1a] taxol,[1b] or brevetoxin.[1c] These landmarks, along with other accomplishments, have to some extent generated the impression that organic synthesis has reached such a degree of maturity that, given enough time, money and effort, any small, reasonably stable molecule can be synthesized.[2] However, at the dawn of the 21st century, obtaining milligram quantities of the required target may not be the principal priority, and issues such as brevity, economy, and ecology can occupy a prominent position in the design of a synthetic plan.[3] For this reason, reduction in the number of synthetic steps, mainly those that use unusual conditions, minimization of waste production, and maximization of efficiency are some of the more relevant challenges for the new age of organic synthesis. One of the best ways to address these challenges relies on the development of methods that allow a maximum increase in target-relevant molecular complexity per synthetic operation while generating minimal amounts of by-products.[4] Undoubtedly, cycloaddition reactions, by virtue of allowing the regio-and stereoselective construction of new rings by simple addition of two or more molecules,[5] occupy a leading position among the tools available to the synthetic chemist that best meet the above requirements. It is not necessary to discuss the enormous impact of the Diels–Alder[6] and 1,3-dipolar[7] cycloaddition reactions in modern organic synthesis. However, the application of these reactions is mainly restricted to the construction of six and five-membered rings. The growing awareness of bioactive natural products that contain larger rings, and in particular the great number of these bearing seven-membered carbocycles,[8] makes the development of cycloaddition methods to assemble these type of rings a truly relevant task. Of the annulation approaches that have already been described for assembling cyclic systems of this size, those that lead to bridged frameworks of type **1** (Figure 1) are among the most attractive because the conformational rigidity imposed by the bridging atom paves the way for the stereoselective introduction of new functionality into the carbocycle. Furthermore, the utility of these bicyclic systems may not be restricted to the synthesis of seven-membered carbocycles that could be unmasked upon appropriate cleavage of the bridge (Figure 1), but they can also serve as building units for other types of structures. The most common methods for the construction of these frameworks are based on [4+3] cycloadditions between allyl cations and pentacyclic dienes,[9] although other [4+3] strategies relying on sequential annulation reactions have also been described recently.[10]

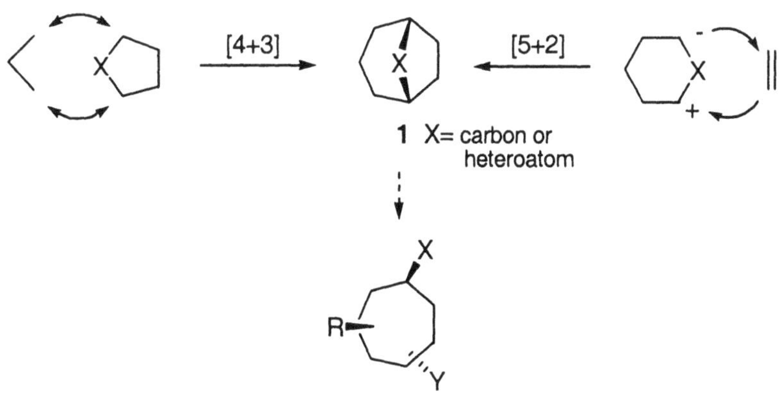

Figure 1.

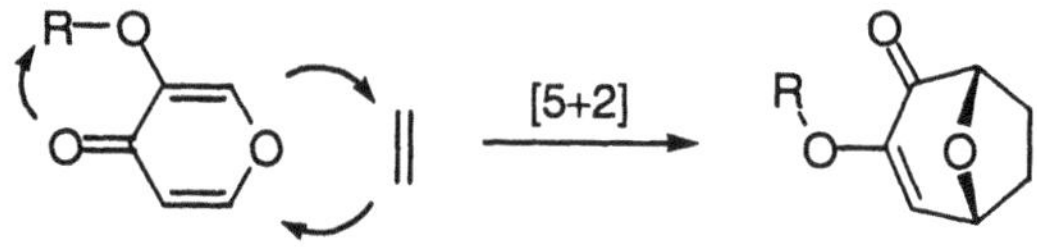

Figure 2.

Alternatively, these bicyclic skeletons (**1**) might be assembled by means of a [5+2] annulation between alkenes or alkynes and appropriate five-carbon cycloaddition partners (Figure 1). Most of the [5C+2C] approaches studied to date rely on the use of 2-alkoxy-1,4-benzoquinone derivatives, or 3-oxidopyrylium or pyridinium betaines as C_5 components of the reaction, reagents whose preparation from available precursors may involve several steps. A preliminary report by Garst in 1983 indicated that thermolysis of 2-(ω-alkenyl)-5-hydroxy-4-pyrones promotes an internal [5+2] cycloaddition.[11] This provided the first compelling evidence that β-alkoxy-γ-pyrones could serve as an alternative source of five-carbon cycloaddition synthons, with the annulation process involving the O-3 → O-4 migration of a suitable R group (Figure 2). The ready availability of several β-hydroxy-γ-pyrones, apparent simplicity and atom-economy of the transformation, and rich but biased functionality of the resulting 8-oxabicyclo[3.2.1]octane adducts led us to embark on a program to study the synthetic scope and potential of this type of promising, yet scarcely studied, annulation. In this review the progress made within our group will be discussed, along with pertinent research from other laboratories. First, however, we consider it convenient to summarize related [5+2] cycloaddition studies that might help to put our work into proper perspective.

II. BACKGROUND: [5+2] CYCLOADDITION APPROACHES TO BICYCLO[3.2.1]OCTANE SYSTEMS

A. Pentadienyl Cation–Alkene Cycloadditions

One of the first examples of a [5+2] cycloaddition reaction was reported in 1965 by the group of Joseph-Nathan[12] who found that thermolysis of perezone (**2**) affords an equimolecular mixture of the α and β-pipitzols (**3a** and **3b**, Scheme 1). Subsequent studies by the same group provided experimental evidence that the reaction proceeds by a concerted process that can be understood in terms of a thermally allowed [π4s+π2s] cycloaddition of a pentadienyl cation intermediate (**4**) to the olefin (Figure 3).[12b] It was later found that the use of BF_3•Et_2O (8 equiv) promotes the cycloaddition at room temperature, which allows diastereoselection in favor of the α-isomer (**3a/3b**, 8:1). The increase in reactivity was interpreted by assuming a change to a stepwise mechanism, presumably involving the formation of an intermediate like **5**.[12c]

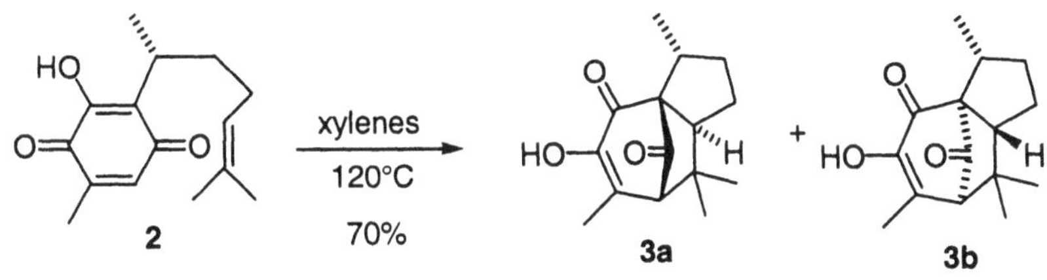

Scheme 1.

Figure 3.

Recognition of this type of cycloaddition as a potentially effective way to assemble bicyclo[3.2.1]octane systems from relatively simple starting materials led Büchi and coworkers to study the application of this method to intermolecular cases, as an entry to several neolignanes and sesquiterpenes that contain this type of structural core.[13] Although attempts to thermally cyclize isosafrole (**6**) to benzoquinone **7** failed, it was found that an alternative, acid-promoted generation of the reactive pentadienyl cation from the quinone ketal **8** allowed its cycloaddition with the alkene **6** to afford the cycloadduct **9**, albeit in only 20% yield (Scheme 2).[13a] The reaction also produced the ketone **10**, which presumably arose from rearrangement of the initially formed cycloadduct **11**. Adduct **9** was easily elaborated to give the natural product guianin by O-methylation followed by selective sodium borohydride reduction of the bridging carbonyl.[13a]

In pursuing the preparation of colchicine analogues, Büchi demonstrated that it is possible to efficiently cleave the bridging carbonyl of cycloadduct **13** by simple treatment with base (Scheme 3). The resulting highly substituted seven-membered carbocycle **14** was further elaborated to the tropolone derivative **15**.[14] Despite the low to moderate yields obtained in these cycloadditions, Büchi's research nicely illustrates how suitable designed complexity-increasing reactions inspired by natural product chemistry can be successfully exported to the synthetic arena.

Several more efficient alternatives to promote related [5+2] cycloadditions which differ in the method used to generate the required pentadienyl cation intermediate have since been developed. Thus, Yamamura and coworkers have shown that anodic oxidation of 3,4-dimethoxyphenol **16** in the presence of excess of isosafrole (**6**)

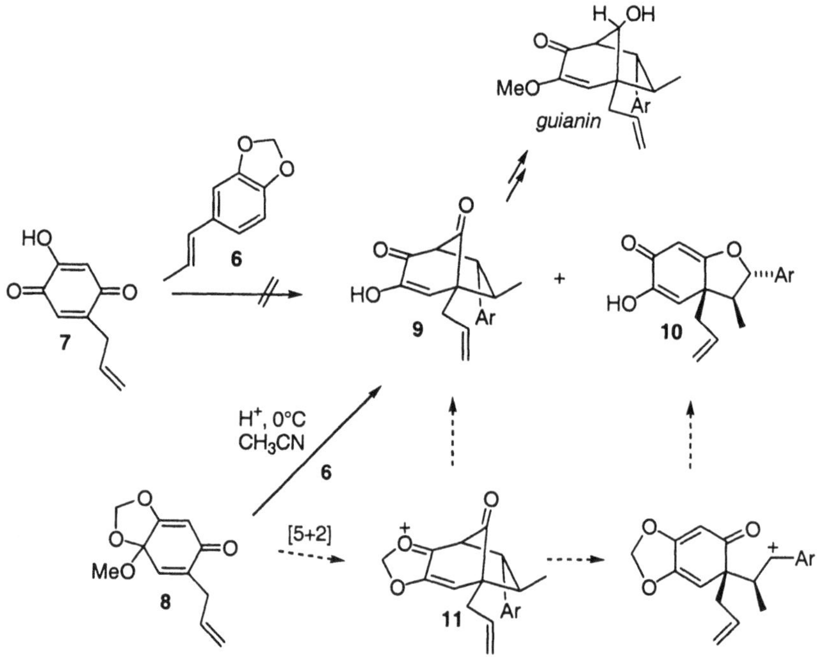

Scheme 2.

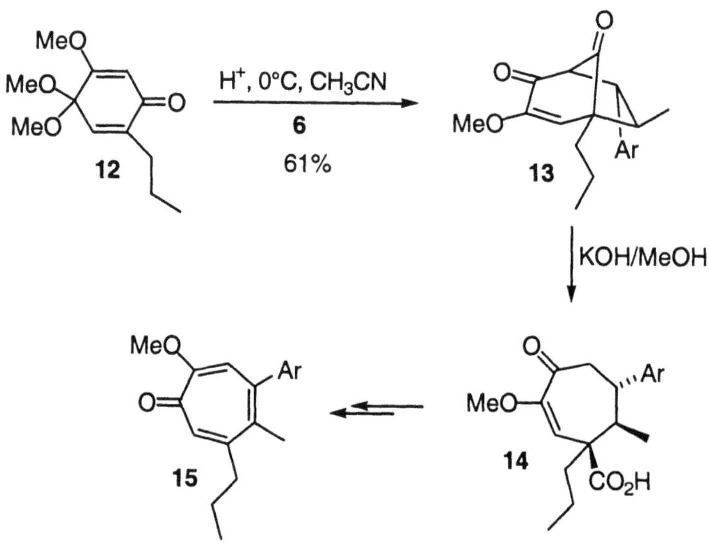

Scheme 3.

gives good yields of the [5+2] adduct **17**, with the reaction presumably occurring by the intermediacy of cation **18** (Scheme 4).[15a] The same group has successfully extended the methodology to intramolecular cases, which allowed the synthesis of several diterpenes containing a tricyclic cedrane-type skeleton.[15b]

Engler and coworkers found that addition of one equiv of $TiCl_4/Ti(O^iPr_4)$ (1:1) to a mixture of *trans*-β-methylstyrene and 2-methoxy-1,4-benzoquinone induces their [5+2] annulation at −78 °C.[16] In addition to the expected [5+2] bicyclic adduct **19**, the reaction produced benzofuran **20** and cyclobutane **21**, compounds that presumably arise from divergent rearrangements of the initially formed adduct **22** (Scheme 5). Despite the fact that the reaction gives mixtures of products, the authors devised elegant alternatives to direct the selective formation of some of them. Particularly significant was the finding that the use of a labile 4-methoxybenzyl (PMB) group as alkoxy substituent in the quinone favors the formation of the [5+2] adduct **19**.[16b]

More recently, Grieco has reported novel experimental conditions for promoting efficient cycloadditions of Büchi's quinone ketals.[17] The method involves the use of trimethylsilyl triflate as an activating agent, but only works successfully if the reaction is carried out in a highly polar medium. Several olefins, such as styrene, cyclopentene, and vinyl acetate, have been demonstrated to participate in the intermolecular annulation with **23** to give moderate to good yields of the bicyclic cycloadducts (Scheme 6). The method has been recently extended to encompass intramolecular cases, and was applied to the formal synthesis of the triquinane isocomene.[17b]

Notwithstanding the notable improvements in the cycloaddition methods for assembling bicyclo[3.2.1]octan-8-ones, the full synthetic power and versatility of

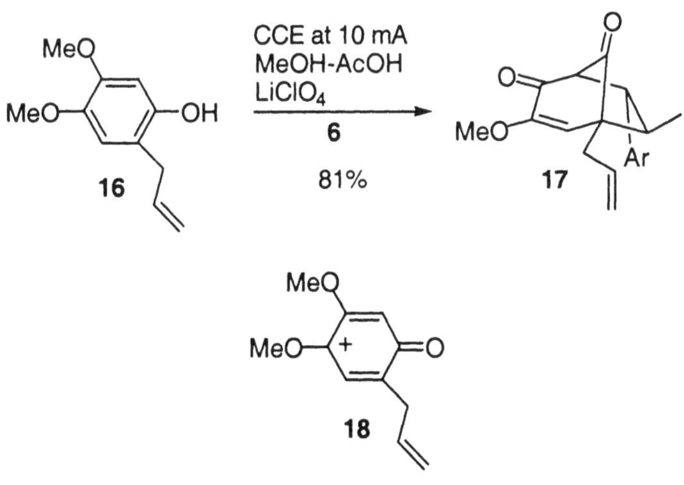

Scheme 4.

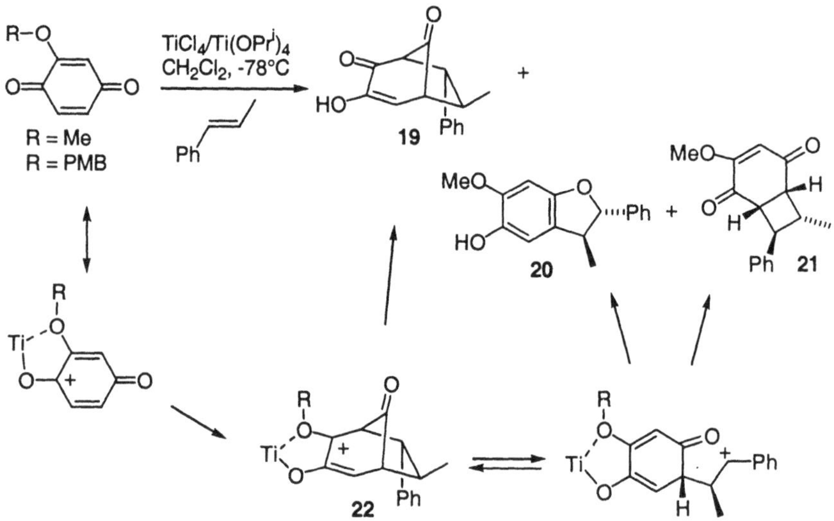

Scheme 5.

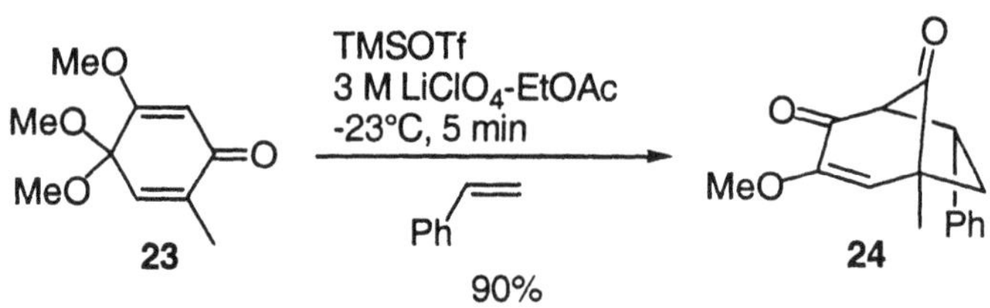

Scheme 6.

the resulting adducts remains to be determined. It is noteworthy that despite the apparent feasibility of breaking the keto bridge of the adducts to give seven-membered carbocycles, pentadienyl cation-alkene approaches to natural products containing this type of cyclic structure have not really been explored.

It would not be appropriate to end this discussion on the different [5+2] techniques for assembling bicyclo[3.2.1]octanes without mentioning the landmark work of Wender in the photochemical arene–alkene *meta*-photocycloadditions.[18] Wender's group has used this reaction as a key step to achieve the syntheses of several triquinane natural products such as cedrene (Scheme 7).

B. Oxidopyrylium and Oxidopyridinium–Alkene Cycloadditions

Among the different bicyclo[3.2.1]octane skeletons, those in which the bridging element is a heteroatom can certainly be considered as the more synthetically

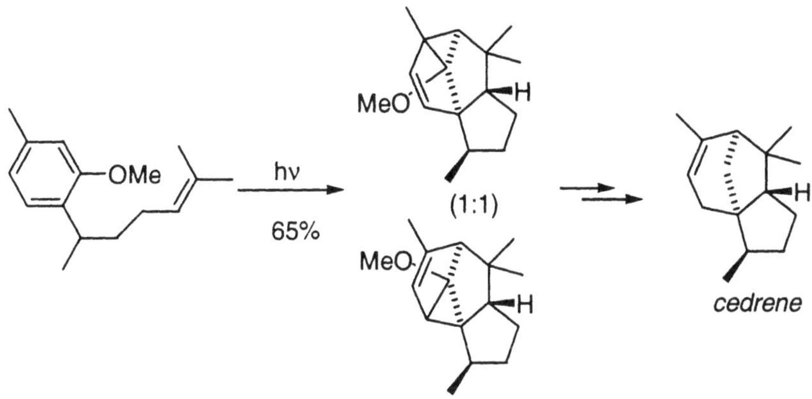

Scheme 7.

attractive. This is particularly the case when the heteroatom is oxygen, since the oxa-bridge serves as a latent hydroxyl substituent of defined stereochemistry, thereby providing for immediate access to hydroxylated cycloheptanoids. The most widely studied [5+2] disconnection for assembling 8-aza and 8-oxabicyclo[3.2.1]octanes involves the cycloaddition of 3-oxidopyridinium (**25a**) or 3-oxidopyrylium (**25b**) betaines to alkenes (Figure 4). Although these zwitterions are heteroaromatic compounds, their cycloaddition chemistry is best understood in terms of an azomethine or carbonyl ylide behavior. Therefore, strictly speaking, these [5C+2C] processes are really 1,3-dipolar reactions.[7]

The chemistry of 3-oxidopyridinium betaines has been extensively studied most notably by Katritzky in the late 1970s and early 1980s.[19] A major advantage of their cycloaddition reactions results from the commercial availability of 3-hydroxypyridine, a direct precursor of unsubstituted systems such as **26** (Scheme 8). Upon thermal activation this and related ylides react with electron-deficient dipolarophiles to give moderate yields of the corresponding azabicyclic adducts, with variable degrees of regio- and stereoselectivity.[19a]

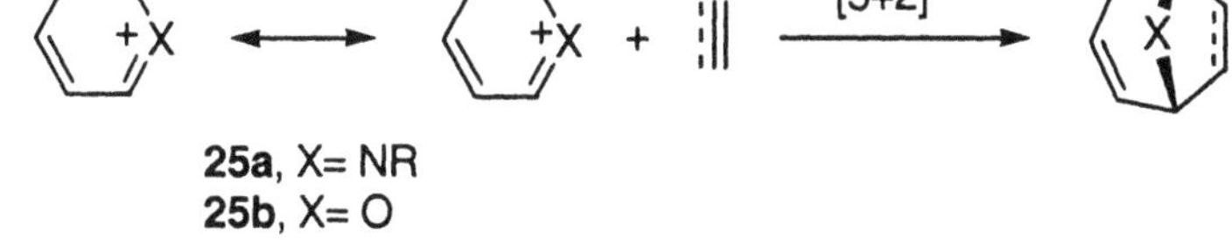

25a, X= NR
25b, X= O

Figure 4.

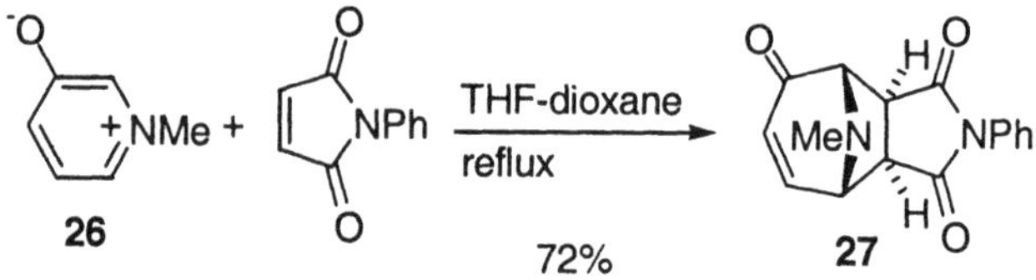

Scheme 8.

In contrast to the abundant coverage of the intermolecular cycloaddition of 3-oxidopyridinium betaines, very few intramolecular applications have been described, which is most probably due to the difficulties associated with obtaining the appropriate precursors. Sammes has reported the cycloaddition of betaines **28** and **29**, reactions that require heating at a minimum of 160 °C (Scheme 9).[20]

Despite the fact that 3-oxidopyrylium betaines are more reactive than the nitrogen homologues, their cycloaddition chemistry has received less attention.[21,19b] The first example of this type of reaction was reported as early as 1960 by Ullman and coworkers, who discovered that the indenone epoxide **30**, upon thermolysis or photolysis, produces the red-colored benzopyrylium oxide **31** which can be trapped with a variety of 2π-addends (e.g. DMAD) to give cycloadducts like **32** (Scheme 10).[22]

In 1980 Hendrickson and Farina reported a truly practical method for generating the unstable parent 3-oxidopyrylium zwitterion **34** by thermolysis of acetate **33**, itself obtained by oxidation of furfuryl alcohol.[23a] Sammes later showed that the

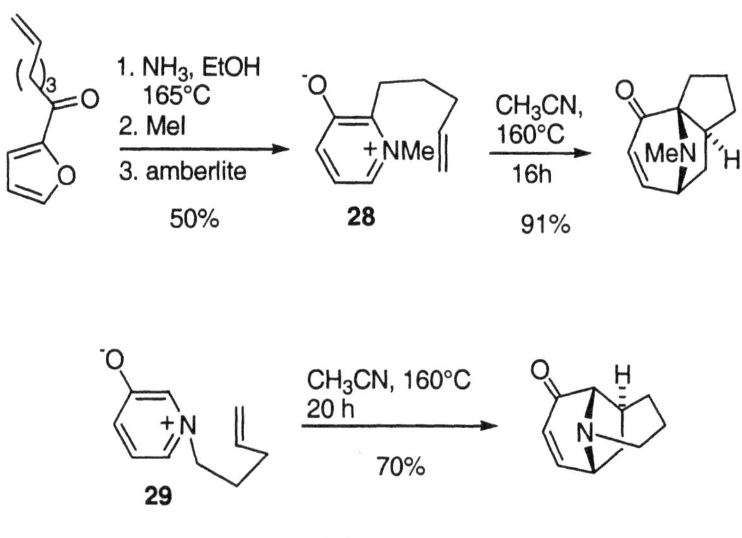

Scheme 9.

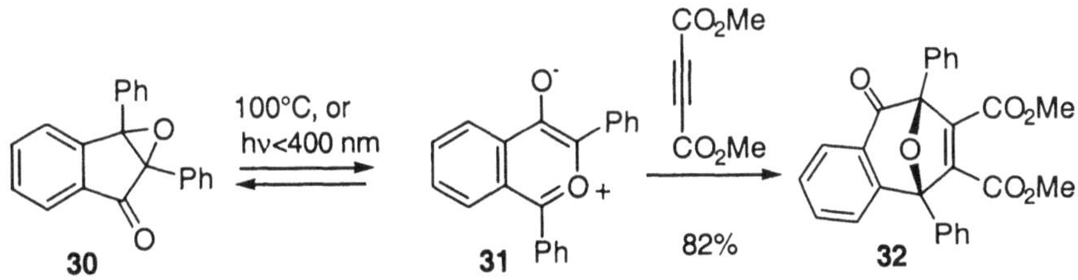

Scheme 10.

oxidopyrylium zwitterion can be generated at room temperature by reaction of **33** with a tertiary base such as Et_3N, and that this ylide adds at that temperature to either electron-rich or electron-deficient dipolarophiles (Scheme 11).[23b]

In contrast to the oxidopyridinium case, there have been relatively numerous studies on intramolecular versions of the oxidopyrylium–alkene cycloaddition.[21,24] Sammes has applied the method for the synthesis of guaiane- and valerane type of sesquiterpenes such as β-bulnesene (**35**) and fauronyl acetate (**36**) (Scheme 12).[24] The latter synthesis involved a Lewis acid-induced rearrangement of the perhydroazulene skeleton to give a decalin system, which can be mechanistically interpreted to occur via the intermediate **37**. More recently, Wender has used an intramolecular oxidopyrylium–alkene cycloaddition as key step in masterly total syntheses of phorbol[25a] and resiniferatoxin[25b] (Scheme 13).

A remarkable alternative to generate 3-oxidobenzopyrylium ylides was disclosed two decades ago by Ibata and coworkers,[26] and is based on a metal-catalyzed intramolecular carbenoid cyclization of diazo compounds like **38** which, when carried out in the presence of a dipolarophile, leads to the corresponding [5+2] cycloadducts (Scheme 14). This type of tandem cyclization–cycloaddition methodology was later studied extensively by the group of Padwa.[27] For instance, this group demonstrated that three-atom tethered carbonyl–diazo substrates like **39**

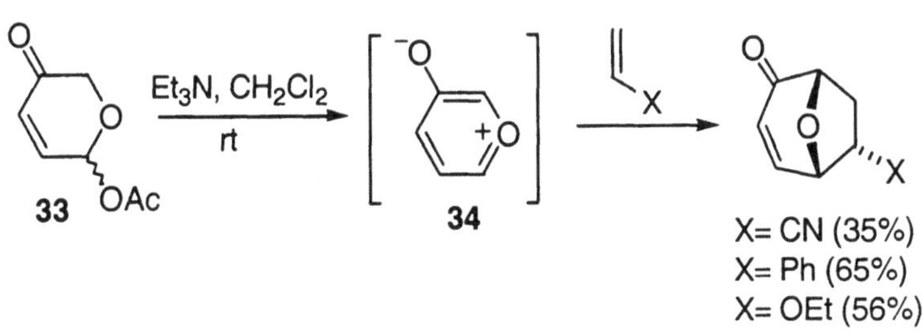

Scheme 11.

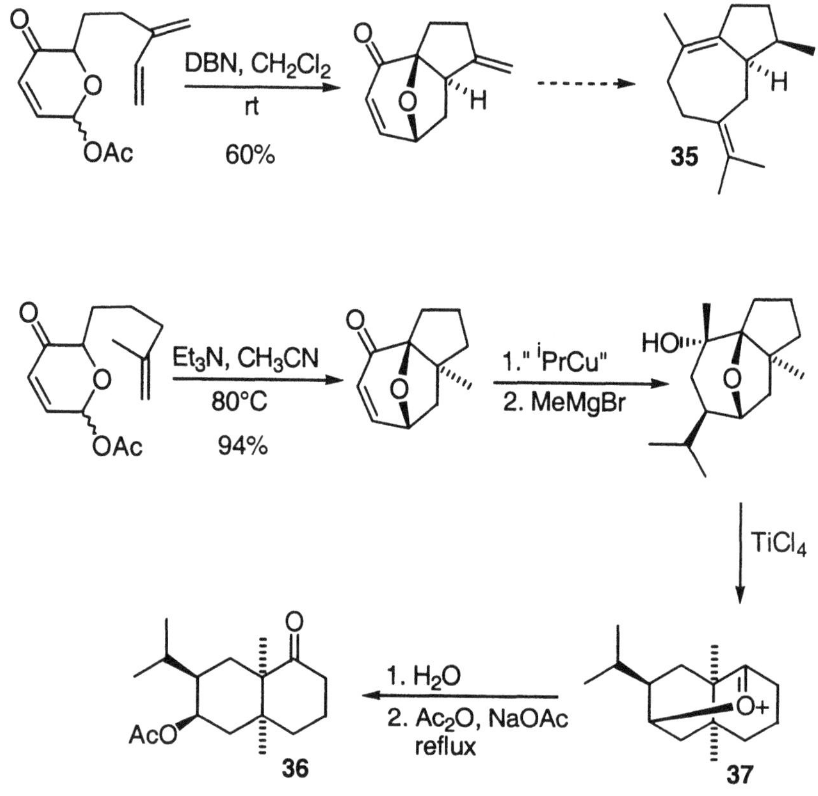

Scheme 12.

easily cyclize to generate highly reactive five-atom carbonyl ylides, which add to several dipolarophiles in a formal [5C+2C] sense.[27b]

C. Base-Promoted Reactions of β-Hydroxy-γ-pyrones with Electron-Deficient Alkenes

Undoubtedly, [5+2] cycloaddition methods described above constitute excellent alternatives for obtaining relatively complex bicyclo[3.2.1]octane frameworks from simpler precursors, and it can be safely stated than their synthetic potential is far from fully realized. However, in many cases, setting up the required reactive C_5 components is not straightforward. This makes worthwhile the pursuit of alternative, readily available molecules that can directly participate as five-carbon partners in the cycloaddition. β-Hydroxy-γ-pyrones can be considered as potential candidates for such a role since they might formally be viewed as five-carbon dipoles by virtue of the simultaneous presence in their skeleton of an α,β-unsaturated ketone and an enol (Figure 5).

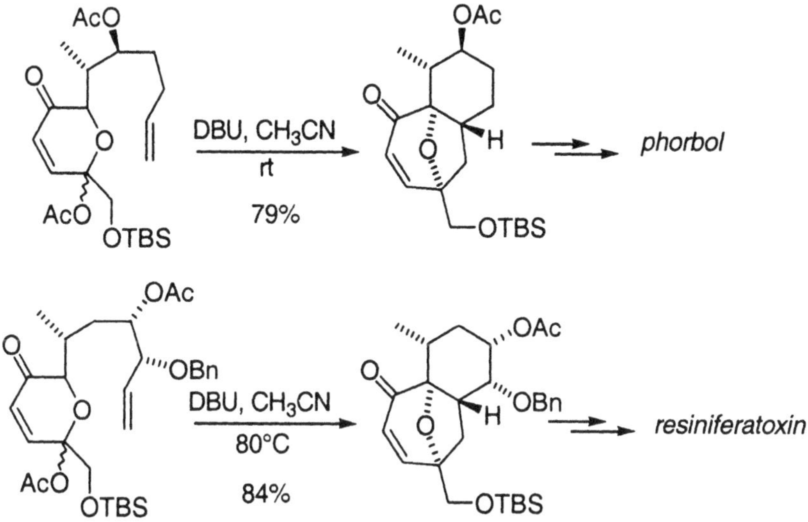

Scheme 13.

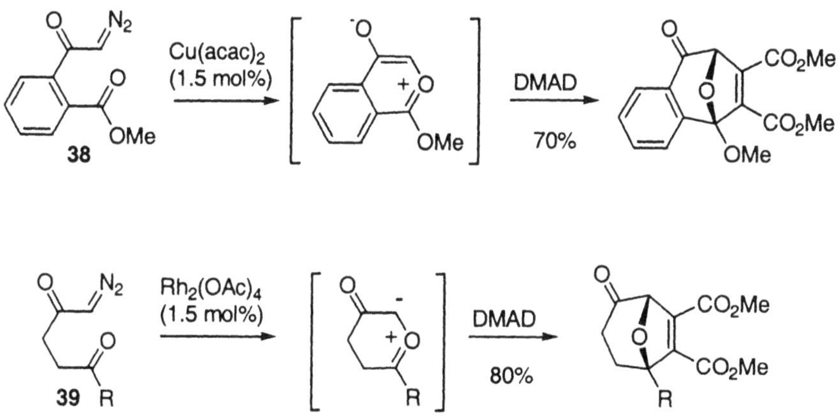

Scheme 14.

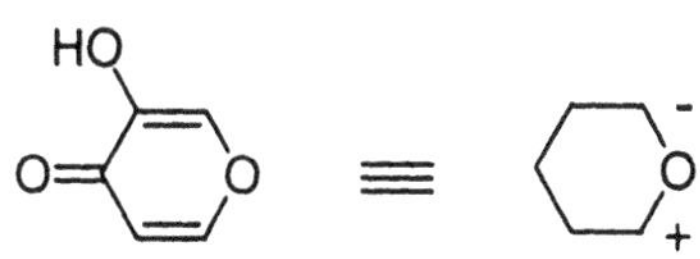

Figure 5.

Initial studies on the chemical behavior of these types of pyrones towards alkenes date from the early 1950s when Woods reported that the base-promoted reaction of kojic acid (**40**) with acrylonitrile gives the conjugate addition product **41**.[28] Six years later, Hurd reexamined this reaction, as well as those of other β-hydroxy-γ-pyrones with acrylonitrile and acrylate esters, and concluded that the assignment made by Woods was erroneous since the major products of the reactions contained two molecules of the hydroxypyrone and one of the acrylic counterpart.[29] However, it was not until the more detailed studies by Volkmann and coworkers, almost two decades later,[30] that the structure of the major adduct of the reaction of kojic acid with acrylonitrile was finally determined to be **42** (Scheme 15). The formation of this compound was rationalized to proceed through an initial tautomerism of kojic

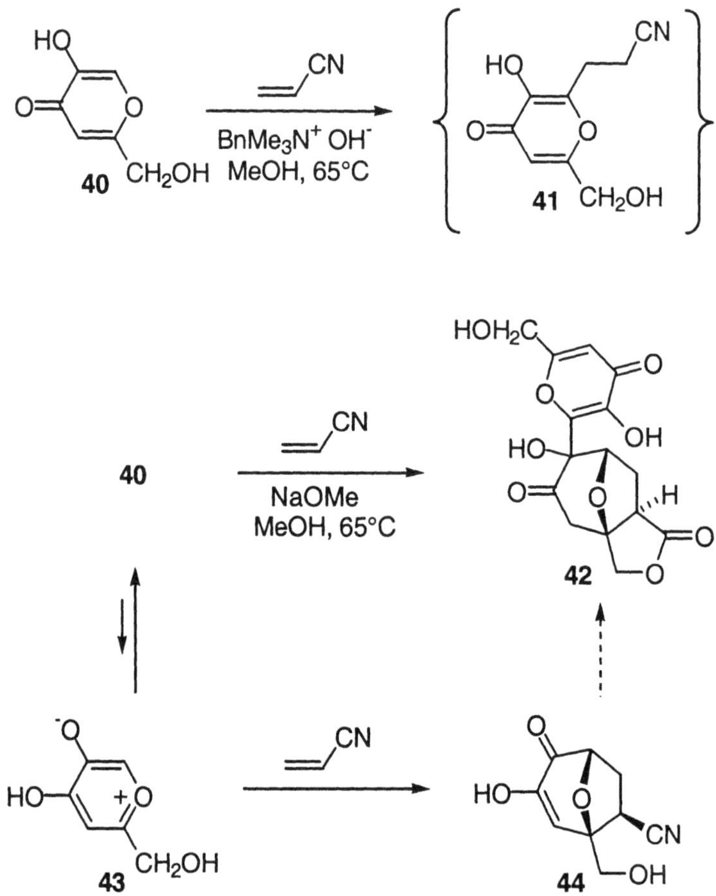

Scheme 15.

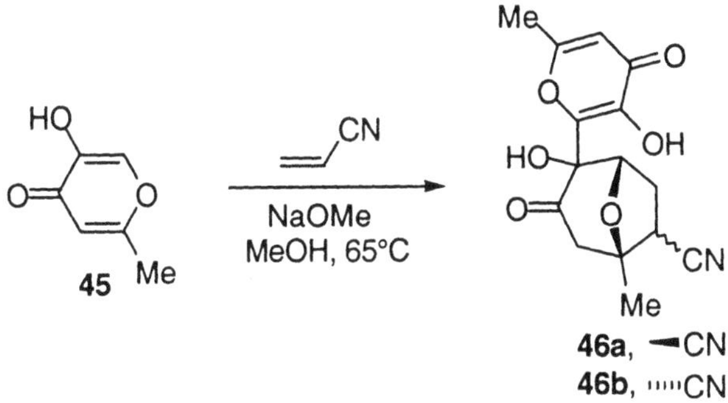

Scheme 16.

acid to the 3-oxidopyrylium zwitterion **43**, which undergoes a [5+2] cycloaddition to the electron-deficient alkene to give adduct **44**. This initial adduct was not observed, but under the conditions of the reaction it presumably underwent an aldol-type addition of a second kojic acid unit and an intramolecular hydroxynitrile condensation to finally give, after hydrolysis, the lactone **42**.

Other β-hydroxy-γ-pyrones, such as α-deoxykojic acid (**45**), reacted with acrylonitrile to give the expected products in a 10:1 *exo* (**46a**)/*endo* (**46b**) ratio (Scheme 16). The absence of an explicit indication of a yield for this reaction led us to repeat the experiment in our laboratory under the conditions described by Hurd (MeO^-, MeOH, 60 °C). Although we did detect the cycloadducts in the ^{1}H NMR spectrum of the reaction residue, they were formed as part of a relatively complex mixture of products, and therefore we didn't pursue their isolation. However, it can be stated that the yield was rather modest. In any event, Volkmann's studies confirmed the formal five-carbon dipole character of kojic acid and related γ-pyrones, opening the door for its possible use in [5+2] cycloaddition reactions to alkenes.

III. THERMAL [5+2] PYRONE–ALKENE CYCLOADDITIONS

A. Intramolecular Precedents

The key article that confirmed the feasibility of achieving [5+2] hydroxypyrone-alkene cycloadditions was published by Garst and coworkers in 1983.[11] Curiously, these authors referred to the intramolecular thermal perezone–pipitzol rearrangement as the enlightening precedent for their studies, rather than the aforementioned intermolecular base-catalyzed reactions of kojic acid and derivatives with electron-deficient alkenes. In any case, they showed that simple heating of amides **47** and **48** in refluxing benzene provided the [5+2] internal adducts **49** and **50** in moderate yields (Scheme 17).

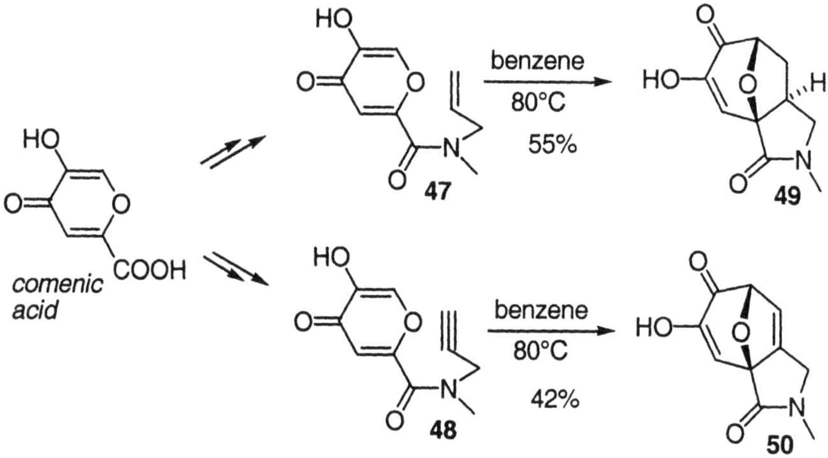

Scheme 17.

They also reported that the cycloaddition can be accomplished when the pyrone and the alkene are linked by carbon tethers. Hence, pyrolysis of **51a** in refluxing benzene for 12 h, followed by acetylation, afforded the expected adduct **52a** (Scheme 18). Enlarging the tether by one carbon has a noticeable influence on the rate of the reaction, although the expected cycloadduct **52b** was obtained after heating a benzene solution of pyrone **51b** at 110 °C for 48 h. A two-atom enlargement, however, led to a complete inhibition of the process, most probably due to the remarkable increase in the entropic component of the activation energy.

In a more recent article,[31] the authors described attempts to use an intramolecular pyrone–alkene cycloaddition of substrates like **53** as the key step to develop a potentially rapid route to colchicine, given that this reaction would set up the basic skeleton and oxygenation of the target (Scheme 19). To this end they prepared precursors **54** and **55**, and although **54** failed to cyclize under simple thermal activation, the pyrone **55a** underwent a [5+2] cycloaddition reaction which conveys

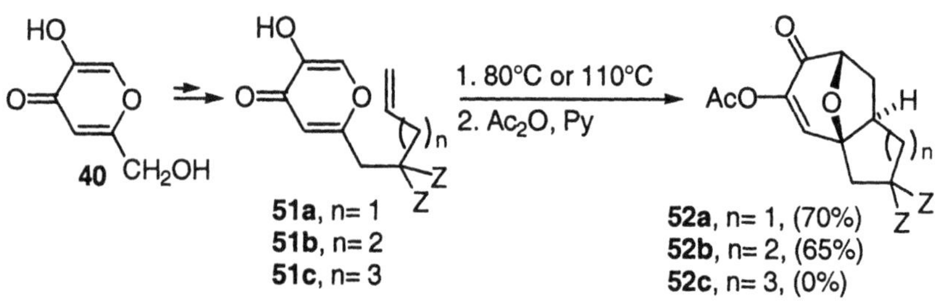

Scheme 18.

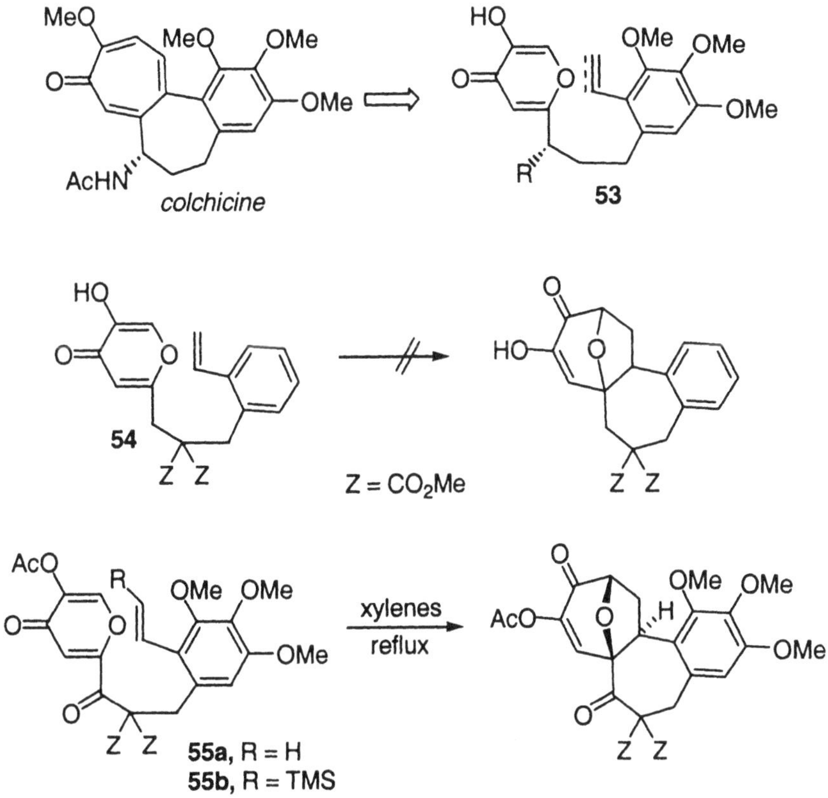

Scheme 19.

a concomitant internal transfer of the acetate group (Scheme 19). Compound **55b**, bearing a TMS group at the terminal position of the alkene, failed to give the cycloaddition, probably for steric reasons. The propensity of **55a** to undergo cycloaddition in comparison to **54** was justified by the authors by the favorable combination of a more electrophilic pyrone due to the presence of the adjacent keto group and an electron-rich styrene. A definitive explanation would, however, require a more detailed study.

In spite of the apparently attractive characteristics of the above [5+2] cycloaddition, the reaction was not really considered by the synthetic community, with the exception being a clever application by Wender and McDonald to the synthesis of phorbol.[32] These authors envisioned that the high oxygenation present in the seven-membered carbocycle of the pyrone–alkene cycloadducts might allow a streamlining of the route to phorbol. To this end, they assembled the precursor **57**, using a Claisen rearrangement of substrate **56** as the key step to attach the required

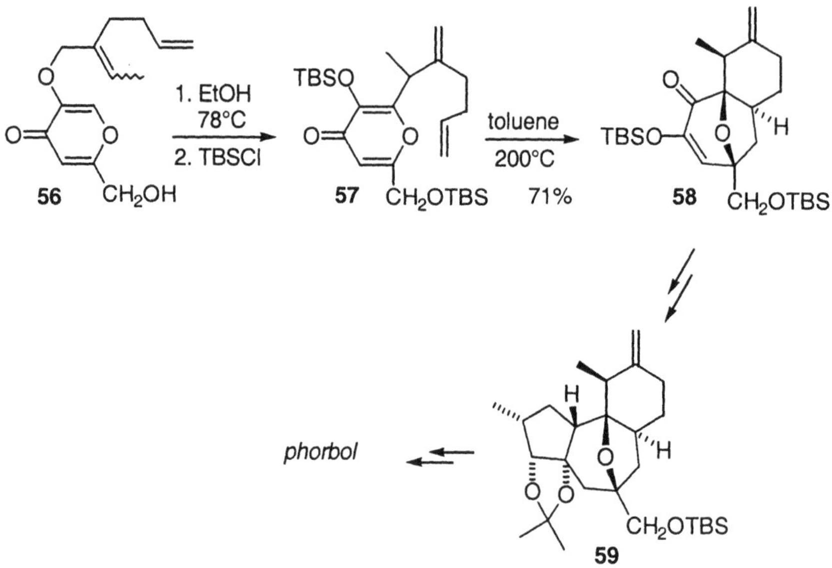

Scheme 20.

carbon chain to the pyrone (Scheme 20). Although the cycloaddition did not take place under the thermal conditions used for the Claisen reaction, they found that heating a toluene solution of silylated derivative **57** at 200 °C for 48 h leads to the *exo*-cycloadduct **58** as a single diastereoisomer. This compound was efficiently elaborated into the polycycle **59**, a known precursor of phorbol. The eventual discovery that the *t*-butyldimethylsilyl group can work as transferable agent in the cycloaddition was of interest owing to its concurrent protecting-group capabilities, which can facilitate the preparation of other precursors.

It should also be pointed out that inducing the cycloaddition at *room temperature* by photochemical activation was feasible, although this led to a very low yield of the product (Scheme 21). The fact that the photochemical annulation proceeded at room temperature suggests a mechanism involving the generation of a highly reactive

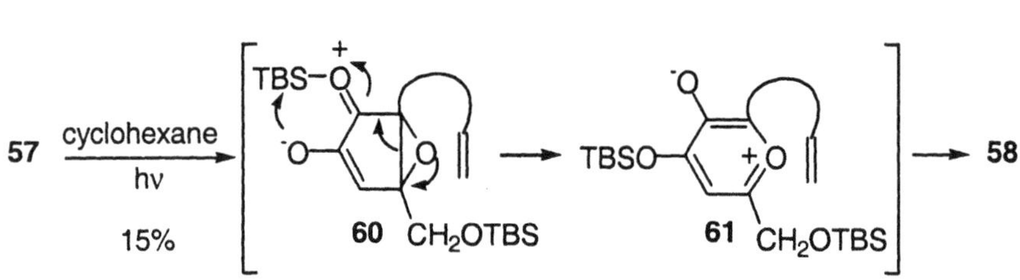

Scheme 21.

3-oxidopyrylium ylide intermediate **61** (see Section IV.A for the room temperature cycloaddition of related ylides). This intermediate may arise by rearrangement of the bicyclic zwitterion **60**, which is presumably formed initially.[33]

B. Temporary Tethered Processes

The aforementioned results of Wender and McDonald confirmed the intramolecular thermally promoted pyrone–alkene cycloaddition to be a simple and practical method for assembling 1,4-oxygen-bridged seven-membered carbocycles, and therefore we anticipated that this transformation might have a wider scope in terms of its synthetic potential. Our initial efforts in the area were influenced by the apparently intrinsic limitation of the success of the method to intramolecular cases since, as discussed earlier in this chapter, the base-promoted intermolecular reaction of kojic acid or its derivatives with electron-deficient alkenes failed to provide the desired cycloadducts.[30] Notwithstanding, it was not known what would happen if the bimolecular process was carried out under exclusive thermal activation (in absence of base). Although the lack of reports on the outcome of this reaction could have been interpreted as a symptom of failure, we found that heating a toluene solution of silylated α-deoxykojic acid **62** with excess of acrylonitrile, for three days at 180 °C in a sealed tube, produces the *exo*-cycloadduct **63**, albeit in only 15% yield (80% based on recovered starting material, Scheme 22).[34] Cycloadducts were not detected in the same reaction with nonactivated alkenes or styrene, although the latter alkene did polymerize at this higher temperature. Despite the fact that the result of the intermolecular reaction with acrylonitrile is not synthetically significant, it is important from a mechanistic point of view in the sense that this alkene must act as electrophilic partner in the reaction. It is therefore highly probable that the reaction proceeds via the intermediacy of the 3-oxidopy-

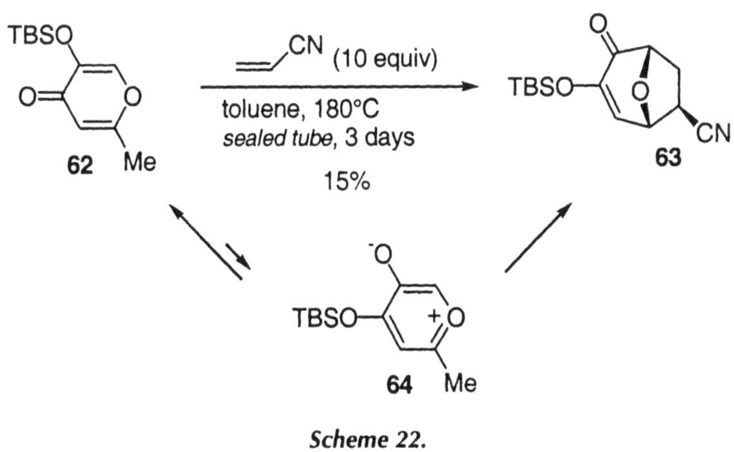

Scheme 22.

rylium ylide **64**, that might exist in a very low proportion in equilibrium with the starting pyrone.

The low efficiency of the intermolecular thermal pyrone–alkene [5+2] process makes it synthetically useless, and therefore we sought to develop strategies that would allow the assembly of the potentially useful bimolecular cycloadducts. We envisaged that connecting the alkene to the pyrone by means of a tether which could be removed after the cycloaddition might provide an acceptable solution not only for inducing the reaction, but also for controlling the regio- and stereochemistry of the process (Scheme 23).

Though connections via sulfur atoms have not been extensively used in temporary intramolecular processes, the prospects of easy assembly of the required precursors, and likelihood of facile post-cycloaddition desulfurization prompted us to explore this alternative first. Indeed, an appropriate substrate (**65**) was rapidly prepared from kojic acid (**40**) in three steps. Although heating a toluene solution of **65** under reflux proved sufficient to induce the cycloaddition, the rate of the process suggested that the reaction be run at higher temperature in a sealed tube. Indeed, heating **65** at 145 °C for 40 h allowed the expected *exo*-cycloadduct **66** to be isolated in good yield (Scheme 24).[35] Somewhat surprisingly, Raney nickel-induced desulfurization of **66** was accompanied by a concurrent reduction–rearrangement reaction at the α-silyloxyenone moiety to give a mixture of ketones **67a** and **67b**. When the reaction was performed in THF instead of ethanol, **67b** was produced exclusively. The formation of this compound can be explained assuming that, in addition to the desulfurization reaction, the Raney nickel promotes the reduction of the carbonyl, which is followed by migration of the silyl group. Overall, the strategy outlined in Scheme 24 led to the achievement of a [5C+2C] formal *intermolecular* cycloaddition between a nonactivated alkene (propene) and α-deoxykojic acid, a transformation that cannot be carried out in the *bimolecular* mode.

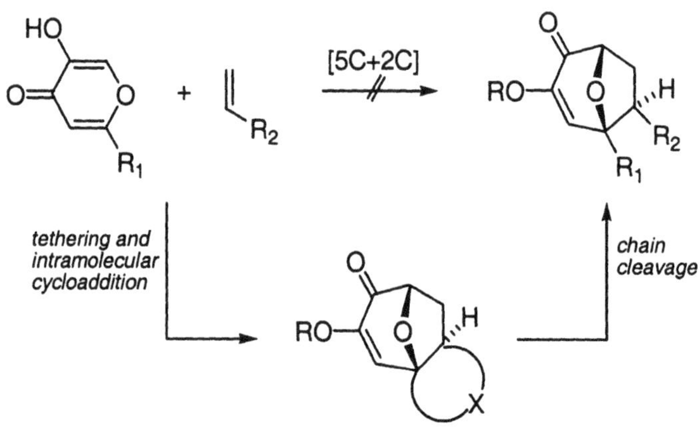

Scheme 23.

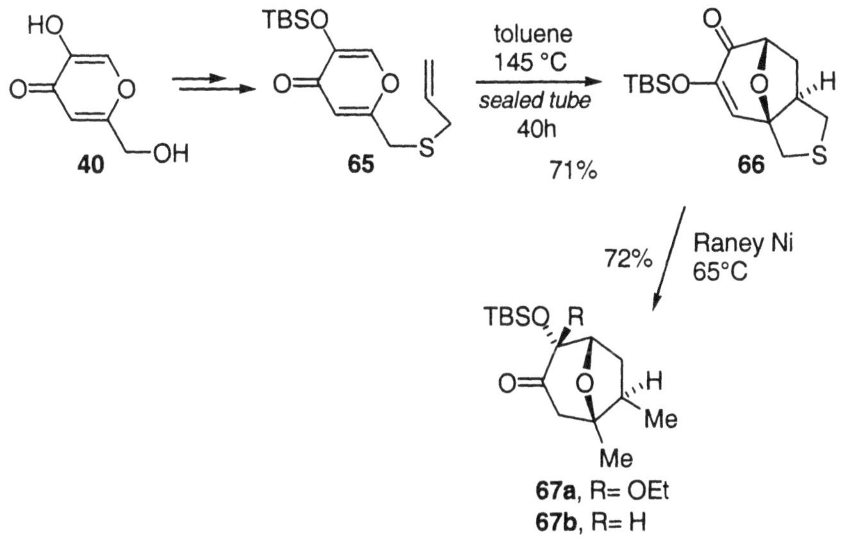

Scheme 24.

With the sulfide **65** in hand, we undertook a brief survey on the influence of the nature of the solvent and the migrating group in the cycloaddition reaction. We found that the cycloaddition of **65** can also be carried out in 1,2-dichloroethane, although it is slower and gives poorer yields than in toluene. Using chloroform as solvent we observed the rapid appearance of insoluble polymeric materials upon heating, and the desired product was isolated in less than 15% yield. The cycloaddition does not work in protic solvents such as MeOH, with the rapid decomposition of the starting pyrone observed when the temperature was increased above 100 °C. Hence, toluene seems to be among the best solvents to carry out the reaction. On the other hand, we observed that the reaction works efficiently even in relatively concentrated solutions, although the best results were obtained using concentrations lower than 0.15 M.[36]

As expected, the cycloaddition can also be accomplished using acyl esters as transferable groups, although this required heating at higher temperatures than with the corresponding silyl derivatives (Scheme 25). Among the acyl groups tested we found that the reaction of the *p*-nitrobenzoate (R = pNO$_2$Bz, Scheme 25) is slightly faster than that of the benzoate (R = Bz), and this is, in turn, faster than that of the acetate (R = Ac) (approx. rate ratio 5:4:3). These rates indicate that the presence of electron-withdrawing groups on the ester activates the migration and facilitates the cycloaddition. We also found that when R is *p*-toluensulfonyl or methyl, the reaction does not take place and the starting compound is completely recovered.[36]

On the other hand, we observed that increasing the oxidation level of the sulfide led to a substantial decrease in the thermal requirements of the reaction, thus the

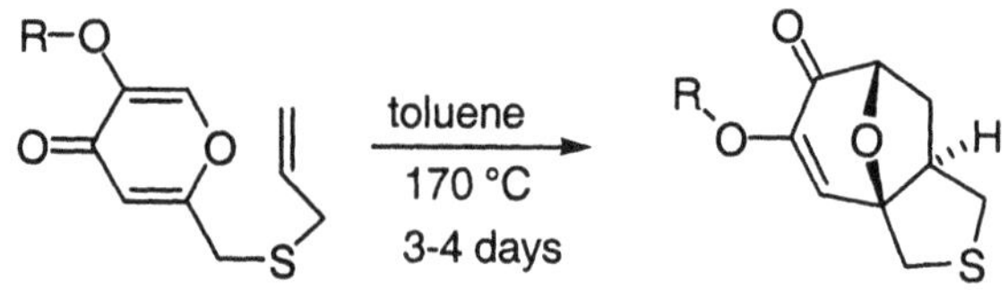

R= p-NO_2Bz, Bz or Ac

Scheme 25.

cycloaddition of the sulfone **68** can be carried out by simple heating at 90 °C for 18 h (Scheme 26).[37] This result may be explained in terms of the Thorpe–Ingold or related effects.[38]

Overall, our own observations, along with the previous results from the laboratories of Garst and Wender, seem to indicate that the most reasonable mechanism for the silyl-transfer-induced reaction involves the formation at equilibrium of a small proportion of an asynchronically developed 4-silyloxy-3-oxidopyrylium ylide **A,** which undergoes a 1,3-dipolar reaction with the alkene (Figure 6). In the case of the ester-induced reaction, a related hemiorthoester intermediate (**B**) can be postulated. Factors that favor the migration of the transferable group, such as the presence of electron-withdrawing substituents in the ester, may accelerate the cycloaddition by increasing the concentration of the reactive ylide intermediate.

The commercial availability of maltol (**69**) provided a possibility to extend the temporary sulfur strategy to prepare regioisomeric adducts (Scheme 27), and hence for eventually expanding the range of ring substitution patterns attainable.[35] On the other hand, the presence of a suitably located primary hydroxyl group in kojic acid pointed to the application of a temporary silicon connection to the pyrone–alkene cycloaddition.[39] Application of this device to our pyrone–alkene cycloaddition demanded the use of an O-5 protecting group that would not interfere in the post-cycloaddition removal of the silicon tether. We found that the benzoyl group can be selectively introduced onto the phenolic hydroxyl of kojic acid and that it fulfills the necessary requirements as a protecting–activating group. Remarkably,

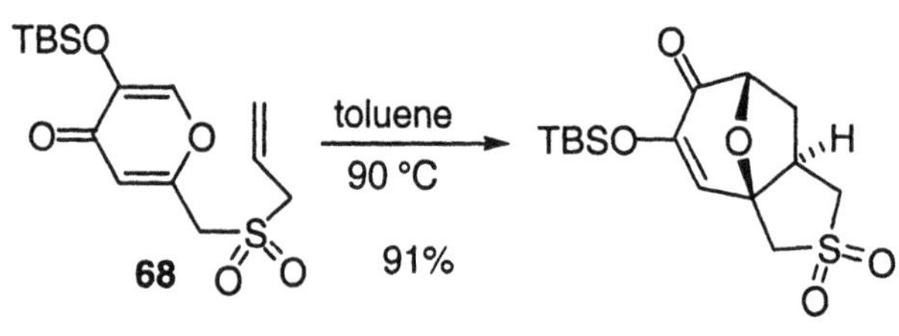

Scheme 26.

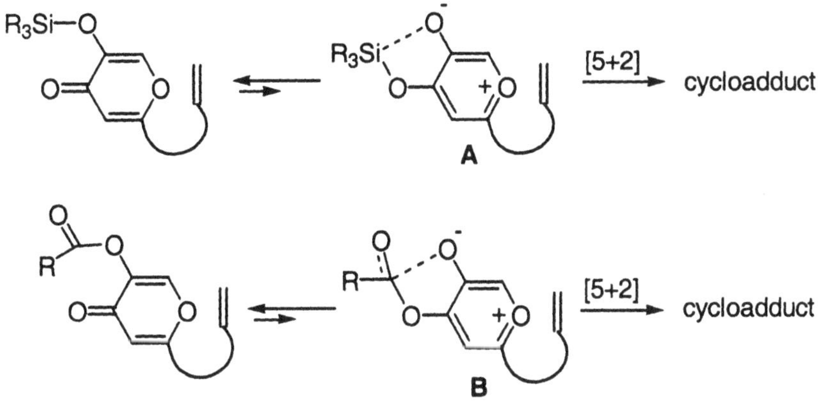

Figure 6.

it was found that the entire cycloaddition sequence, including oxidative workup of the carbon–silicon bond of the cycloadduct, can be carried out in a "one pot" protocol: (i) formation of the silyl ether **71**; (ii) replacement of the solvent by toluene and heating in a sealed tube at 170 °C, and (iii) replacement of the solvent by DMF and treatment of the residue with potassium fluoride and *m*-chloroperbenzoic acid (Scheme 28).[36] It should be noted that initial attempts to oxidize the silanoxy bond of the cycloadduct under typical Tamao conditions (H_2O_2, K_2CO_3, DMF) failed, leading to complex mixtures of products. Overall, the transformation **70** → **72** formally constitutes a completely regio- and stereoselective *intermolecular* [5+2] cycloaddition between kojic acid and the enolic tautomer of acetaldehyde. In conclusion, linking the reacting partners via sulfur or silicon tethers allowed the achievement of a thermal [5+2] cycloaddition between nonactivated alkenes and 3-alkoxy-4-pyrones with complete regio- and stereocontrol.

C. Pyrone–Benzyne Cycloadditions

It is well known that benzyne is an extremely reactive alkyne and it was therefore suspected that it could participate in a [5+2] cycloaddition process with β-hydroxy-γ-pyrones without the need of strong thermal activation. Indeed, this hypothesis

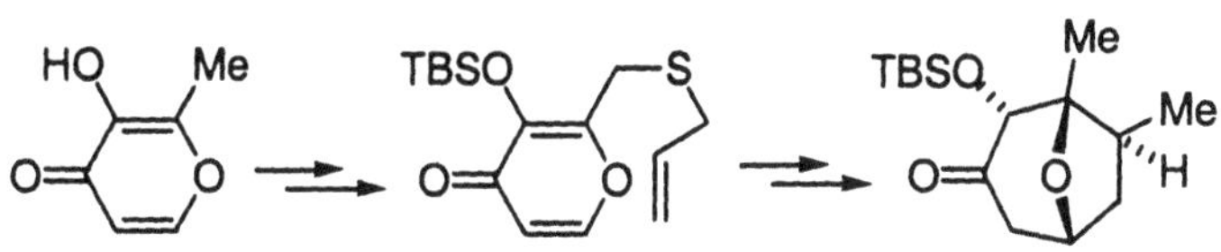

69

Scheme 27.

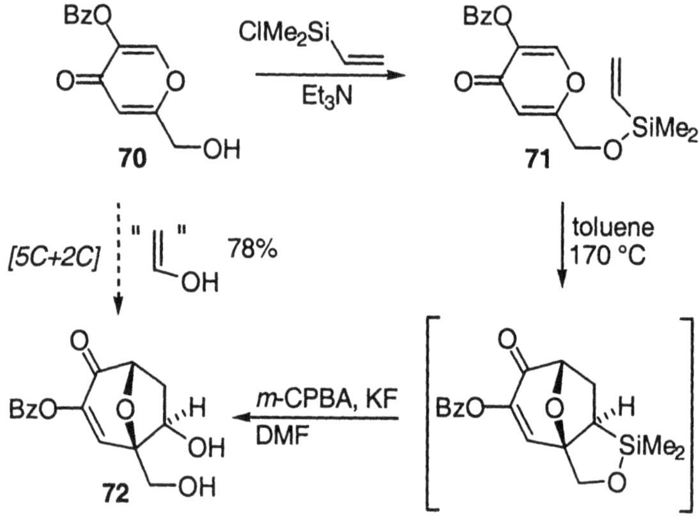

Scheme 28.

was demonstrated by Guitián and coworkers,[40] who showed that generating benzyne from benzenediazonium-2-carboxylate in the presence of α-deoxykojic acid leads, among other products, to the [5+2] adduct **73**. The formation of this compound was explained according to the mechanistic pathway indicated in Scheme 29.

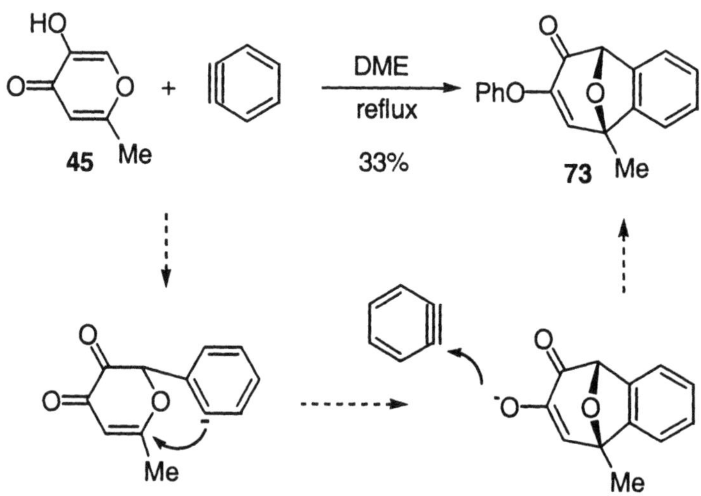

Scheme 29.

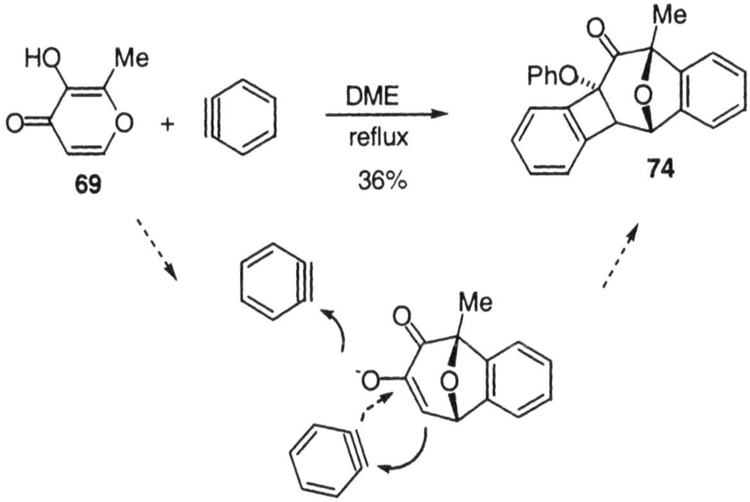

Scheme 30.

Curiously, the use of maltol (**69**) as pyrone partner led to the major isolated product being the polycycle **74**. The formation of this compound was explained assuming a carbon rather than an oxygen trapping by benzyne of the initial [5+2] adduct, followed by intramolecular cyclization to give a cyclobutane (Scheme 30).[40] An alternative scenario involving an initial O-arylation followed by [2+2] cycloaddition of benzyne to the alkene cannot be discarded.

IV. HYDROXYPYRONES AS PRECURSORS OF REACTIVE OXIDOPYRYLIUM AND OXIDOPYRIDINIUM YLIDES

A. Preparation and Reactions of 4-Methoxy-3-oxidopyrylium Ylides

In research aimed at the preparation of advanced bicarbocyclic precursors of tigliane and daphnane diterpenes, Wender and Mascareñas found that pyrone–alkene substrates containing sensitive functionalities do not survive the relatively stringent conditions required to achieve the thermal intramolecular [5+2] cycloaddition.[41] This was the case for compound **75b** (with a third double bond in the chain), which, in contrast to its homologue **75a** (without such double bond), decomposed on heating at the temperatures required for the cycloaddition (Scheme 31).

To overcome this problem, a practical alternative that allowed the cycloaddition to be carried out at room temperature was devised. The strategy consisted of converting pyrone **76a** into a highly reactive 4-methoxy-3-oxidopyrylium ylide by sequential O-4 methylation (with MeOTf) and O-3 desilylation (with fluoride). As

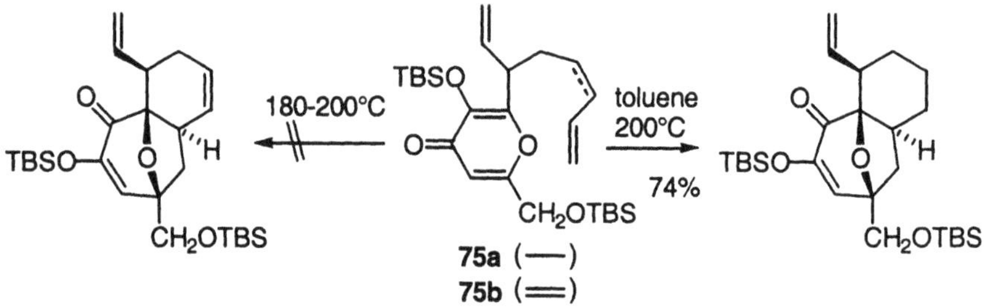

Scheme 31.

might be expected from the precedents with the parent 3-oxidopyrylium zwitterions, this ylide underwent a smooth internal [5+2] cycloaddition, even at room temperature (Scheme 32). The strategy can also be implemented starting from the hydroxypyrone **76b**, although in this case the generation of the ylide from the corresponding 3-hydroxy-4-methoxypyrylium salt was best carried out using a non-nucleophilic base such as 2,2,6,6-tetramethylpiperidine (TMP).[41]

The discovery of this new cycloaddition-inducing protocol prompted us to investigate whether it could be extended to intermolecular examples, owing to the aforementioned difficulties associated with the thermal bimolecular pyrone–alkene reaction. The 4-methoxypyrylium salt of α-deoxykojic acid (**45**) was readily prepared by mild heating of a chloroform solution of the pyrone with MeOTf. Upon treatment with a hindered base, this salt (**77**) evolved to the dimer **78**, suggesting the formation of the desired oxidopyrylium zwitterion (Scheme 33).[42] Performing the deprotonation in the presence of a dipolarophile such as acrylonitrile afforded the expected cycloadduct, albeit in very low yield, with the major product being the dimer. This result suggested that an efficient [5+2] annulation would require the development of experimental conditions that would keep the concentration of the reactive oxidopyrylium betaine to a minimum. The best result was achieved using a relatively weak, non-nucleophilic base, such as *N*,*N*-dimethylaniline (DMA), as

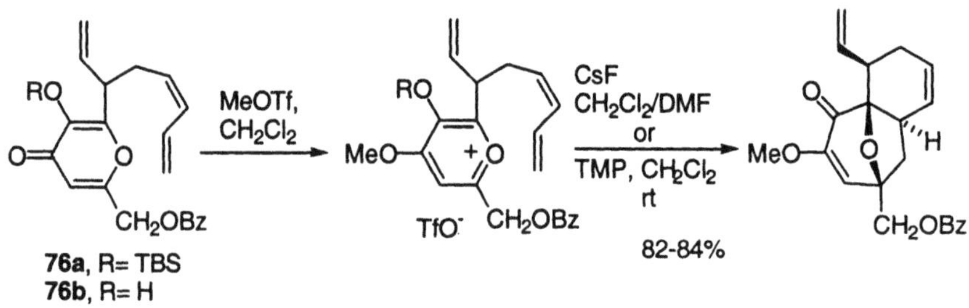

Scheme 32.

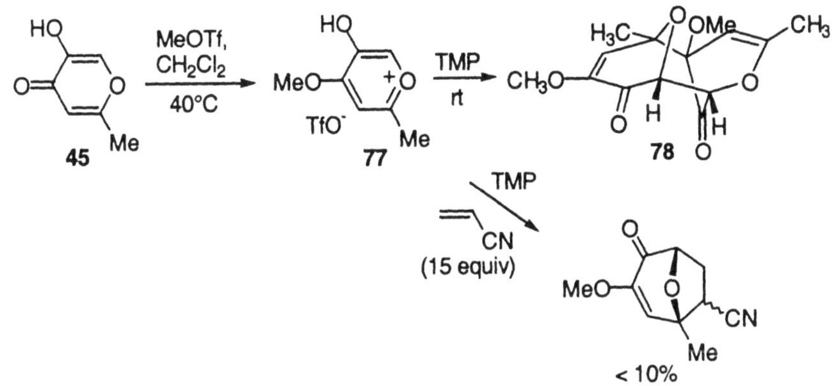

Scheme 33.

deprotonating agent. This variation led to excellent yields of the cycloadducts with a range of dipolarophiles (Table 1).[42] The regio- and stereoselectivity of the cycloaddition is in keeping with that predicted from the application of the perturbation molecular orbital theory, a technique that had already shown its utility for explaining the results obtained with the parent 3-oxidopyrylium systems.[21]

Table 1.

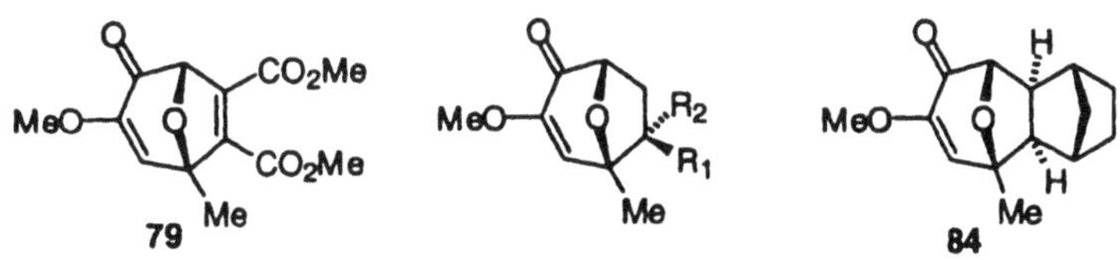

80, R_1= CN, R_2= H
81, R_1= H, R_2= CN
82, R_1= Ph, R_2= H
83, R_1= H, R_2= Ph

Dipolarophiles	*Isolated Yield*	*Products*	*(exo : endo)*
DMAD	73%	**79**	—
acrylonitrile	78%	**80/81**	(2.3 : 1)
styrene	58%	**82/83**	(1 : 2.2)
norbornene	73%	**84**	(1 : 0)

B. Formation and Reactions of 4-Methoxy-3-oxidopyridinium Ylides

Having learned that β-alkoxy-γ-pyrones can be used as five-carbon components in [5+2] annulations to alkenes, it was of interest to determine if the homologous pyridones could serve the same role, since the resulting 8-azabicyclo[3.2.1]octane adducts might constitute useful frameworks to construct a variety of functionalized tropane derivatives (Figure 7).[43]

The presumption that the required pyridones could be readily prepared from the corresponding pyrones by treatment with primary amines was confirmed by synthesizing pyridone **86** from a suitable protected hydroxypyrone (**85**, Scheme 34). Remarkably, while pyrone **65** gave the corresponding cycloadduct **66** by thermolysis at 145 °C, pyridone **86** remained unchanged after heating in toluene for several hours, even at 190 °C. Although the reasons for this lack of reactivity have not been established, they are probably related to the greater aromatic character of the pyridone and pyridinium systems with respect to their oxygen-containing analogues.

At this point it was envisaged that a viable alternative to induce the pyridone–alkene cycloaddition might be based on the conversion of the pyridone into a 4-methoxy-3-oxidopyridinium zwitterion, which should add to alkenes in a similar fashion than its unsubstituted parent analogue. This hypothesis was first tested intermolecularly owing to the easy preparation of the pyridinium salt **88**. Treatment of this salt with *N*-phenylmaleimide in refluxing acetonitrile in presence of 1.5 equiv of TMP gave the expected *exo*-cycloadduct in good yield (Scheme 35).[44] The reaction also worked with other electron-deficient dipolarophiles, such as acrylonitrile, methyl acrylate, and phenyl vinyl sulfone, leading to mixtures of *endo*- and *exo*-stereoisomers (Table 2). Comparison of the coefficients at the reactive centers of the HOMO of our betaine with those of the parent unsubstituted 3-oxidopyrid-

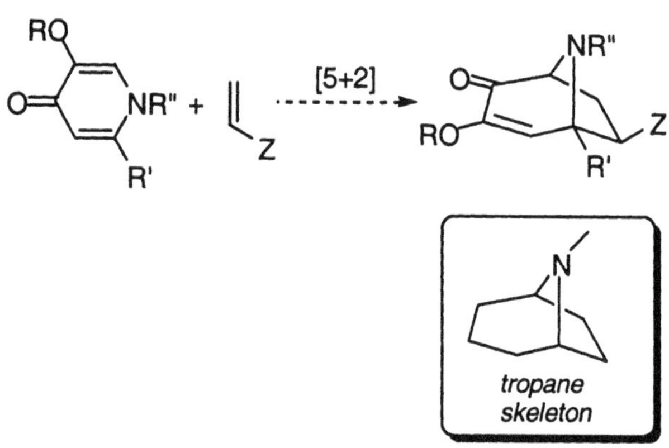

Figure 7.

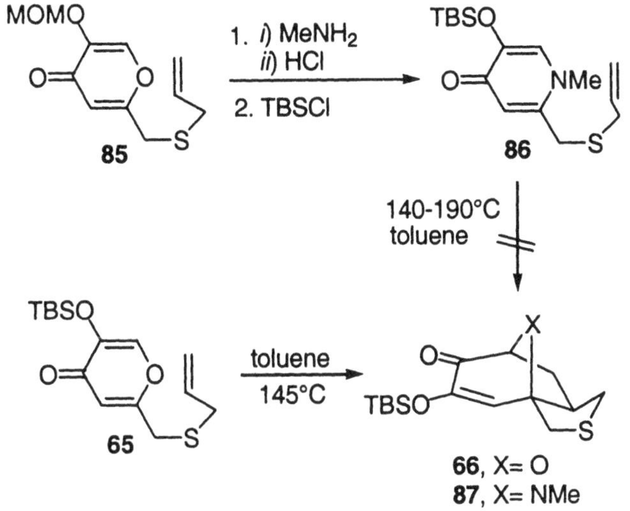

Scheme 34.

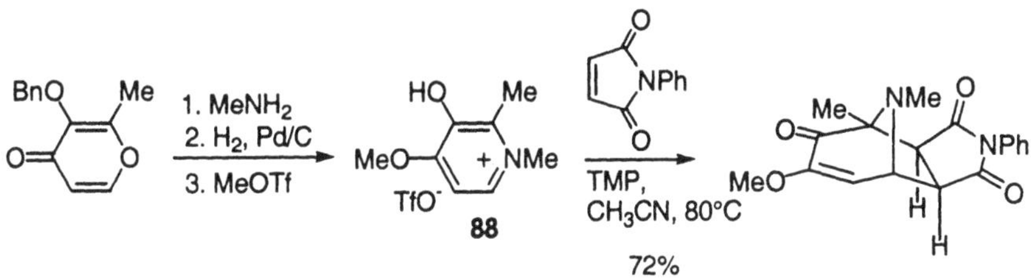

Scheme 35.

Table 2.

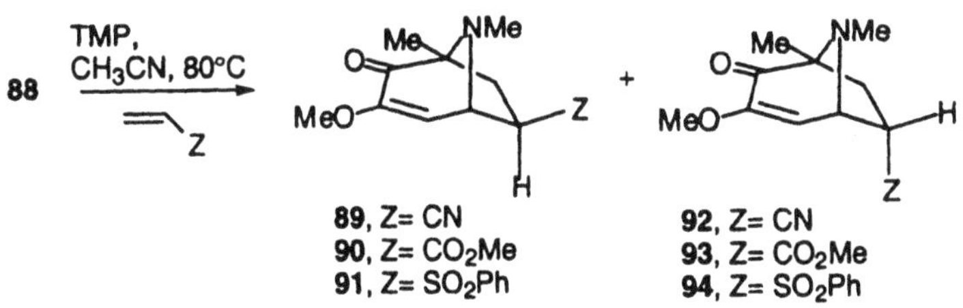

Dipolarophiles	*Isolated Yield*	*Products*	*(exo : endo)*
acrylonitrile	68%	**89/92**	(5.6 : 4.4)
methyl acrylate	64%	**90/93**	(5.8 : 4.2)
phenyl vinyl sulfone	58%	**91/94**	(8 : 2)

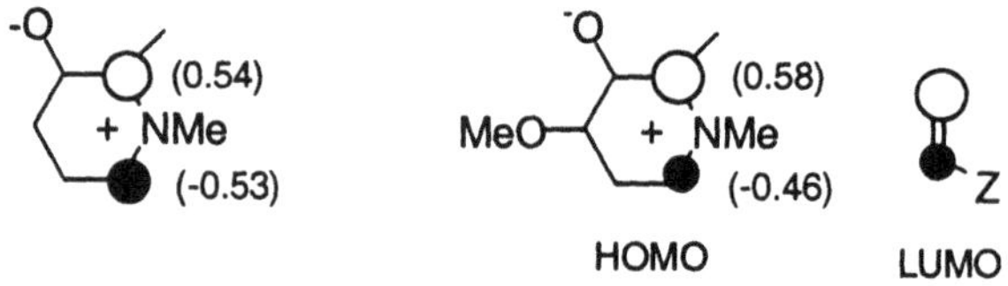

Figure 8.

inium dipole (using PM3 MOPAC) revealed that the presence of the 4-methoxy group induces a slight but significant increase in their size difference. This is in consonance with the entire regioselectivity of our reactions in comparison with that reported for cycloadditions of 2-methyl-4-unsubstituted oxidopyridinium ylides (Figure 8).[45]

This cycloaddition protocol can be extended to intramolecular cases as was demonstrated for the pyridone **95** which bears an ether-linked pendant alkene. As expected, thermolysis of this pyridone in toluene for several hours at 190 °C didn't promote the cycloaddition, but prior conversion to the 4-methoxypyridinium salt, followed by deprotonation and heating of the mixture at 100 °C, provided the desired *exo*-cycloadduct **96** in excellent yield (Scheme 36).[44]

It can be concluded that although β-alkoxy-γ-pyridones failed to undergo a direct thermal [5C+2C] cycloaddition to alkenes, the annulation can be achieved by driving the reaction through a dipolar mechanism by converting the pyridones into 4-methoxy-3-oxidopyrylium zwitterions. Thus, although β-hydroxy-γ-pyrones do not have an obvious structural resemblance to the tropane azabicyclic skeleton, viewing the former as oxidopyridinium ylide precursors led to a practical entry to the latter.

C. Preliminary Studies on the Generation and Cycloaddition Properties of 4-Alkyl-3-oxidopyrylium Ylides

While the reduced electrophilic character of the carbonyl group of γ-pyrones[46] hinted at the potential difficulties of achieving nucleophilic additions to this group,

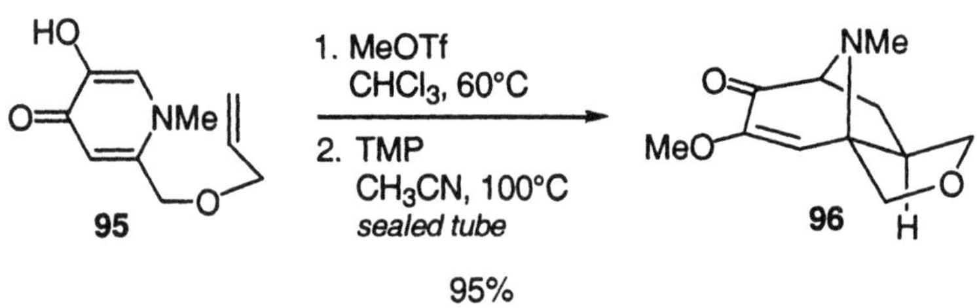

Scheme 36.

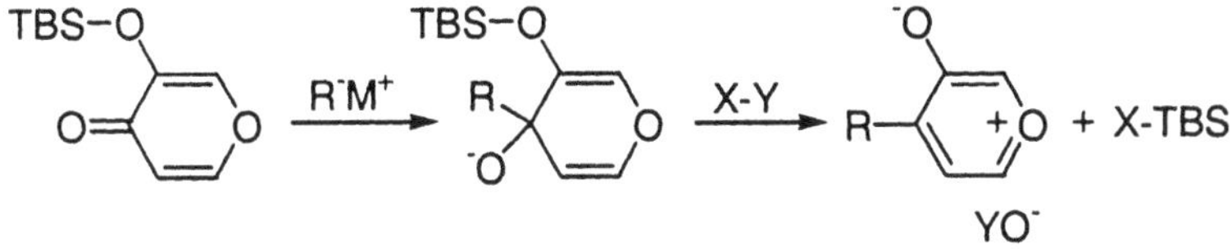

Figure 9.

several precedents indicating that Grignard reagents do add to some substituted γ-pyrones[47] led us to envisage the possibility of obtaining 4-alkyl-3-oxidopyrylium zwitterions by promoting a sequence of reactions similar to that outlined in the Figure 9.

We preferred to test the feasibility of this protocol on an intramolecular example since the potential generation of the dipole could be rapidly detected by its presumably immediate trapping by the internal alkene. Unfortunately, the best results obtained to date involving the addition of MeLi to the pyrone **65**, followed by trifluoromethanesulfonic acid (TfOH, approx. 2 equiv) and an excess of TBAF, gave the desired product **97** in only 15% yield (Scheme 37).[48] Given that the process seems to offer a versatile entry to a range of substituted adducts, further research on the development of an improved protocol is warranted. Preliminary observations point to the organometallic addition reaction as the problematic step, and we are therefore currently investigating alternatives to optimize this reaction. Somewhat surprisingly, initial assays involving the sequential addition of BF_3•Et_2O and MeLi to the pyrone **65** led, at –78 °C, to the conjugate addition products **98** as an approximately 1:1 mixture of *cis*- and *trans*-diastereoisomers.[34]

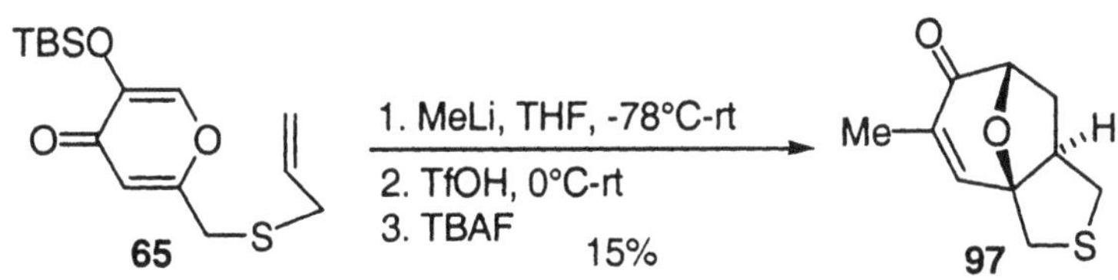

Scheme 37.

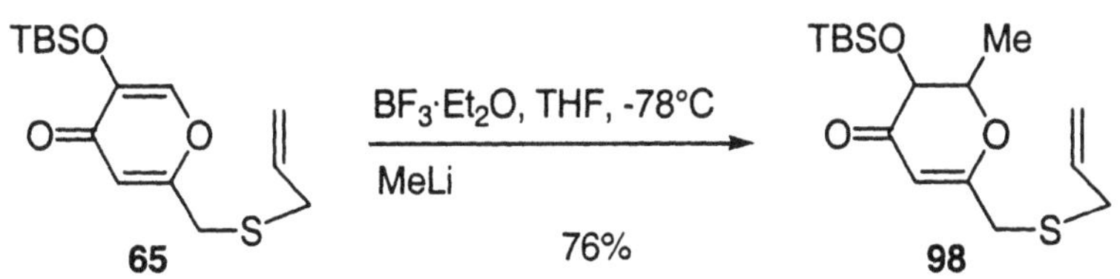

Scheme 38.

V. ACID-INDUCED [5+2] CYCLOADDITIONS OF HYDROXYPYRONES

We have already discussed several alternatives to promote the [5+2] cycloadddition of β-alkoxy-γ-pyrones to alkenes. In the initial report by Garst on the intramolecular thermally induced pyrone–alkene cycloadditions, it is mentioned that substrate **54**, which failed to cyclize under simple thermolysis, did undergo cycloaddition to give **99** when refluxed in MeOH in the presence of methanesulfonic acid (Scheme 39).[11] This isolated finding, which is reminiscent of the acid-induced reaction of Büchi's quinone ketals (see Section II.A), attracted our attention as presumably entailing a different mechanistic pathway than the simple thermal cycloaddition, given that under the acidic conditions of the reaction, the formation of a 3-oxidopyrylium ylide would certainly not be feasible. We therefore turned our attention to this process, finding that this type of acidic activation can indeed be used to induce intramolecular reactions of substrates like **100a** or **100b** (Scheme 40).[49] Since for **100b** proton NMR monitoring of the reaction course revealed the probable intermediacy of the 3-hydroxy-4-methoxypyrylium salt **102**, we attempted its deliberate preparation by O-4 methylation of the pyrone with MeOTf. Interestingly, heating a solution of **100b** in $CHCl_3$ with MeOTf provided the cycloadduct **103** in a 53% yield. This enol ether can be completely converted into the dimethylketal **101b** upon treatment with TfOH in refluxing methanol. These results seem to confirm that the acid-mediated reaction involves the formation of a 4-methoxypyrylium ion intermediate.

With this information in hand it was of interest to find out whether the above acidic conditions could be applied to intermolecular examples. Indeed, heating maltol (**69**) with β-methylstyrene under such conditions gave the cycloadduct **104** as a single diastereoisomer, albeit in a modest 38% yield (Scheme 41).[49] It is noteworthy that in this case the enol ether did not evolve to the dimethylketal as it did in the intramolecular case, perhaps because the absence of the fused five-membered ring allows for better conjugation of the double bond with the ketone. The regio- and stereochemistry of **104** is similar to that observed for the major product

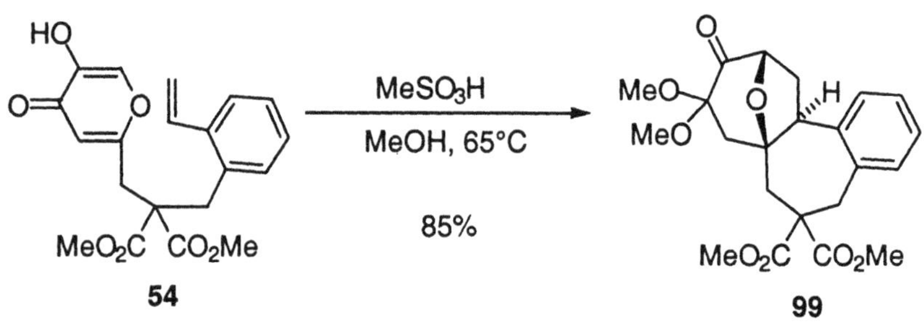

Scheme 39.

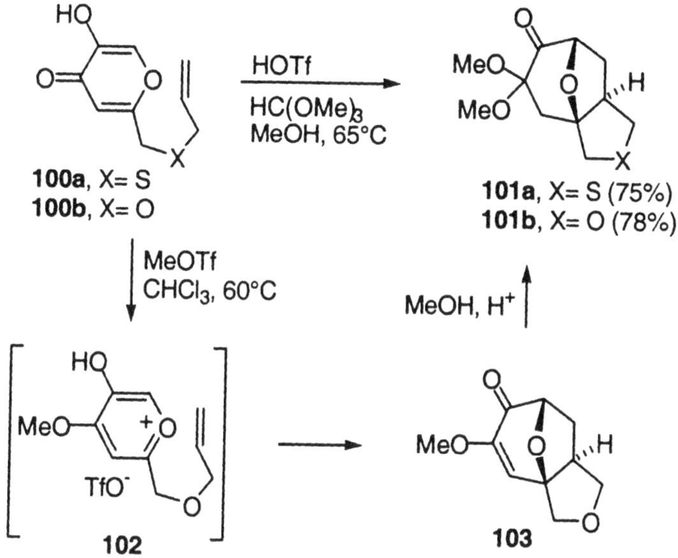

Scheme 40.

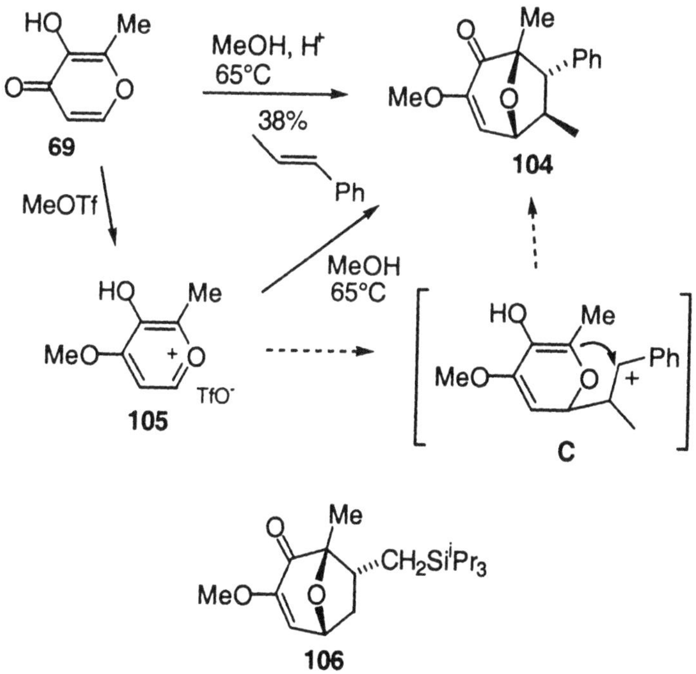

Scheme 41.

in the reaction of 3-alkoxy-1,4-benzoquinones to β-methylstyrene.[16] Despite the low yield of the annulation, the considerable increase in skeletal and stereochemical complexity achieved in one single reaction from commercially available compounds must be noted.

In consonance with the above proposed intermediacy of a 4-methoxypyrylium cation in the acid-induced reaction, we found that cycloadduct **104** can also be obtained (52% isolated yield) by direct heating of pyrylium salt **105** with an excess of β-methylstyrene in MeOH. The intermolecular annulation was also successful with other electron-rich olefins, such as allyltriisopropylsilane (which gave exclusively the *endo*-cycloadduct **106** in 35% yield), but failed with electron-deficient alkenes such as acrylonitrile or methyl acrylate. This failure, and the fact that the acidic conditions must prevent the formation of a 3-oxidopyrylium ylide, rules out the possibility that the pyrylium salt–alkene annulation occurs through a 1,3-dipolar mechanism, and suggests that the hydroxypyrylium ion is working as an electrophile in an apparently stepwise process that might imply an intermediate such as **C** (Scheme 41).

VI. CHEMISTRY OF THE OXABICYCLIC PYRONE–ALKENE ADDUCTS

A. Access to Seven-Membered Carbocycles by Opening of the Oxa-Bridge

Certainly the synthetic value of the above cycloaddition strategies is intimately ligated to the feasibility of rapid elaboration of the resulting oxabicyclic adducts into more advanced precursors for relevant products. The first and most appealing structural element embedded in the oxabicyclic frame is the seven-membered carbocycle, a type of medium-sized ring that occurs in a large number of bioactive natural products[8] and that is not easily constructed by the usual cyclization methods. Therefore, a major challenge in manipulating the pyrone–alkene cycloadducts consists of removing the bridging ether to unmask the carbocycle itself. Opening of this type of bridge in oxabicyclic[2.2.1] and [3.2.1] fragments has been the subject of intensive research by several groups, and a number of efficient methods have been developed.[50] In the case of the [3.2.1] architectures, most of these

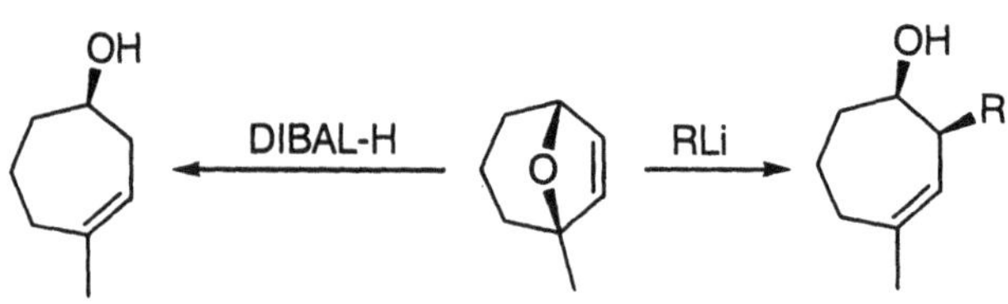

Scheme 42.

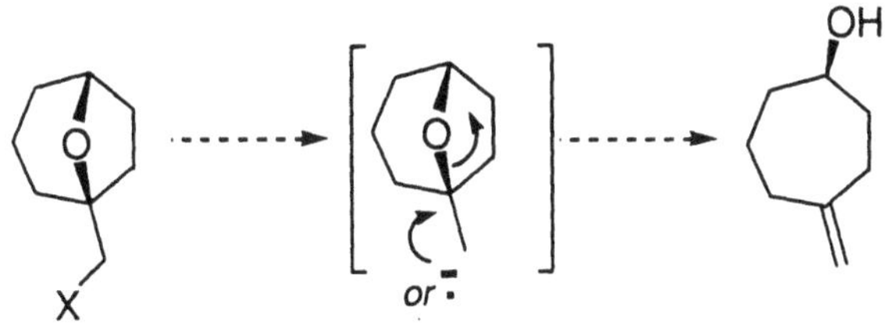

Figure 10.

methods require the presence of a C–C double bond in the furan unit of the oxabicycle in order to trigger the opening (Scheme 42),[50] although other tactics that rely on the generation of a radical or carbanion in the position α to the bridgehead carbon have also been used (Figure 10).[51,32]

The presence of an electron-donating enol ether in the pyran portion of our [5+2] pyrone–alkene cycloadducts seemed to offer a particularly attractive site for launching the bridge opening, one that might be more effective after eliminating the conjugation with the ketone (Figure 11). This deconjugation was easily carried out in the substrate **66** by addition of a Grignard reagent, such as MeMgBr, that exclusively gave the *exo*-addition product **107** (Scheme 43). To our surprise, when the same reaction was carried out with MeLi instead of the Grignard reagent, the rearranged product **108a** was obtained exclusively. This compound must arise from migration of the TBS group to the lithium alkoxide generated in the addition reaction, with the concomitantly formed enolate becoming protonated upon work-up. Interestingly, this enolate can be trapped in situ with an alkylating agent such as MeI to obtain the *exo*-methylated product **108b**.[52] The tandem *addition–migration–alkylation* sequence was found to be general for a variety of different organolithium and alkylating agents, and hence constitutes a versatile method for the stereoselective introduction of two alkyl groups onto the seven-membered carbocycle. The complete facial selectivity demonstrates the stereochemical directing potential of the bridged frame.

Initial attempts to induce the bridge-opening were carried out on the enol ether **107** by attempting to enhance the leaving group ability of the bridging oxygen with Lewis acids such as $TiCl_4$ or $BF_3{\bullet}Et_2O$. However, the unique products detected arose from simple hydrolysis of the silyl ether, while the bridge remained intact.

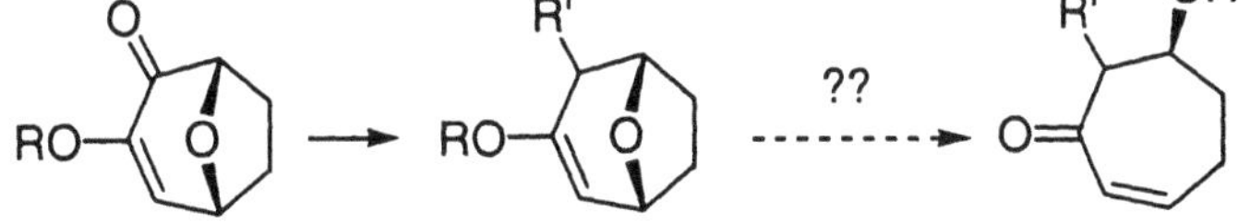

Figure 11.

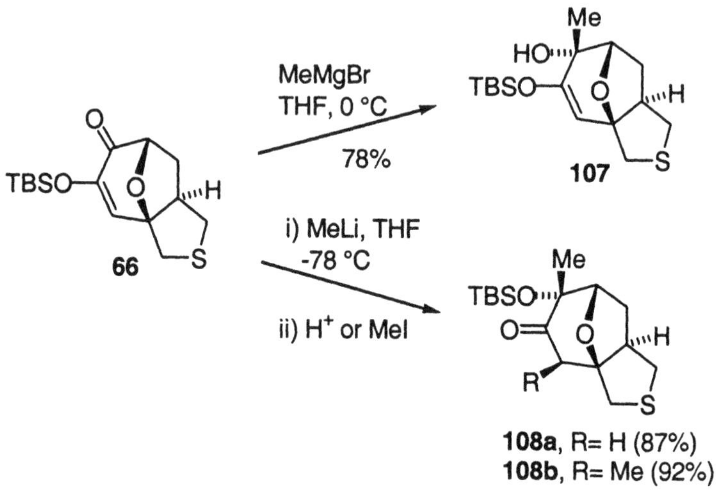

Scheme 43.

On the other hand, it was relatively surprising to find that although **108a** was unaltered upon treatment with $BF_3{\bullet}Et_2O$, its reaction with BBr_3 led to the rearranged oxabicycle **109** (Scheme 44). The formation of this compound can be explained in terms of an oxygen-bridge-promoted expulsion of the leaving group

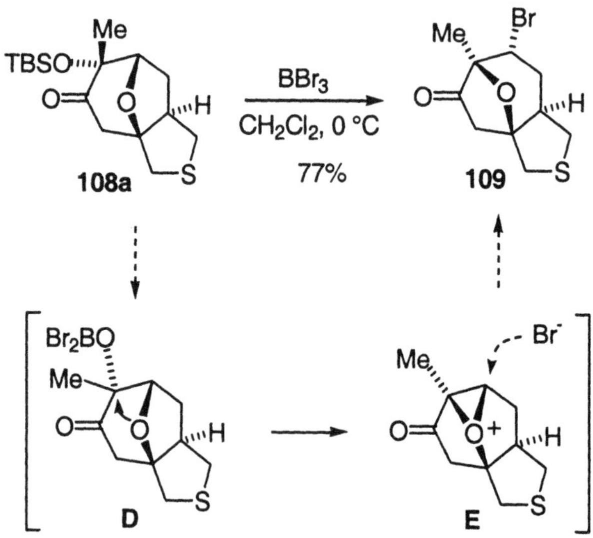

Scheme 44.

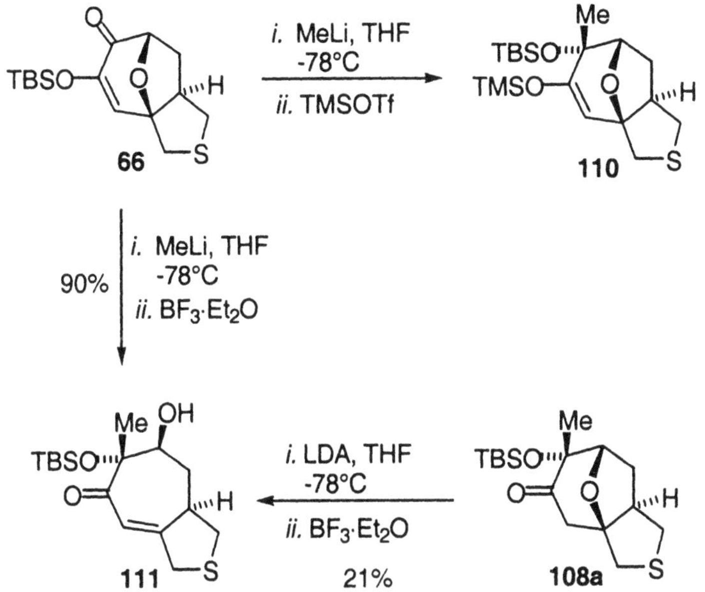

Scheme 45.

at C-8 with concurrent back-side attack of the bromide on the presumed oxonium intermediate **E.**[53]

The failure to open the bridge with Lewis acids led us to test the feasibility of a base-induced β-elimination reaction, although we were concerned about foreseeable difficulties associated with this process due to the imperfect stereoelectronic alignment for an E_2 elimination reaction. In keeping with this presumption, the lithium enolate resulting from the addition of MeLi to **66** did not evolve to the opened system even after several hours in refluxing THF. These results prompted us to attempt other strategies based on the simultaneous action of a strong electron-donating group (enolate) and an oxophilic agent (TMSOTf), conditions that have been reported to work in related systems, albeit in rather poor yields.[54] Sequential treatment of **66** with MeLi and TMSOTf gave the silyl enol ether derivative **110** as the only product. However, the use of BF_3•Et_2O smoothly produced the desired hydroxyenone **111** in excellent yield, even at −78 °C (Scheme 45).[53] Curiously, generation of the same enolate by deprotonation of **108a** with LDA in THF, followed by addition of 5 equiv of BF•Et_2O led to the same cycloheptenol **111**, albeit in a modest 21% yield with most of the recovered compound being starting material. This lower yield could be due to the existence of an equilibrium between the bridged and opened structures, being completely shifted to the desired cycloheptenols when secondary reagents such as amines are absent from the medium.

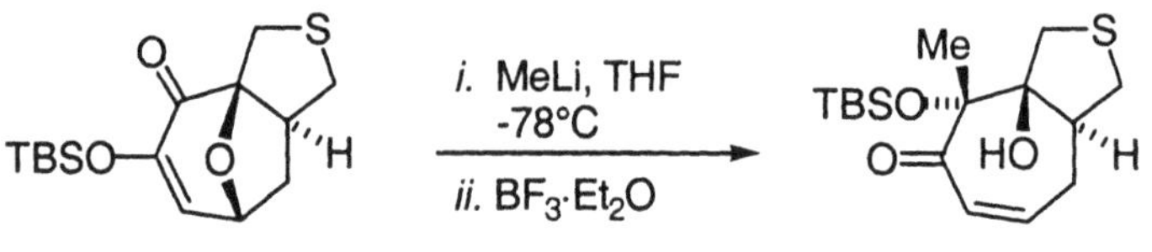

Scheme 46.

At this stage, it was of interest to assess whether the tandem addition–migration–opening sequence could be extended to obtain regioisomeric adducts such as **113** due to the fact that a large number of natural products incorporate this type of junction between a seven- and a five- or six-membered ring.[55] Satisfyingly, treatment of **112** with MeLi followed by the addition of BF_3•Et_2O gave the expected opened derivative **113** in good yield (Scheme 46).[53] This result confirmed the oxabicycle opening protocol as a mild, efficient, and general method for removing the oxa-bridge, and therefore further validates the pyrone–alkene cycloaddition as a practical route to highly substituted seven-membered carbocycles containing several interesting sites for subsequent manipulations.

B. Conversion into Highly Functionalized Tetrahydrofurans

One of the advantages of oxabicyclic frameworks is that their chemistry is not limited to the construction of carbocycles, but they can also serve as useful building blocks to prepare other structures such as tetrahydrofurans.[56,50] This goal was particularly easy to fulfill in our pyrone–alkene cycloadducts since their oxygenation pattern sets the stage for a direct oxidative cleavage of the carbocycle. In this way, **108a** and **108b** could be readily transformed into the *cis*-2,5-tetrahydrofurans **114a** and **114b** by desilylation and treatment of the resulting hydroxyketone with $Pb(OAc)_4$.[52] Remarkably, the oxidative cleavage can be carried out in a single pot by stirring the oxabicycles with TBAF and $Pb(OAc)_4$ in MeOH (Scheme 47). Since the oxabicyclic adducts are rapidly obtained from kojic acid as described above,

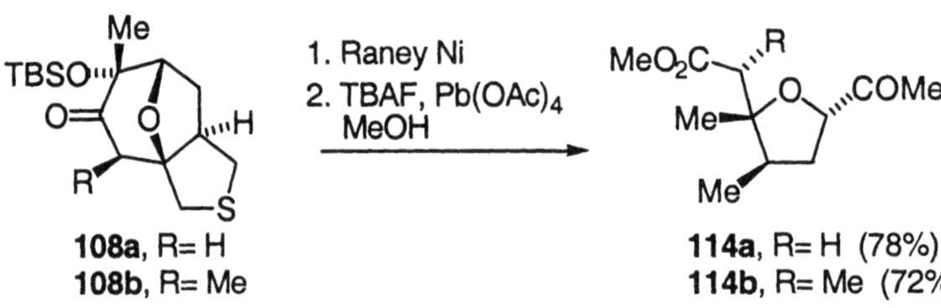

108a, R= H
108b, R= Me

114a, R= H (78%)
114b, R= Me (72%)

Scheme 47.

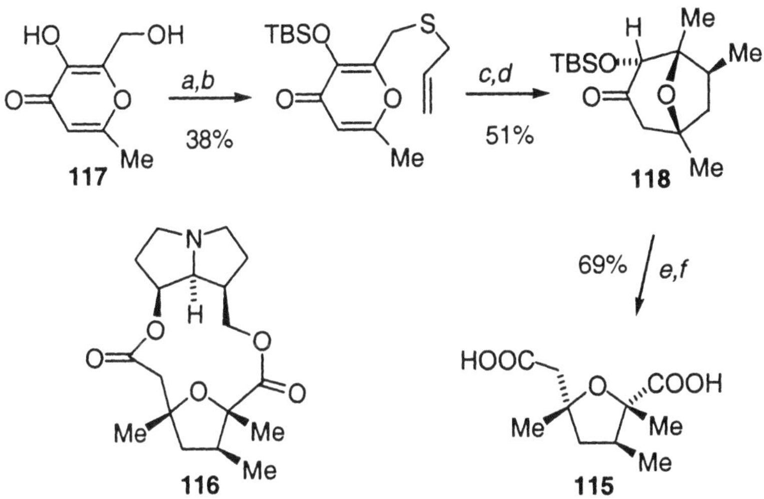

(a)*i.* $SOBr_2$, $CHCl_3$, rt; *ii.* Et_3N, $HSCH_2CH{=}CH_2$, THF, rt. (b) Et_3N, TBSCl, CH_2Cl_2 (c) Toluene, 175°C. (d) Raney Ni, THF, 65°C. (e) *i.* TBAF; *ii.* $Pb(OAc)_4$, MeOH. (f) *i.* CrO_3, H_2SO_4, acetone, rt; *ii.* H_2O, OH^-

Scheme 48.

the sequence provides a succinct route to *cis*-2,5-disubstituted tetrahydrofurans with up to four stereocenters.

The synthetic value of the approach was recently proved by its application to a short synthesis of (±)-nemorensic acid (**115**), the diacid portion of nemorensine (**116**), a prominent member of the Senecio alkaloid family. Implementation of the strategy required preliminary conversion of kojic acid into **117**, which was achieved following known deoxygenation and hydroxymethylation procedures. This pyrone was elaborated to give the cycloadduct **118**, which bears the correct stereochemical array for the target, following the steps indicated in Scheme 48.[57] Oxidative cleavage of the carbocycle gave the desired tetrahydrofuran, which was synthesized in roughly half the number of steps necessary in the previously published syntheses.[58]

C. Conversion into 1,4-O-Bridged Nine- and Ten-Membered Carbocycles

Since the combined structural effects of the α-silyloxyenone functionality and the rigidifying bridging ether of the pyrone–alkene cycloadducts facilitated the one-step introduction of two *exo*-alkyl groups into the oxabicyclic frame, it was reasoned that if both alkyl chains were connected, subsequent oxidative fragmen-

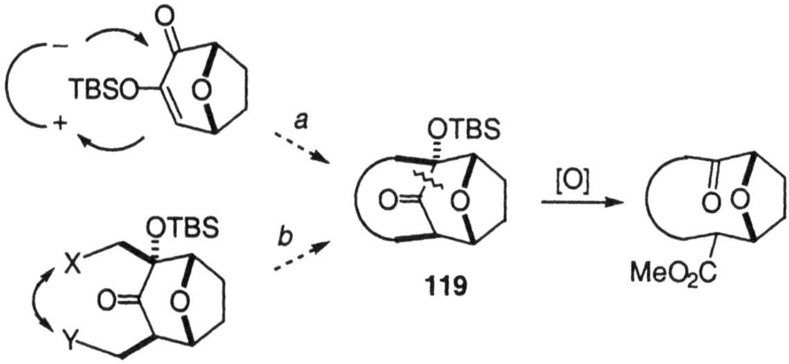

Figure 12.

tation would provide access to oxygen-bridged medium-sized carbocycles (Figure 12). This was of interest owing to the well-known synthetic challenges associated to the construction of rings of this size. The required tricyclic system **119** could, in principle, be prepared by an annulation reaction (*route a*), or a *bis*-alkylation followed by ring closure (*route b*). Reasoning that introduction of chains bearing terminal alkenyl functionalities (X and Y equal to double bonds) would be straightforward, we decided to follow *route b* using a ring-closing metathesis C–C bond-forming reaction (RCM) to link the two *exo*-alkenyl chains.[52]

The strategy was first evaluated for the synthesis of nine-membered carbocycles, a ring type present in a large number of naturally occurring terpenoids and for which methods of assembly remain particularly scarce.[59] Addition of vinyllithium to cycloadduct **66** and alkylation of the resulting enolate with allyl bromide gave the expected α,α′-dialkylated ketone **120** in good yield. MM2 calculations showed that the conformation **120a**, with a pseudoaxial disposition of the alkyl substituents, is energetically favored over **120b**, and served to predict that this system should be particularly prone to ring-closing metathesis, since the olefins are constrained to be in proximity (Scheme 49).

However, attempts to induce the reaction using Grubbs ruthenium catalyst **121** in several solvents failed, leading to recovery of the starting material. On the assumption that this failure could have been due to poisoning of the catalyst by the sulfur atom, and given that removal of this atom was complicated by concurrent reduction of the terminal alkenes, the reaction was evaluated on the analogous adduct **122**, which bears an oxygen rather than a sulfur atom in the accessory five-membered heterocycle. In this case, refluxing a dichloromethane solution of **122** in presence of catalyst **121** (4 mol%) provided the expected tetracycle **123** in excellent yield (Scheme 50). The oxidative ring-enlargement reaction, that required more severe conditions than those used previously for the fragmentation to tetrahydrofurans, was best executed after reducing the C–C double bond.[52]

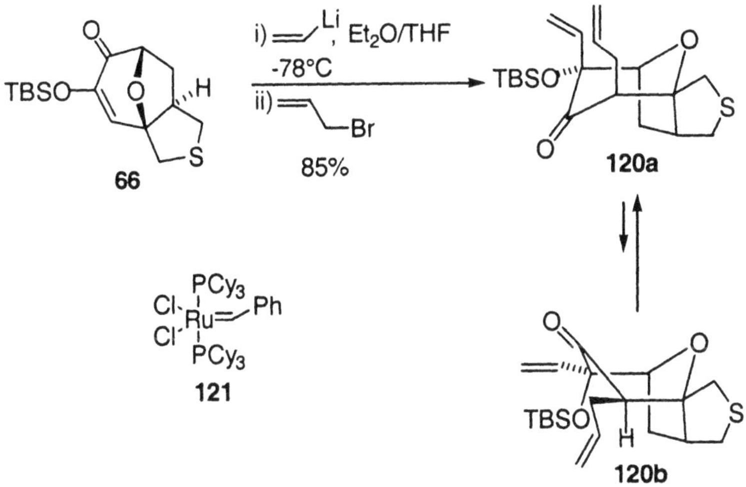

Scheme 49.

In order to evaluate the versatility of this strategy we next targeted the homologous 10-membered carbocycle. This was especially appealing because the resultant 11-oxabicyclo[6.2.1]undecane architecture is the main structural motif of a large number of diterpenes of the eunicillane and cladiellane families,[60] as well as of the recently discovered potent cytotoxic agents eleutherobins and sarcodictyins.[61]

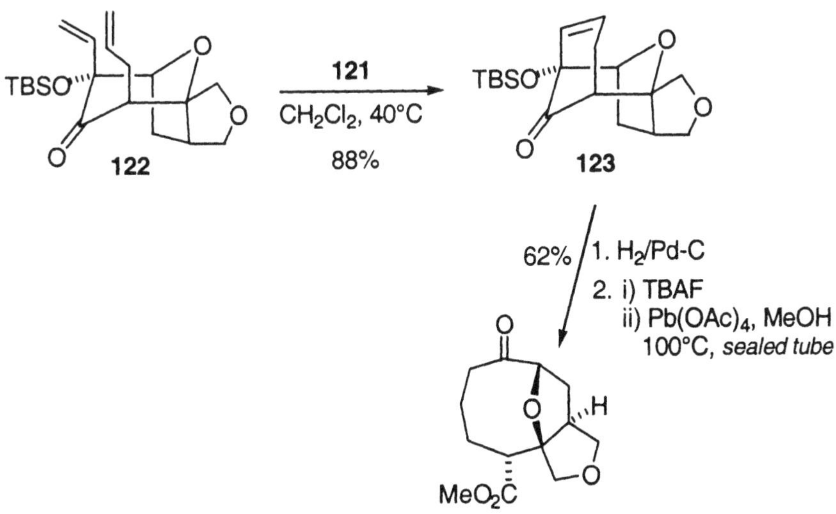

Scheme 50.

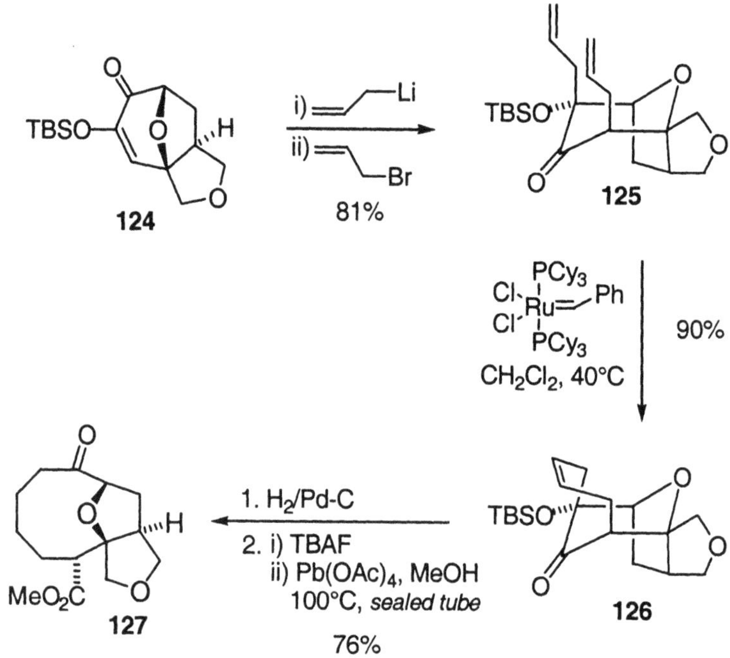

Scheme 51.

Implementation of the approach required only a minor change in the synthetic sequence; the use of allyllithium instead of vinyllithium in the addition reaction to the enone (Scheme 51). It is worth noting that the *addition–alkylation/RCM* can be formally viewed as a two-step method to achieve [3+3] and [4+3] annulation processes.

D. Construction of Fused 6,7,5-Tricarbocyclic Systems by Tandem [5+2]/[4+2] Cycloadditions

The recognition of a relatively low-field ^{1}H NMR chemical shift for the alkenyl hydrogen of the oxabicyclic [5+2] pyrone–alkene adducts, which is suggestive of substantial conjugation of the double bond with the carbonyl group, prompted us to examine the ability of these compounds to participate as dienophiles in a Diels–Alder type of reaction. This would provide a way to fuse a six-membered ring to the seven-membered carbocycle, which was of interest owing to the considerable amount of natural terpenoids that contain this type of ring system.[62] Furthermore, the relative reluctance of 4-pyrones to undergo Diels–Alder reactions with dienes[63] suggested that both the [5+2] and the [4+2] processes could possibly be coupled in a single step. If cycloaddition reactions are among those methods that best fit with the demands of modern organic synthesis (see INTRODUCTION,

Section I), needless to say that tandem processes that integrate several cycloadditions in a single operation are very appealing.[64]

The feasibility of the tandem process was demonstrated by heating a toluene solution of sulfide **128** in presence of 5 equiv of 2,3-dimethylbutadiene. This reaction revealed that while the [5+2] process could be carried out at 145 °C at a moderate rate, the Diels–Alder reaction required heating to a minimum of 160 °C (Scheme 52). In any case, the tandem process could be efficiently carried out by heating at this higher temperature in a sealed tube. Once again the facial bias of the oxabicycle forces a complete diastereoselectivity in the reaction, with the diene approaching from the more accessible *exo*-face of the dienophile.

The abundance of natural products having fused 6,7,5-tricarbocyclic skeletons as their structural core[62] prompted us to investigate the extension of the above tandem strategy to include these types of compounds. Pyrones **130** and **132,** bearing carbon-tethered alkenes, were rapidly assembled from maltol and kojic acid in moderate yields. The tandem process was efficiently carried out by heating a 1:5 mixture of the pyrones and the diene at 160 °C in toluene for 30 h (for **130**) or 60 h (for **132**, Scheme 53).[65] Therefore, relatively complex tetracycles embodying a fused 6,7,5-tricarbocyclic system which has a noticeable relationship to the basic polycyclic skeleton of dolastane and sphaeroane diterpenes (Figure 13) could be assembled in a single step from readily available precursors by simple thermal activation. It should also be noted that the tandem processes described above can be considered thoroughly atom-economical since they take place without the need for the addition of any external reagents, and the products result from a simple summation of the reactant atoms.[3g]

Certainly, adjustment of these strategies for the synthesis of natural diterpenes and their analogues requires appropriate processing of the tetracyclic adducts, with the opening of the oxa-bridge being a particularly challenging step. Although the attempts carried out to date in our laboratories to address this challenge have not

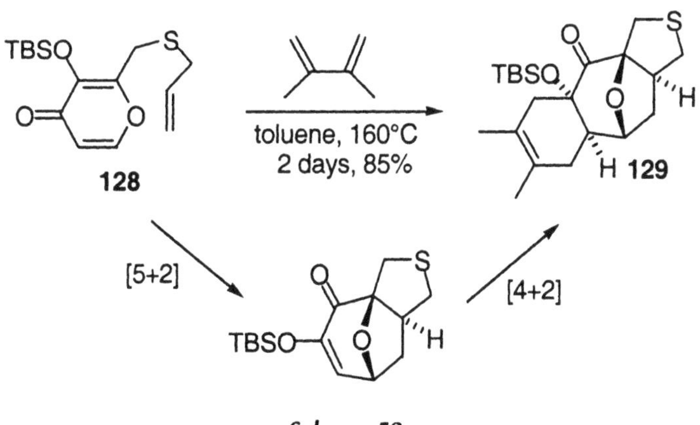

Scheme 52.

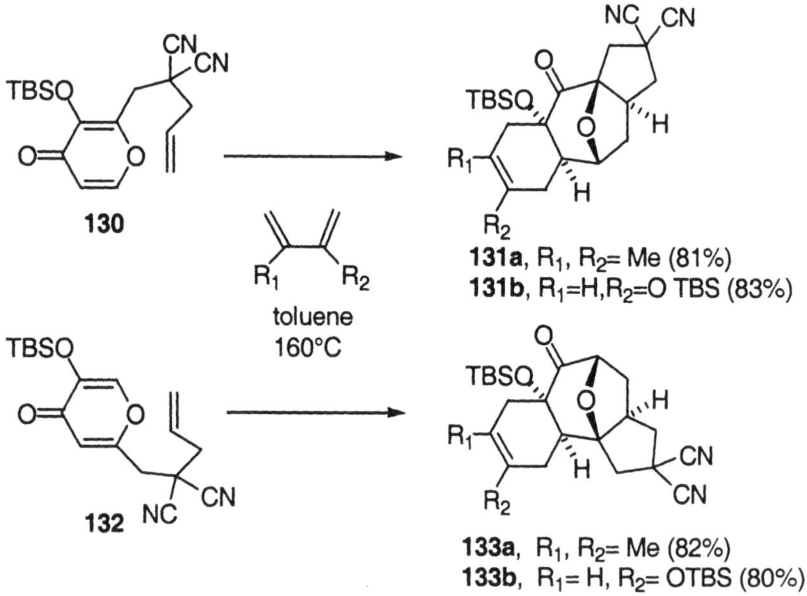

Scheme 53.

succeeded, some interesting reactions were nonetheless uncovered. Hence, in attempting to achieve a reductive opening of the oxabicycle by treatment of **131a** with SmI_2, we found the exclusive reaction product to be the deoxygenated derivative **134** (Scheme 54). This reaction is extremely fast in THF/MeOH, but does not take place if THF is used as single solvent. Other electron transfer reducing agents, such as Na/Bu_2O, gave mixtures of products as a consequence of concurrent reactions with the nitrile groups. We also found that **134** remained unchanged upon treatment with SmI_2 even after mild heating. In light of previous observations by Williams in related systems,[51] we thought that transformation of the ketone into an

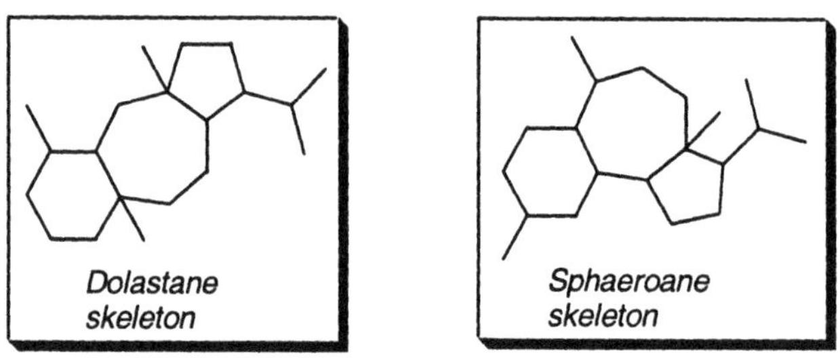

Figure 13.

exocyclic alkene could lead to the generation of a more reactive radical that might then induce the opening. Unfortunately, we were unable to achieve the methylenation, presumably due to steric constraints for approaching the carbonyl, and we observed only products arising from methylation of one of the nitriles. We were also unable to induce the opening of **131a** by using standard Wharton conditions (hydrazine and acetic acid), with the corresponding hydrazones being the major products.

On the other hand, reaction of **131a** with Lewis acids such as BBr_3 or BF_3•Et_2O led to complex reaction mixtures. Curiously, however, treatment of **131a** with excess TMSOTf in refluxing benzene gave the aromatic derivative **135** as the major product.[65] Although we could not determine the mechanistic path of the whole process, we did find that the transformation involves the initial desilylation of the tertiary hydroxy group. This desilylation is most probably caused by the presence of traces of HOTf in the medium, since the reaction is inhibited by the presence of Et_3N. Subsequent formation of the observed product could then be explained in

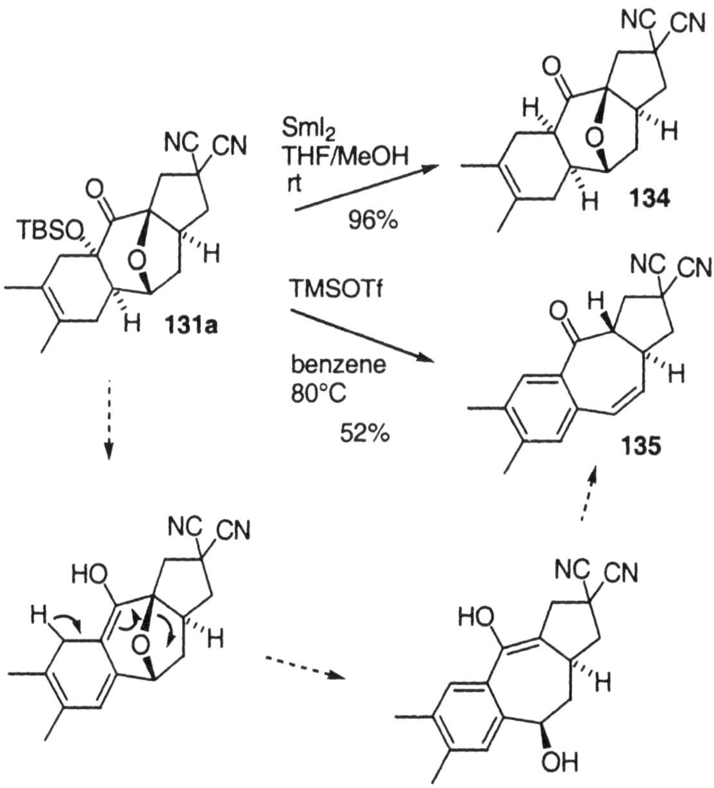

Scheme 54.

terms of an initial elimination of the hydroxy group followed by γ-enolization of the ketone, aromatization-driven opening of the oxa-bridge, and final dehydration of the resulting benzylic alcohol (Scheme 54).

Undoubtedly, further exploration of these processes is needed in order to gain more insights that might help to discover appropriate conditions for controlling the opening of the oxa-bridge. We are also currently investigating other alternatives based on incorporating a double bond within the furan-portion of the oxabicycles in order to use it as open-triggering device.

VII. ASYMMETRIC INDUCTION IN [5+2] PYRONE–ALKENE CYCLOADDITIONS

It is without doubt that one of the major current and future challenges in the construction of 8-heterobicyclo[3.2.1]octanes using annulation methods is related to the development of asymmetric versions that can allow access to enantiopure compounds. Unfortunately, not much progress has been made in this regard,[66] particularly in the field of [5+2] cycloadditions. The introduction of chiral auxiliaries in the dipolarophile has brought about modest levels of facial selectivity in cycloaddition with 3-oxidopyridinium betaines.[67] The most recent report by Kozikowski shows that moderate levels of facial diastereoselectivity can be achieved in the cycloaddition of oxidopyridinium ylide **135** with (*R*)-*p*-tolyl vinyl sulfoxide (Scheme 55).[67b]

We briefly investigated an alternative approach to achieve diastereoselectivity in oxidopyridinium-alkene cycloadditions based on the introduction of the chiral auxiliary in the pyridinium dipole instead of the dipolarophile, by taking advantage of the relative ease of preparing pyridones from pyrones. To that end the pyridone **136**, which bears a chiral group attached to the nitrogen, was readily assembled from maltol. The reaction of its 4-methoxypyridinium ylide derivative with *N*-

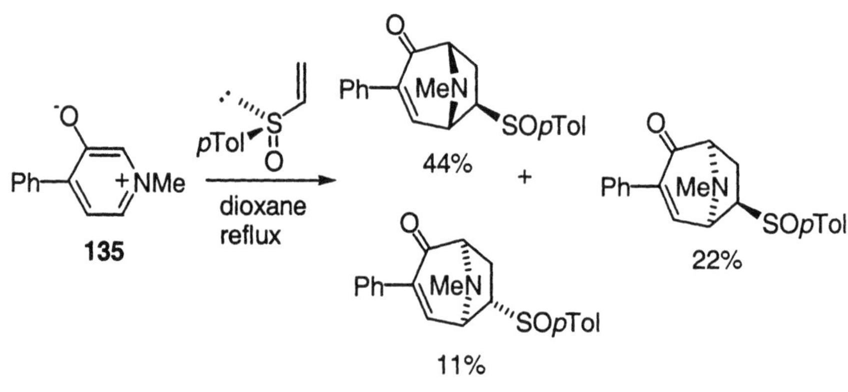

Scheme 55.

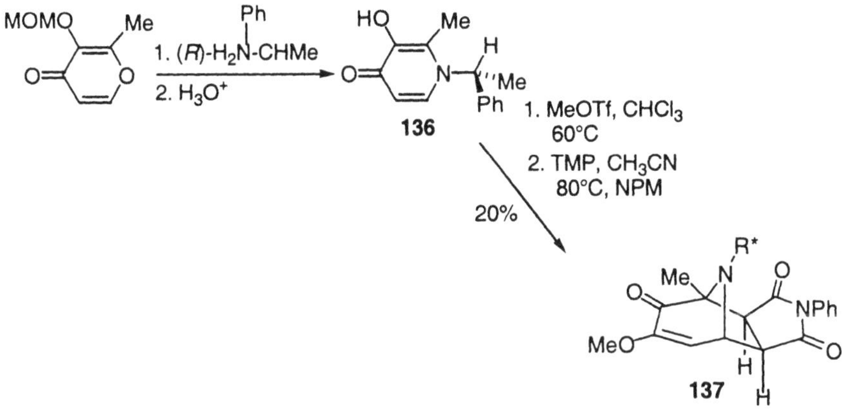

Scheme 56.

phenylmaleimide (NPM) led to a single *exo*-cycloadduct, although it could only be isolated in a 20% yield (Scheme 56).[36] The structure of the diastereoisomer obtained has not yet been confirmed, but most probably is that arising from attack of the dipolarophile from the less hindered face of the more stable conformer of the betaine (Figure 14). In any event, a useful implementation of this approach for asymmetric induction will require the development of more efficient alternatives, which will probably rely on incrementing the reactivity of the system by, for instance, carrying out the reaction intramolecularly with activated dipolarophiles.

Several precedents indicating that the presence of stereocenters in the tether connecting the reactants induces high diasterofacial selectivity in intramolecular oxidopyrylium–alkene annulations,[21,25] led us to consider an asymmetric extension of the temporary tethering strategies for [5+2] pyrone–alkene cycloadditions based on the introduction of a stereogenic element (X = chiral auxiliary) on the disposable

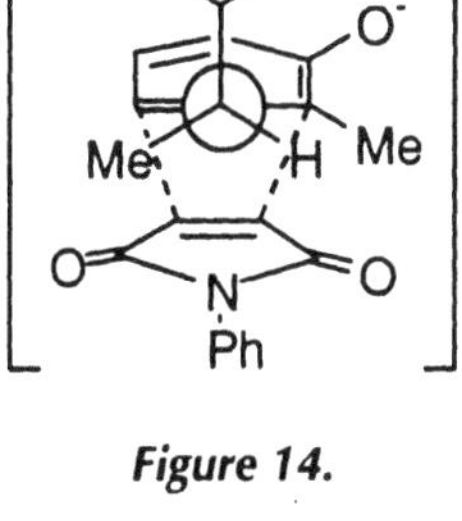

Figure 14.

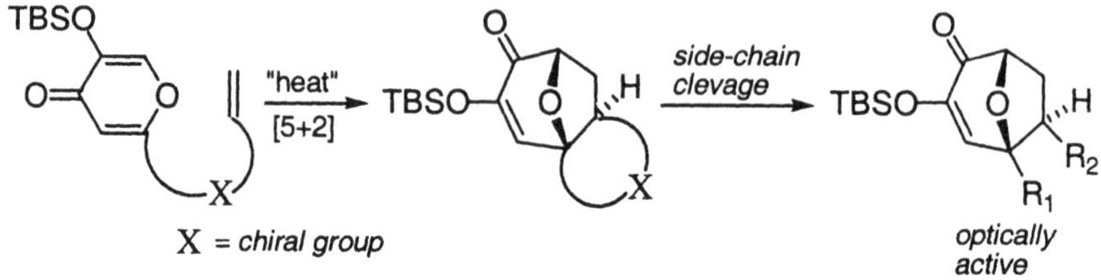

Figure 15.

tether (Figure 15). Since the linker connecting the reactants is removed after the cycloaddition, the approach would potentially provide a direct route to enantiomerically pure oxabicycles.

The availability of thioether **66** prompted us to examine whether the corresponding sulfoxide could participate in the cycloaddition and induce diastereofacial selectivity. Although sulfoxides have been widely used as asymmetric inducers, particularly in intermolecular Diels–Alder reactions, to the best of our knowledge incorporation of the sulfinyl group in the tether connecting the reacting partners was unprecedented. We found that heating a solution of **138** at 120 °C in a sealed tube gave an excellent yield of a 3:1 mixture of the expected diastereoisomeric cycloadducts **139** and **140** (Scheme 57).[37] A possible explanation for the stereochemical outcome of the annulation is that formation of the minor product (**140**) entails a disfavored interaction between the sulfinyl oxygen and the oxygen bridge of the pyrone (Figure 16). Unfortunately, all attempts to improve the diastereoselectivity by changing the solvent or using additives proved unsuccessful.

Although the degree of stereochemical induction achieved was modest, it was our expectation that altering the electronic or steric features of the disposable stereocenter may provide better results. Unfortunately, our initial investigations addressing the introduction of a larger *N*-tosylsulfimide group were unsuccessful as a consequence of an extremely easy 2,3-sigmatropic rearrangement occurring on the presumably formed sulfimide to give the corresponding sulfonamide **141** (Scheme 58).[68]

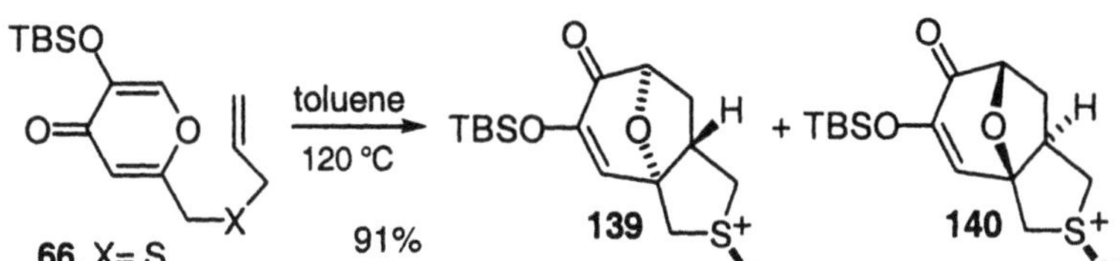

66, X= S
138, X= S→O
139
140

Figure 57.

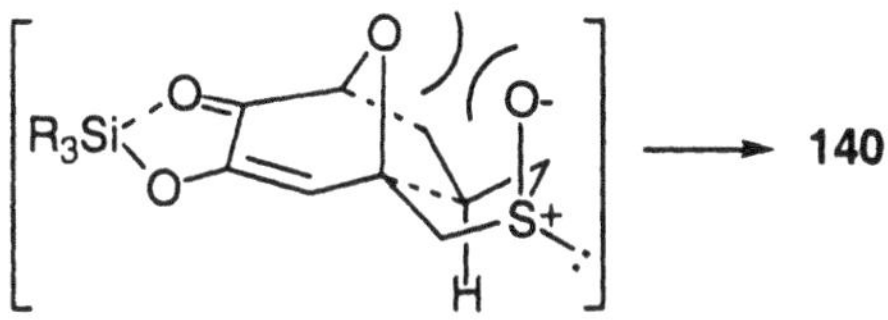

Figure 16.

Other approaches currently under scrutiny in our laboratory are focused on placement of the inducing stereocenter in a different position on the tether and on the use of chiral migrating groups.

Finally, a recent report by Ohkata and coworkers describing moderate levels of diastereofacial induction in the intramolecular cycloaddition of pyrone **142** by placing chiral auxiliaries in the side-chain esters is worthy of mention (Scheme 59).[69] These authors were able to induce the cycloaddition at room temperature by using TBSOTf and 2,6-lutidine as promoting agents, with the reaction most probably occurring via a 3-oxidopyrylium ylide mechanism similar to that operating when MeOTf is used (see Section IV.A).

A considerable amount of research effort still needs to be undertaken before a general, practical asymmetric [5+2] cycloaddition methodology can be implemented.

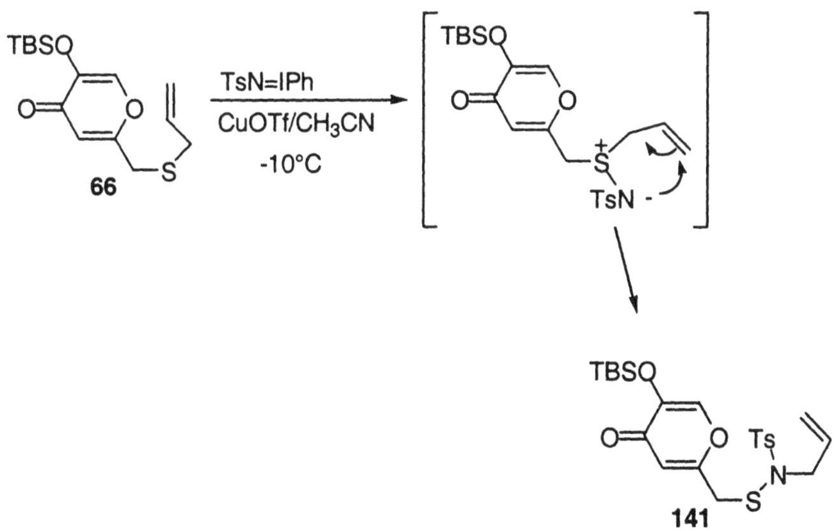

Scheme 58.

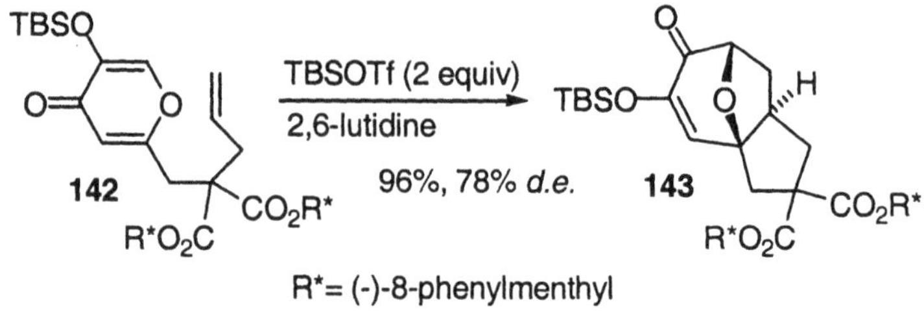

Scheme 59.

VIII. CONCLUSIONS

As postulated at the outset of our research, β-hydroxy-γ-pyrones did have a rich synthetic potential as a result of their capability to participate as five-carbon partners in [5+2] cycloaddition with alkenes. We have already demonstrated that the high functionalization of the resulting adducts offers unique possibilities for divergent

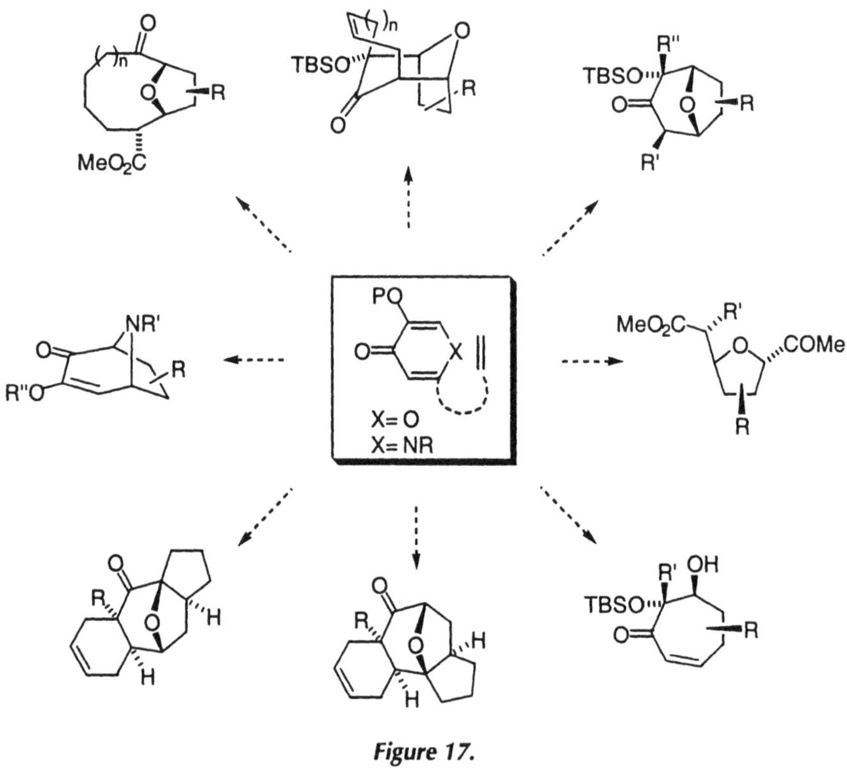

Figure 17.

manipulation, and hence for gaining practical access to a variety of valuable multifunctional cyclic skeletons (Figure 17).

Further studies addressing the development of efficient asymmetric versions of the reaction, the feasibility of using catalysis, and the implementation of specific synthetic applications are in progress.

ACKNOWLEDGMENTS

A great percentage of the work described above, developed in our laboratories in Santiago de Compostela, was carried out by a single Ph.D. student, my former coworker Antonio Rumbo, who is currently in a postdoctoral position in Germany. I am very grateful to him as well as to my present Ph.D. students J. Ramón Rodríguez and Fernando López, who are responsible for the more recent developments. Ignacio Pérez and Delia Muñoz also made a significant contribution to some aspects of the work during their brief stays in the laboratory. I would also like to thank Professor Luis Castedo and Antonio Mouriño, my Ph.D. mentors, for their teaching and continuous support. Finally, I must say that most of the synthetic work carried out in the group arose from my postdoctoral stay at Stanford in the group led by Professor Paul Wender, to whom I am deeply indebted, among other things, for being the main influence responsible for my current attitude toward organic synthesis. Finally I would like to thank the Spanish Ministry of Education and the Government of Galicia for providing the funding for the research.

REFERENCES AND NOTES

1. (a) Acher, T. D.; Buszek, K. R.; Fang, F. G.; Forsyth, C. J.; Jung, S. H.; Kishi, Y.; Matelich, M. J.; Scola, P. M.; Spero, D. M.; Yoon, S. K. *J. Am. Chem. Soc.* **1992**, *114*, 3162. (b) Wender, P. A.; Badham, N. F.; Conway, S. P.; Floreancig, P. E.; Glass, T. E.; Houze, J. B.; Krauss, N. E.; Lee, D.; Marquess, D. G.; McGrane, P. L.; Meng, W.; Natchus, M. G.; Shuker, A. J.; Sutton, J. C.; Taylor, R. E. *J. Am. Chem. Soc.* **1997**, *119*, 2757. (c) Nicolaou, K. C.; Yang, Z.; Shi, G.-Q.; Gunzner, J. L.; Agrios, K. A.; Gärtner, P. *Nature* **1998**, *392*, 265.
2. (a) Seebach, D. *Angew. Chem., Int. Ed. Engl.* **1990**, *29*, 1320. (b) Hanessian, S. *Pure Appl. Chem.* **1993**, *65*, 1189. (c) Cornforth, J. W. *Aldrichimica Acta* **1994**, *27*, 71.
3. (a) Hudlicky, T.; Natchus, M. G. In *Organic Synthesis: Theory and Applications*; Hudlicky, T., Ed.; JAI: Greenwich, CT, 1993, Vol. 2, p. 1. (b) Wender, P. A.; Miller, B. L. In *Organic Synthesis: Theory and Applications*; Hudlicky, T., Ed.; JAI: Grennwich, CT, 1993, Vol. 2, p. 27. (c) Wender, P. A.; Handy, S. T.; Wright, D. L. *Chem. Ind.* **1997**, 766. (d) Hendrickson, J. B. *Chem. Tech.* **1998**, September, p. 35. (e) Sheldon, R. A. *Chem. Tech.* **1994**, March, p. 38. (f) Hall, N. *Science* **1994**, *266*, 32. For a discussion on atom-economy, see: (g) Trost, B. M. *Science* **1991**, *254*, 1471. (h) Trost, B. M. *Angew. Chem., Int. Ed. Engl.* **1995**, *34*, 259.
4. Bertz, S. H.; Sommer, T. J. In *Organic Synthesis: Theory and Applications*; Hudlicky, T., Ed.; JAI Press: Greenwich, CT, 1993, Vol. 2, p. 67.
5. (a) Carruthers, W. *Cycloaddition Reactions in Organic Synthesis;* Pergamon: Oxford, 1990. (b) Ghosez, L. In *Stereocontrolled Organic Synthesis* Trost, B. M., Ed.; Blackwell Science: Oxford, 1994, p. 193. (c) Dell, C. P. *J. Chem. Soc., Perkin Trans. 1* **1998**, 3873 and references therein.
6. Oppolzer, W. In *Comprehensive Organic Synthesis*; Trost, B. M.; Fleming, I. A., Eds.; Pergamon: New York, 1991, Vol. 5, Chap. 4.1.
7. *1,3-Dipolar Cycloaddition Chemistry*; Padwa, A., Ed.; Wiley: New York, 1984.

8. For reviews and leading references, see: (a) Fisher, N. H.; Olivier, E. J.; Fisher, H. D. *Fortschr. Chem. Org. Naturst.* **1979**, *38*, 47. (b) Heathcock, C. H.; Graham, S. L.; Pirrung, M. C.; Pavoac, F.; White, C. T. In *The Total Synthesis of Natural Products*; ApSimon, J., Ed.; Wiley: New York, 1983, Vol. 5, p. 333. (c) Rigby, J. H. In *Studies in Natural Products Chemistry*; Rahman, A., Ed.; Elsevier: Amsterdam, 1988, Vol. 12, p. 233. (d) Tanis, S. P.; Robinson, E. D.; McMills, M. C.; Watt, W. *J. Am. Chem. Soc.* **1992**, *114*, 8349. (e) Fraga, B. M. *Nat. Prod. Rep.* **1996**, *13*, 307. (f) Lee, E.; Lim, J. W.; Yoon, C. H.; Sung, Y.; Kim, Y. K.; Yun, M.; Kim, S. *J. Am. Chem. Soc.* **1997**, *119*, 8391 and references cited therein.
9. (a) Desimoni, G.; Tacconi, G.; Barco, A.; Pollini, G. P. In *Natural Product Synthesis through Pericyclic Reactions*; Marjorie, C., Ed.; ACS Monograph: Washington, DC, 1983, p. 255. (b) Noyori, R.; Hayakawa, Y. *Org. React.* **1983**, *29*, 163. (c) Hoffman, H. M. R. *Angew. Chem., Int. Ed. Engl.* **1984**, *23*, 1. (d) Mann, J. *Tetrahedron* **1986**, *42*, 4611. (e) Föhlisch, B.; Krimmer, D.; Gehrlach, E.; Käshammer, D. *Chem. Ber.* **1988**, *121*, 1585. (f) Hoffman, H. M. R.; Wagner, D.; Wartchow, R. *Chem. Ber.* **1990**, *123*, 2131. (g) Hosomi, A.; Tominaga, Y. In *Comprehensive Organic Synthesis*; Trost, B. M.; Fleming, I. A., Eds.; Pergamon: New York, 1991, Vol. 5, Chap. 5.1. (h) Hoffman, H. M. R. In *Encyclopedia of Reagents for Organic Synthesis*; Paquette, L. A., Ed.; Wiley: 1995, Vol. 7, p. 4591. (i) Walters, M. A.; Arcand, H. R.; Lawrie, D. J. *Tetrahedron Lett.* **1995**, *36*, 23. (j) Harmata, M.; Gamlath, C. B.; Barnes, C. L.; Jones, D. E. *J. Org. Chem.* **1995**, *60*, 5077. (k) Harmata, M. In *Advances in Cycloaddition*; Lautens, M., Ed.; JAI: Greenwich, CT, 1996, Vol. 4, p. 41. (l) Harmata, M.; Elomari, S.; Barnes, C. L. *J. Am. Chem. Soc.* **1996**, *118*, 2860. (m) Jin, S.; Choi, J.-R.; Oh, J.; Lee, D.; Cha, J. K. *J. Am. Chem. Soc.* **1995**, *117*, 10914. (n) Rigby, J. H.; Pigge, F. C. *Org. React.* **1997**, *51*, 351. (o) Harmata, M. *Tetrahedron* **1997**, *53*, 6235.
10. (a) Molander, G. A.; Cameron, K. O. *J. Am. Chem. Soc.* **1993**, *115*, 830. (b) Molander, G. A.; Eastwood, P. R. *J. Org. Chem.* **1995**, *60*, 8382. (c) Davies, H. M. L.; Clark, D. M.; Alligood, D. B.; Eiband, G. R. *Tetrahedron* **1987**, *43*, 4265. (d) Davies, H. M. L. *Tetrahedron* **1993**, *49*, 5203. (e) Marson, C. M.; Khan, A.; McGregor, J.; Grinter, T. J. *Tetrahedron Lett.* **1995**, *36*, 7145. (f) Marson, C. A.; Campbell, J.; Hursthouse, M. B.; Malik, K. M. A. *Angew. Chem., Int. Ed. Engl.* **1998**, *37*, 1122.
11. Garst, M. E.; McBride, B. J.; Douglass, J. G. *Tetrahedron Lett.* **1983**, *24*, 1675.
12. (a) Walls, F.; Padilla, P.; Joseph-Nathan, P.; Giral, F.; Romo, J. *Tetrahedron Lett.* **1965**, 1577. (b) Joseph-Nathan, P.; Mendoza, V.; Garcia, E. *Tetrahedron* **1977**, *33*, 1573. (c) Sanchez, I. H.; Yañez, R.; Enriquez, R.; Joseph-Nathan, P. *J. Org. Chem.* **1981**, *46*, 2818.
13. (a) Büchi, G.; Mak, C. P. *J. Am. Chem. Soc.* **1977**, *99*, 8073. (b) Büchi, G.; Chu, P. S. *J. Org. Chem.* **1978**, *43*, 3717.
14. Mak, C. P.; Büchi, G. *J. Org. Chem.* **1981**, *46*, 3.
15. (a) Yamamura, S.; Shizuri, H.; Shigemori, Y.; Okuno, Y.; Ohkubo, M. *Tetrahedron* **1991**, *47*, 635. (b) Takakura, H.; Toyoda, K.; Yamamura, S. *Tetrahedron Lett.* **1996**, *37*, 4043.
16. (a) Engler, T. A.; Combrink, K. D.; Takusagawa, F. *J. Chem. Soc., Chem. Commun.* **1989**, 1573. (b) Engler, T. A.; Letavic, M. A.; Combrink, K. D.; Takusagawa, F. *J. Org. Chem.* **1990**, *55*, 5810. (c) Engler, T. A; Combrink, K. D.; Letavic, M. A.; Lynch, K. O.; Ray, J. E. *J. Org. Chem.* **1994**, *59*, 6567.
17. (a) Collins, J. L.; Grieco, P. A.; Walker, J. K. *Tetrahedron Lett.* **1997**, *38*, 1321. (b) Grieco, P. A.; Walker, J. K. *Tetrahedron* **1997**, *53*, 8975.
18. For a recent review, see: Wender, P. A.; Dore, T. M. In *CRC Handbook of Organic Photochemistry and Photobiology*; Horspool, W. M.; Song, P.-S., Eds.; CRC: Boca Raton, 1995, p. 280.
19. (a) Dennis, N.; Katritzky, A. R.; Takeuchi, Y. *Angew. Chem., Int. Ed. Engl.* **1976**, *15*, 1. For a review of cycloaddition reactions of heteroaromatic six-membered rings, including oxidopyridinium and oxidopyrylium systems, see: (b) Katritzky, A. R.; Dennis, N. *Chem. Rev.* **1989**, *89*, 827.
20. (a) Bromidge, S. M.; Archer, D. A.; Sammes, P. G. *J. Chem. Soc., Perkin Trans. 1* **1990**, 353. (b) Bromidge, S. M.; Archer, D. A.; Sammes, P. G. *Synthesis* **1992**, 645.

21. For a review of oxidopyrylium cycloadditions, see: Sammes, P. G. *Gazz. Chim. Ital.* **1986**, *119*, 109.
22. (a) Ullman, E. F.; Milks, J. E. *J. Am. Chem. Soc.* **1962**, *84*, 1315. (b) Ullman, E. F.; Henderson, W. A. *J. Am. Chem. Soc.* **1966**, *88*, 4942.
23. (a) Hendrickson, J. B.; Farina, J. S. *J. Org. Chem.* **1980**, *45*, 3359. (b) Sammes, P. G.; Street., L. J. *J. Chem. Soc., Perkin Trans. 1*, **1983**, 1261.
24. Sammes, P. G.; Street., L. J. *J. Chem. Soc., Chem. Commun.* **1983**, 666.
25. (a) Wender, P. A.; Rice, K. D.; Schnute, M. E. *J. Am. Chem. Soc.* **1997**, *119*, 7897. (b) Wender, P. A.; Jesudason, C. D.; Nakahira, H.; Tamura, N.; Tebbe, A. L.; Ueno, Y. *J. Am. Chem. Soc.* **1997**, *119*, 12976.
26. Ibata, T.; Motoyama, T.; Hamaguchi, M. *Bull. Chem. Soc. Jpn.* **1976**, *49*, 2298.
27. (a) Padwa, A. *Acc. Chem. Res.* **1991**, *24*, 22. (b) Padwa, A.; Hornbuckle, S.; Fryxell, G. E.; Zhang, Z. J. *J. Org. Chem.* **1992**, *57*, 5747.
28. Woods, L. L. *J. Am. Chem. Soc.* **1952**, *74*, 3959.
29. (a) Hurd, C. D.; Trofimenko, S. *J. Am. Chem. Soc.* **1958**, *80*, 2526. (b) Hurd, C. D.; Sims, R. J.; Trofimenko, S. *J. Am. Chem. Soc.* **1959**, *81*, 1684.
30. Volkmann, R. A.; Weeks, P. D.; Kuhla, D. E.; Whipple, E. B.; Chmurny, G. N. *J. Org. Chem.* **1977**, *42*, 3976.
31. McBride, B. J.; Garst, M. E. *Tetrahedron* **1993**, *49*, 2839.
32. Wender, P. A.; McDonald, F. E. *J. Am. Chem. Soc.* **1990**, *112*, 4956.
33. (a) Barton, D. H. R.; Hulshof, L. A. *J. Chem. Soc., Perkin Trans. 1* **1977**, 1103. (b) West, F. G. In *Advances in Cycloaddition*; Lautens, M., Ed.; JAI: Greenwich, CT, 1996, Vol. 4, p. 1.
34. Rodríguez, J. R.; Castedo, L.; Mascareñas, J. L. Unpublished results.
35. Rumbo, A.; Castedo, L.; Mouriño, A.; Mascareñas, J. L. *J. Org. Chem.* **1993**, *58*, 5585.
36. Rumbo, A. Ph.D. Dissertation, Universidad de Santiago, 1996.
37. Rumbo, A.; Castedo, L.; Mascareñas, J. L. *Tetrahedron Lett.* **1997**, *38*, 5885.
38. For a discussion of these effects, see: Sammes, P. G.; Weller, D. J. *Synthesis* **1995**, 1205.
39. For recent reviews in the field of temporary silicon connections, see: (a) Bols, M; Scrydstrup, T. *Chem. Rev.* **1995**, *95*, 1253. (b) Fensterbank, L.; Malacria, M.; Sieburth, S. M. *Synthesis* **1997**, 813. (c) Gauthier, D. R.; Zandi, K. S.; Shea, K. J. *Tetrahedron* **1998**, *54*, 2289.
40. Nogueira, E.; Guitián, E.; Castedo, L.; Castiñeiras, A. *Aust. J. Chem.* **1997**, *50*, 751.
41. Wender, P. A.; Mascareñas, J. L. *J. Org. Chem.* **1991**, *56*, 6267.
42. Wender, P. A.; Mascareñas, J. L. *Tetrahedron Lett.* **1992**, *33*, 2115.
43. For leading references to tropane alkaloids, see: (a) Fodor, G.; Dharanipragada, R. *Nat. Prod. Rep.* **1994**, *11*, 443. (b) Lounasma, M. In *The Alkaloids*; Brossi, A., Ed.; Academic: New York, 1988, Vol. 33, pp. 1–81. (c) Lounasma, M.; Tamminen, T. In *The Alkaloids*; Cordell, G. A., Ed.; Academic: New York, 1993, Vol. 44, p. 1. (d) Stoelwinder, J.; Roberti, M.; Kozikowski, A. P.; Johnson, K. M.; Bergmann, J. S. *Bioorg. and Med. Chem. Lett.* **1994**, *4*, 303.
44. Rumbo, A.; Mouriño, A.; Castedo, L.; Mascareñas, J. L. *J. Org. Chem.* **1996**, *61*, 6114.
45. (a) Katritzky, A.; Dennis, N.; Chaillet, M.; Larrieu, C.; Mouhtadi, M. E. *J. Chem. Soc., Perkin Trans. 1* **1979**, 408. (b) Hamaguchi, M.; Matsuura, H.; Nagai, T. *J. Chem. Soc., Chem. Commun.* **1982**, 263.
46. Kingsbury, C. A.; Cliffton, M.; Looker, J. H. *J. Org. Chem.* **1976**, *41*, 2777.
47. Köbrich, G. *Annalen* **1961**, *648*, 114.
48. Muñoz, D. M. S. Dissertation, Universidad de Santiago, 1997.
49. (a) Mascareñas, J. L.; Pérez, I.; Rumbo, A.; Castedo, L. *Synlett* **1997**, 81. (b) Pérez, I. M. S. Dissertation, Universidad de Santiago, 1996.
50. An exhaustive review of the different techniques to open this type of bridge has recently appeared: Chiu, P.; Lautens, M. In *Topics in Current Chemistry*; Springer-Verlag: Berlin, 1997, Vol. 190, p. 1.

51. (a) Williams, D. R.; Benbow, J. W.; McNutt, J. G.; Allen, E. E. *J. Org. Chem.* **1995**, *60*, 833. (b) Molander, G. A.; Eastwood, P. R. *J. Org. Chem.* **1995**, *60*, 8382.
52. Mascareñas, J. L.; Rumbo, A.; Castedo, L. *J. Org. Chem.* **1997**, *62*, 8620.
53. Rodríguez, J. R.; Rumbo, A.; Castedo, L.; Mascareñas, J. L. Presented at the 10th IUPAC OMCOS Symposium, Versailles (France), July 1999.
54. Barbosa, L. C. A.; Rubinger, M. M. M.; Mann, J.; Mansell, H. L. *Tetrahedron* **1996**, *52*, 11297.
55. (a) Lautens, M.; Kumanovic, S. *J. Am. Chem. Soc.* **1995**, *117*, 1954. (b) Marson, C. M.; McGregor, J.; Khan, A.; Grinter, T. J. *J. Org. Chem.* **1998**, *63*, 7833, and references cited therein.
56. (a) Lautens, M. *Pure Appl. Chem.* **1992**, *64*, 1873. (b) Lautens, M. *Synlett* **1993**, 177.
57. Rodríguez, J. R.; Rumbo, A.; Castedo, L.; Mascareñas, J. L. *J. Org. Chem.* **1999**, *64*, 4560.
58. (a) Klein, L. L. *J. Am. Chem. Soc.* **1985**, *107*, 2573. (b) Klein, L. L.; Shanklin, M. S. *J. Org. Chem.* **1988**, *53*, 5202. (c) Dillon, M. P.; Lee, N. C.; Stappenback, F.; White, J. D. *J. Chem. Soc., Chem. Commun.* **1995**, 1645.
59. Rigby, J. H.; Fales, K. R. *Tetrahedron Lett.* **1998**, *39*, 1525, and references cited therein.
60. (a) Connolly, J. D.; Hill, R. A. *Dictionary of Terpenoids*; Chapman & Hall, London, 1991, Vol. 2, p. 1005. (b) Faulkner, D. J. *Nat. Prod. Rep.* **1996**, 106.
61. Nicolaou, K. C.; Xu, J. Y.; Pfefferkorn, J.; Ohshima, T.; Vourloumis, D.; Hosokawa, S. *J. Am. Chem. Soc.* **1998**, *120*, 8661.
62. For a detailed account, see: Connolly, J. D.; Hill, R. A. *Dictionary of Terpenoids*; Chapman & Hall, London, 1991, Vol. 2.
63. (a) Hirsch, J. *J. Heterocycl. Chem.* **1975**, *12*, 785. (b) Hsung, R. P. *J. Org. Chem.* **1997**, *62*, 7904.
64. For recent reviews on tandem processes, see: (a) Ho, T.-L. *Tandem Organic Reactions*; Wiley: New York, 1992. (b) Tietze, L. F.; Beifuss, U. *Angew. Chem., Int. Ed. Engl.* **1993**, *32*, 131. (c) Tietze, L. F. *Chem. Rev.* **1996**, *96*, 115. (d) Grigg, R. *Tetrahedron Symposia-In-Print Number 62*, **1996**, *52*, 11385. (e) Bunce, R. A. *Tetrahedron* **1995**, *51*, 13103. (f) Parsons, P. J.; Penkett, C. S.; Shell, A. J. *Chem. Rev.* **1996**, *96*, 195. (g) Neuschutz, K.; Velker, J.; Neier, R. *Synthesis* **1998**, 227.
65. Rodríguez, J. R.; Rumbo, A.; Castedo, L.; Mascareñas, J. L. *J. Org. Chem.* **1999**, *64*, 966.
66. Several asymmetric [4+3] annulation methods to oxabicyclic compounds have been developed, see: (a) Davies, H. M. L.; Ahmed, G.; Churchill, M. R. *J. Am. Chem. Soc.* **1996**, *118*, 10774. (b) Lautens, M.; Aspiotis, R.; Colucci, J. *J. Am. Chem. Soc.* **1996**, *118*, 10930. (c) Harmata, M.; Jones, D. E. *J. Org. Chem.* **1997**, *62*, 1578. (d) Stark, C. B. W.; Eggert, U.; Hoffmann, H. M. R. *Angew. Chem., Int. Ed. Engl.* **1998**, *37*, 1267. (e) Harmata, M.; Jones, D. E.; Kahraman, M.; Sharma, U.; Barnes, C. L. *Tetrahedron Lett.* **1999**, *40*, 1831.
67. (a) Pham, V. C.; Charlton, J. L. *J. Org. Chem.* **1995**, *60*, 8051. (b) Araldi, G. L.; Prakash, K. R. C.; George, C.; Kozikowski, A. P. *J. Chem. Soc., Chem. Commun.* **1997**, 1875.
68. López, F. M. S. Thesis, Santiago de Compostela, 1998.
69. Ohmori, N.; Yoshimura, M.; Ohkata, K. *Heterocycles* **1997**, *45*, 2097.

METALLOCARBENOID-INDUCED CYCLIZATIONS OF ACETYLENIC DIAZO CARBONYL COMPOUNDS

Albert Padwa and Christopher S. Straub

Advances in Cycloaddition
Volume 6, pages 55–95.

ISBN: 0-7623-0531-2

I. INTRODUCTION

The chemistry of metal carbene complexes has provided chemists with exceptionally fertile ground for the design and development of new stereoselective bond formation processes for application to organic synthesis.[1] Since the pioneering work by Stork and Ficini[2] on the cyclopropanation reaction of α-diazo ketones, intramolecular reactions of metal carbene complexes have been extensively studied from both a mechanistic and synthetic viewpoint.[3] Rhodium(II) carboxylates are particularly effective catalysts for the decomposition of diazo compounds and many chemical syntheses are based on this methodology.[4] Among the more synthetically useful processes of the resulting carbenoid intermediates are intramolecular C–H insertion,[5] cyclopropanation,[6] and ylide generation.[7] Elegant and practical examples of the application of these reactions include the synthesis of gibberelin/gibberellic acid[8] and the triquinane sesquiterpenes.[9] In contrast to these processes, the corresponding reaction of alkynes with metal carbenes has been far less studied. Only in recent years has some attention been focused on the intramolecular cyclization reactions of α-diazo ketones containing tethered alkynes (i.e. **1**) in the presence of transition metal catalysts. The overall reaction observed is believed to proceed via an initial decomposition of the α-diazo ketone to generate a rhodium carbenoid intermediate **2** (Scheme 1). Attack on the carbenoid carbon by the tethered alkyne generates a new intermediate **3** in which carbene-like character has been transferred to the β-carbon of the alkyne. This intermediate vinyl carbenoid may then react further in either an *intramolecular* or *intermolecular* fashion to give novel products. This review will describe the breadth of work carried out in the author's laboratory. The chapter is divided into two parts. The first part covers systems in which the α-diazo acetyl and alkynyl groups are situated *ortho* to each other on a phenyl ring. Much of the early work in this area was carried out on systems such as these, and these studies are used to probe the mechanistic issues of the reaction as well as the scope and limitations of the cyclization process. The

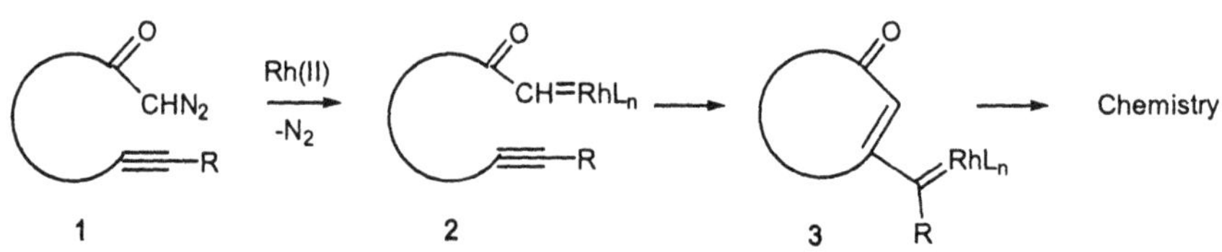

Scheme 1.

second part of the review covers systems where the conformational constriction of *ortho* substitution has been removed.

II. GENERAL MECHANISTIC CONSIDERATIONS

A. The Role of Metal and Solvent on the Cyclization Process

The mechanism of the diazo ketone–alkyne cyclization reaction has been the subject of some study over the past several years. In an early report, Hoye and Dinsmore demonstrated that the distribution of products was markedly dependent on the nature of the metal catalyst used.[10] For example, treatment of α-diazo ketoester **4** with catalytic palladium(II) acetoacetonate produced cyclopropane **5** in 78% yield, while treatment with rhodium(II) acetate provided furan **6** in 56% yield (Scheme 2). Furan **6** arises from a 1,5-electrocyclization of the initially produced vinyl carbenoid intermediate onto the adjacent carbonyl group (vide infra).[10] The fact that the chemistry of **4** is catalyst-dependent suggests that a metalated species is involved in the product-determining step. One possible mechanism to explain the products involves the initial decomposition of the α-diazo moiety to give the metal carbenoid **7**. In the next step, the rhodium metal migrates from the original diazo carbon to the alkynyl carbon via a metathesis reaction and ultimately produces metallocyclobutene **8** (Scheme 3). This intermediate could then ring-open to furnish the vinyl carbenoid **10** which goes on to afford the observed products. Another possible variation would be formation of the highly strained cyclopropene **9**. This intermediate could then be rapidly converted into the 5-*exo*-vinyl carbenoid **10** or the 6-*endo*-carbenoid **11**, both of which can undergo further chemistry.[11] This pathway has precedent from the known metal-catalyzed ring opening of cyclo-

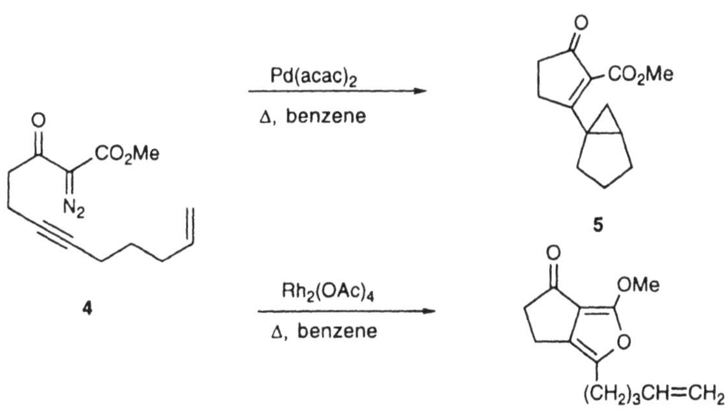

Scheme 2.

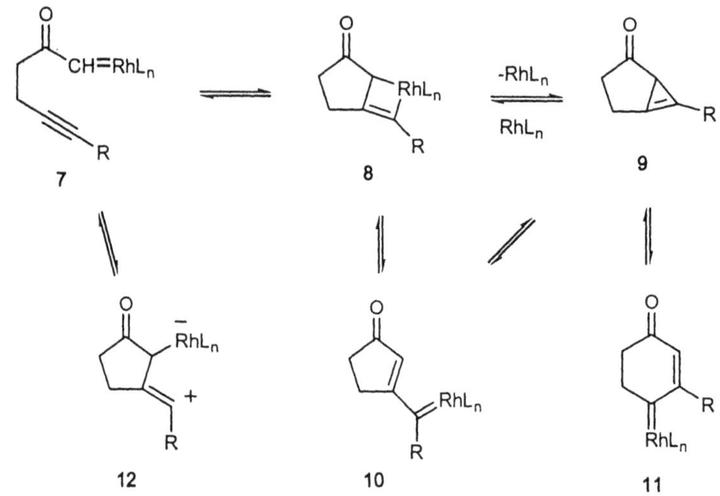

Scheme 3.

propenes to vinyl carbenes.[12] It should be noted, however, that Hoye and Dinsmore have dismissed this sequence and have suggested, instead, the formation of a zwitterionic intermediate **12** which proceeds on to the observed products (vide infra).[10,13]

In contrast to the Hoye mechanism, the intermediacy of metallocyclobutene **8** and the transfer of rhodium to the alkyne carbon to form vinyl carbenoid **10** was initially postulated in the authors' laboratory[14] and is related to work described by Jones in 1975.[15] Vinyl carbenes have been of considerable interest in recent years,[16] and these species are typically generated via pyrolysis of cyclopropenes[17] or vinyl diazomethanes[18] and have been used in a variety of reactions. The rhodium-catalyzed methodology described above represents an alternative method for generating these interesting reactive intermediates for study.[19] As was pointed out above, the major pathway according to Hoye and Dinsmore does not proceed via a metallocyclobutene intermediate. This conclusion was based upon a comparison of the product distribution obtained from the decomposition of α-diazo ketone **13** with that obtained from the isomeric diazo cyclopentenone **14** (Scheme 4). Hoye argued that the rhodium carbenoid intermediate formed from **13** should correspond to the same species as that derived from **14** if a metallocyclobutene intermediate is involved. Consequently, one would expect identical product ratios. The distinct difference in product distribution obtained from the two isomeric diazo alkynes **13** and **14** does not seem to be compatible with a common intermediate such as **10**. This observation led these authors to propose that the cyclization reaction proceeds in a stepwise fashion (vide infra).[13]

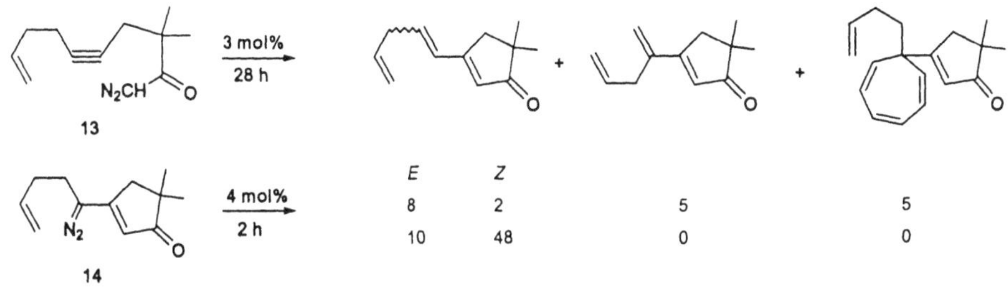

Scheme 4.

More recently, these same authors reported on the Rh(II)-catalyzed *double internal–external alkyne insertion* reaction of an acetylenic α-diazo ketone such as **15** (Scheme 5).[20] The initially formed rhodium carbenoid intermediate was suggested to undergo internal insertion into the tethered alkyne bond and this was followed by bimolecular addition to the external alkyne to produce a cyclopropenyl substituted cyclopentenone derivative **17**. Once again, migration of the rhodium metal to the remote alkyne carbon via a [2+2]-*cycloaddition/cycloreversion* path (i.e. **15** → **19**) was discounted on the basis that the distribution of products derived from **15** differed significantly from those obtained from the rhodium carbenoid

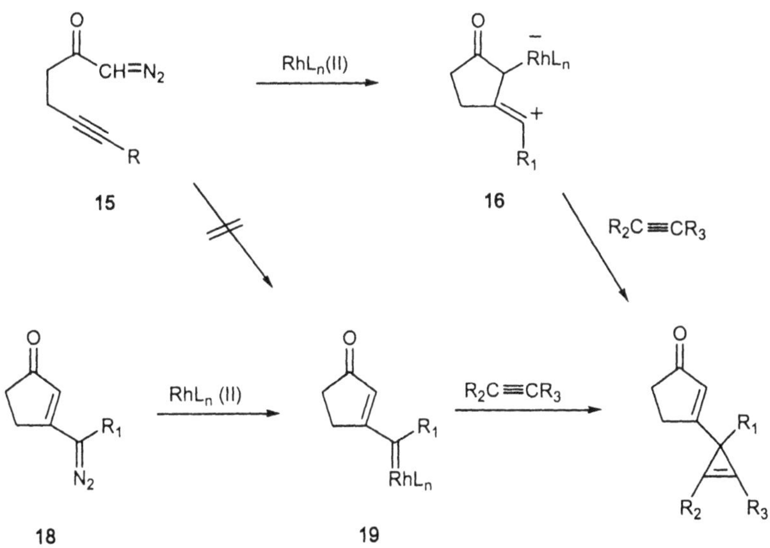

Scheme 5.

species **19**, which was generated from the vinylogous α-diazo ketone precursor **18**. Zwitterion **16** was postulated as the key intermediate in the transformation of **15** to **17**.[20]

More recent results in our laboratory showed, however, that the reaction mechanism is markedly dependent on the solvent employed in these Rh(II)-catalyzed insertion processes. Thus, treatment of **20** with a catalytic amount of rhodium(II) acetate resulted in a 2:1-mixture of the *cis*- and *trans*-alkenyl-substituted indenones **21** (85% combined yield; Scheme 6). No signs of cyclopropene **22** (< 2%) could be detected in the crude reaction mixture by NMR spectroscopy. Interestingly, when pentane was used as the solvent, cyclopropene **22** (80%) was the exclusive product. The degree of chemoselectivity that was achieved in this reaction by simply changing the solvent from dichloromethane to pentane is most remarkable.[21] A reasonable explanation that nicely accounts for the formation of indenone **21** involves stepwise cyclization of the initially formed keto carbenoid in accord with the Hoye–Dinsmore proposal to give **23** (Scheme 7). A 1,2-hydrogen shift results in the formation of allylic cation **24** and this is followed by collapse to **21** and regeneration of the rhodium catalyst. The intermediates involved in the formation of **21** are dipolar, which would explain why the formation of **21** is strongly inhibited in nonpolar solvents. When pentane is used as the solvent, metal migration occurs via the metallocyclobutene intermediates **25** and **26** so as to avoid charge buildup.[21] Thus, it would appear as though the reaction mechanism of these alkyne cyclizations is markedly dependent on the nature of the solvent used.

Hoye and Dinsmore have also dismissed the involvement of a cyclopropene intermediate such as **9**, arguing that the strain inherent in this intermediate would

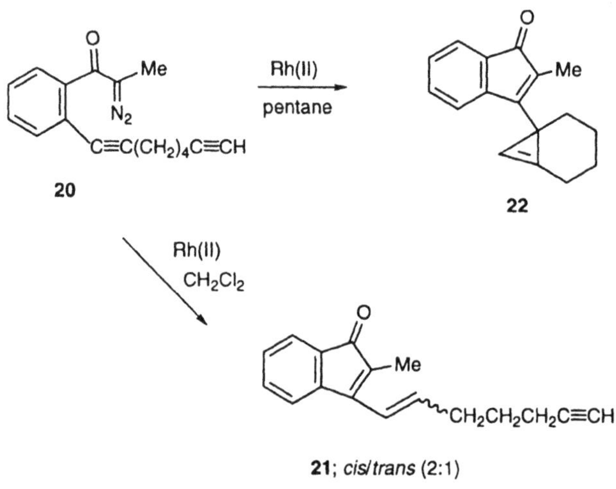

Scheme 6.

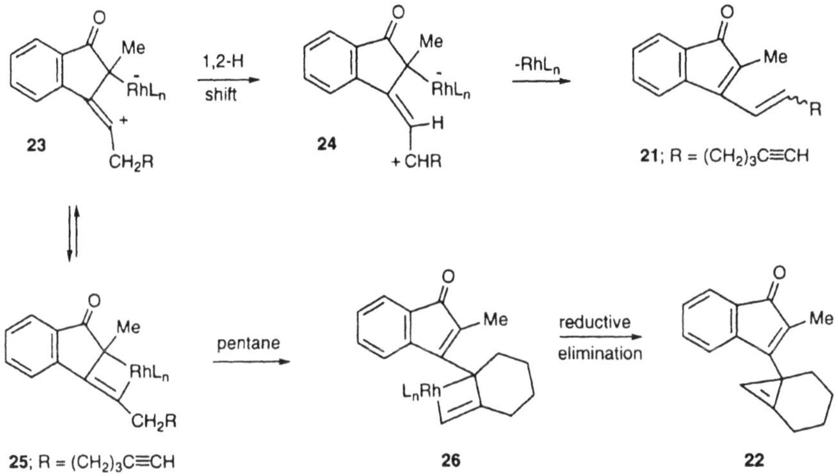

Scheme 7.

preclude its formation.[10] Our own group carried out a number of experiments designed to probe the transient existence of a strained cyclopropene in these reactions. The interception of highly strained and reactive cyclopropenes with furan by a Diels–Alder reaction is well established.[22] Consequently, the Rh(II)-catalyzed reaction of a typical *o*-alkynyl α-diazoacetophenone such as **27** in the presence of furan was carried out with the hope of trapping intermediate **30** as the Diels–Alder adduct **28** (Scheme 8). However, no such trapping product was observed.[23]

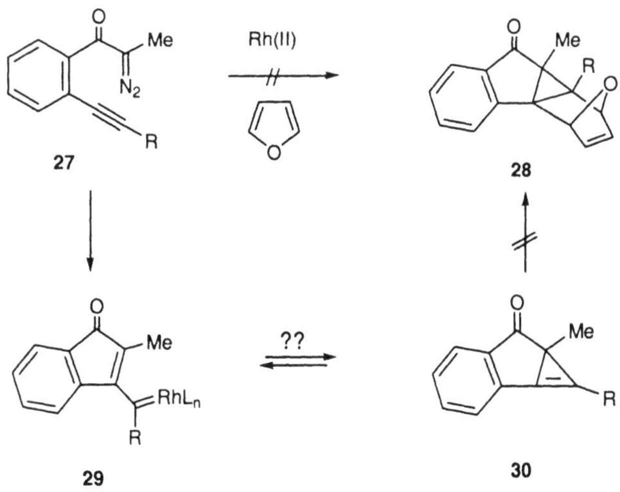

Scheme 8.

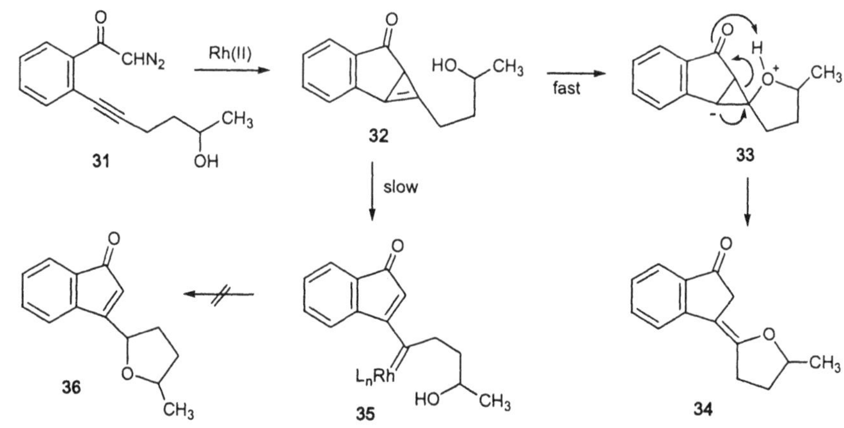

Scheme 9.

The strongest evidence for the involvement of a cyclopropene intermediate in these cyclizations comes from studies which take advantage of the rapid addition of alcohols to these reactive species.[24] Diazo alkyne **31** was allowed to react with rhodium(II) mandelate in benzene at 25 °C (Scheme 9). The only product isolated in 85% yield corresponded to structure **34**. No signs of product **36** derived from vinyl carbenoid insertion into the neighboring OH group could be detected in the crude reaction mixture. Indenone **34** can be accounted for in the following manner. Intramolecular addition of the rhodium-stabilized carbenoid onto the acetylenic

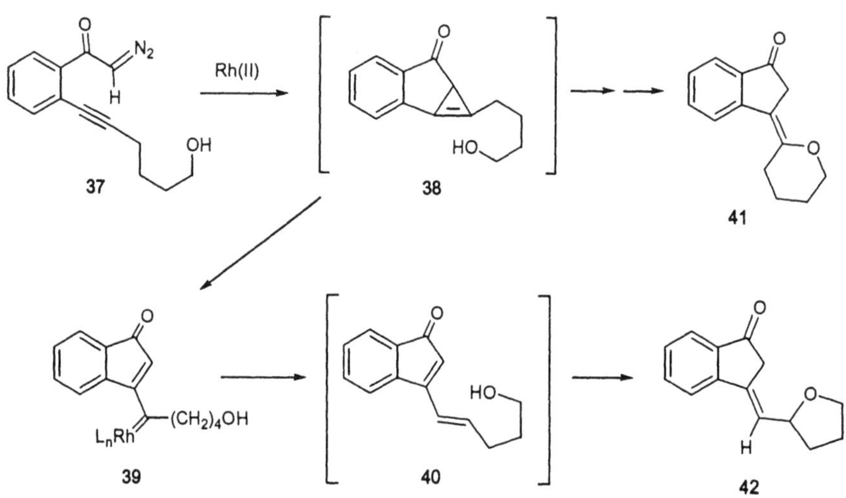

Scheme 10.

π-bond generates the highly strained cyclopropene **32**. This species is too strained to survive at ambient temperature. Instead, attack of the hydroxyl group onto the double bond occurs leading to **33** which rapidly undergoes ring cleavage to produce **34**. In this case, intramolecular nucleophilic addition of the hydroxyl group onto the cyclopropene ring is faster than ring opening to give vinyl carbenoid **35**, which if formed, would have produced indenone **36**.[19]

The behavior of the homologous α-diazo ketone **37** was also studied (Scheme 10). In this case, the rhodium(II)-catalyzed reaction afforded a 2:1-mixture of cyclic ethers **41** and **42**. Apparently, extension of the chain by one methylene unit sufficiently retards alcohol addition to the putative cyclopropene ring, thereby allowing a competing process to occur. In addition to trapping by the adjacent alcohol to form **41**, the initially formed cyclopropene **38** can also undergo a Rh(II)-induced ring opening to produce vinyl carbenoid **39**. Compound **42** is formed from **39** by way of a 1,2-hydrogen shift to first produce **40**. This highly activated dienone undergoes rapid internal conjugate addition with the neighboring alcohol to produce the indenylidene tetrahydrofuran **42**.[11]

B. The Role of the Ligand Groups on the Metal

Further complicating the mechanistic picture is the occasional dependence of the process on the ligand groups attached to the rhodium metal. In certain cases, no ligand dependence was noted as was found in the case of diazo ketone **43** (Scheme 11). Varying the ligands from the electron-withdrawing trifluoroacetate group to the more electron donating mandelate or acetate ligand always afforded a 60% yield of indenone **44** as the only observed product.[11] In other systems, however, the cyclization reaction was found to be ligand-dependent. For example, treatment of α-diazoacetophenone **45** with a Rh(II) catalyst in benzene at 25 °C afforded mixtures of the δ-CH insertion product **46** (1:2-*cis/trans*) as well as the *Z*-substituted alkene **47** derived from 1,2-hydride migration to the carbenoid center (Scheme 12). Varying the catalyst from rhodium(II) octanoate to rhodium(II) perfluorobutyrate caused a significant change in product distribution.[23] The data (Table 1) indicates that the more electron-withdrawing perfluorobutyrate ligand favors

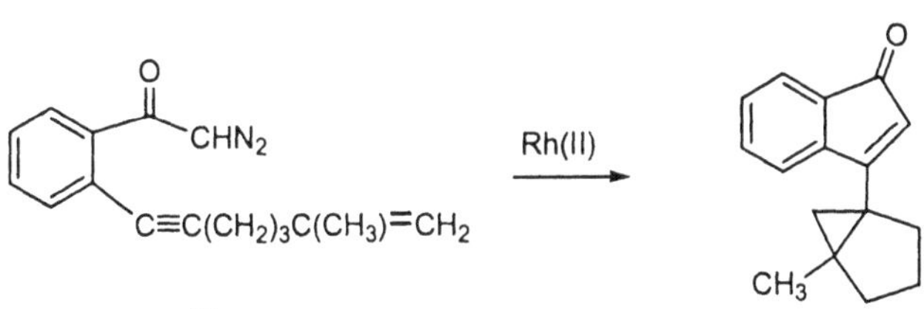

43 **44**

Scheme 11.

Table 1

$Rh_2(oct)_4$	80/20
$Rh_2(OAc)_4$	66/33
$Rh_2(man)_4$	50/50
$Rh_2(pfb)_4$	10/90

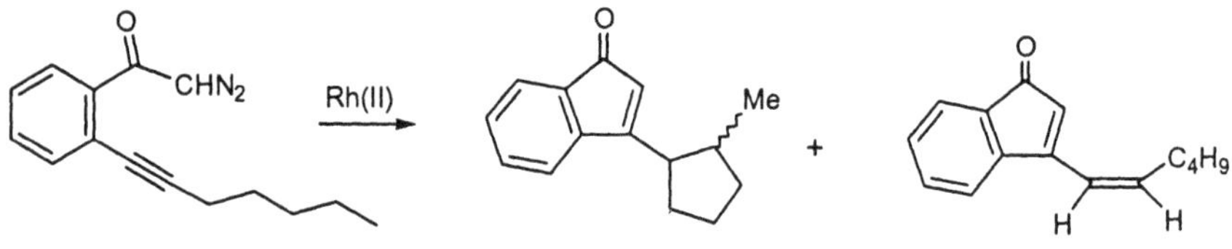

45 46 47

Scheme 12.

1,2-hydride migration, whereas the 1,5-insertion reaction is favored by the more electron-donating ligands. It has been well recognized that rhodium carbenoid intermediates are highly electron deficient at the carbon center and are destabilized by the presence of an electron-withdrawing ligand on the metal. Thus, when the ligand corresponds to an electron-withdrawing group, the entropically less demanding 1,2-hydride migration pathway is favored. Related ligand dependencies have been found with other systems.[21,25] Another point worth noting is the preferential

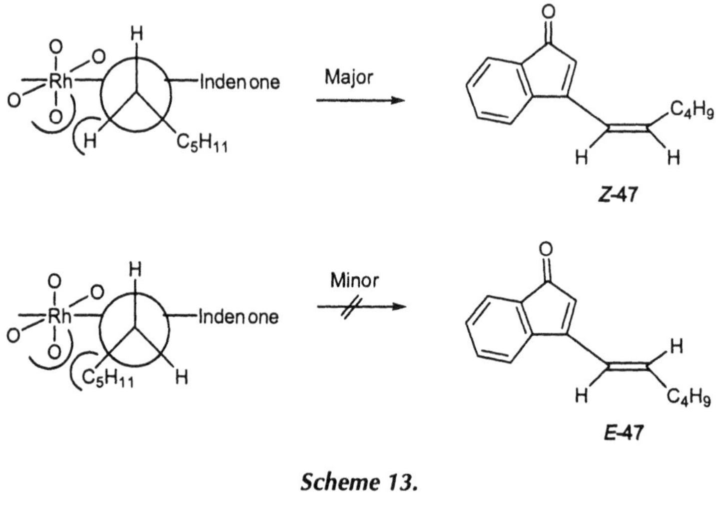

Scheme 13.

formation of the thermodynamically less stable Z-isomer **47** which is formed from the 1,2-hydride shift. This stereochemical result has been attributed to steric constraints associated with the orientation of the alkyl chain for hydride migration in the metal carbene intermediate (Scheme 13).[23,26]

C. 5-*exo* versus 6-*endo* Selectivity

Still an additional point worth mentioning is the preference for formation of the 5-*exo*-vinyl carbenoid **10** over the 6-*endo*-vinyl carbenoid **11**. Although most work published on these systems show a propensity for 5-*exo*-vinyl carbenoid formation, there are some exceptions.[13,20,23,27] In an effort to shed light on the source of this preference, we carried out some studies dealing with the nature of the substituent group on the alkyne and how it effects the mode of cyclization. Our initial efforts focused on the rhodium(II)-catalyzed reaction of 2-ethynyl-α-diazoacetophenone **48** (Scheme 14). The reaction of **48** with rhodium(II) mandelate in methanol or isopropanol afforded naphthols **49** or **50** as the only products formed in good yield. When the reaction was carried out using benzene as the solvent, 4-phenyl-1-naphthol (**51**) was obtained in 70% yield.[23]

Experimental results[28] as well as theoretical MO calculations indicate that thermodynamic factors are not important in influencing the 5-*exo* versus 6-*endo* cyclization selectivity with the *o*-alkynyl-substituted α-diazoacetophenone system.[23] Rather, the regioselectivity of cyclization seems to be highly dependent on the nature of the substituent group attached to the alkyne tether. The 5-*exo* versus 6-*endo* selectivity appears to be primarily due to steric interactions between the substituent group on the alkyne and the ligand groups present on the catalyst. When the keto carbenoid intermediate **52** possesses a terminal hydrogen (R = H), products derived primarily from 6-*endo* closure are observed (Scheme 15). On the other hand, alkynyl-substituted α-diazo ketones, which contain a terminal alkyl group (R = alkyl), produce products derived from 5-*exo* cyclization. Steric factors combined with the ability of the substituent group (R) to stabilize the vinylogous carbenoid intermediate **53** seemingly determines the chemoselectivity.[23,27]

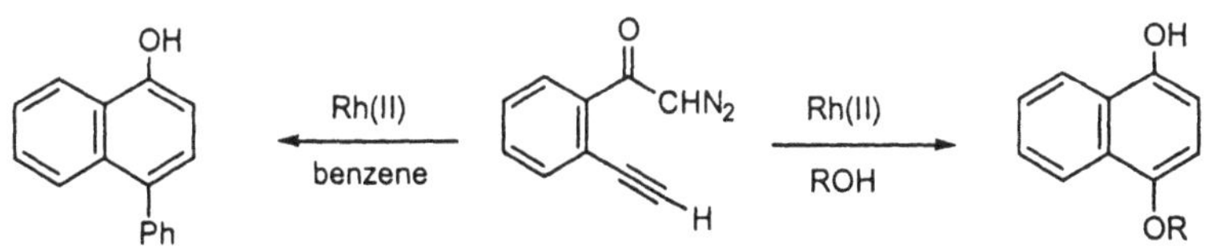

51 48

49; R=Me
50; R=$(CH_3)_2CH$

Scheme 14.

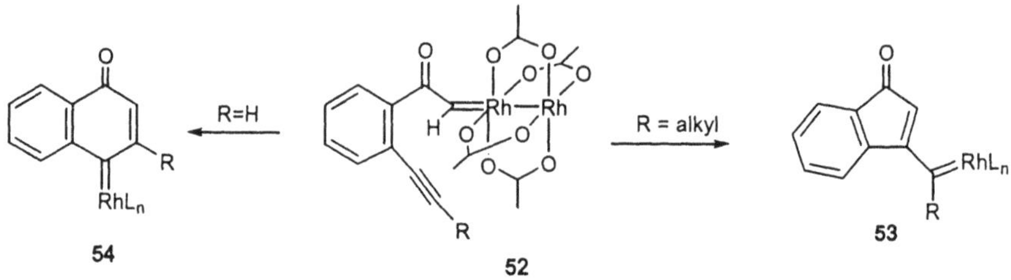

Scheme 15.

III. RHODIUM CATALYSIS VERSUS PHOTOCHEMICAL OR THERMAL CYCLIZATION

Another interesting aspect of these systems is the difference in behavior of the rhodium-catalyzed reaction relative to what happens on thermolysis or photolysis of the same compounds.[29,30] A typical example involves treating diazo ketone **55** with a catalytic quantity of rhodium(II) acetate in the presence of 2 equiv of ethyl vinyl ether (Scheme 16). The major product corresponds to cyclopropane **56** (91%) which is derived from a transient vinyl carbenoid adding to the π-bond of the vinyl ether. The thermal reaction of **55** differs significantly from the transition metal catalyzed process. Thus, heating a sample of **55** in a sealed tube at 130 °C gave naphthol **57** in 69% yield. The same naphthol was also produced from the photolysis of **55**.[30] The formation of the naphthol ring can be explained in terms of a photochemical Wolff rearrangement[31,32] of the initially formed carbene which rearranges to give an *o*-alkynyl-substituted arylketene intermediate **59** (Scheme 17). Cyclization of the ketene leads to diradical **60** which subsequently abstracts hydrogen from the solvent to give the phenol.[30] Supporting evidence for the proposed mechanism was obtained by carrying out the irradiation of diazo ketone **55** in the presence of 2 equiv of methanol. Isolation of methyl ester **61** in high yield from this experiment provides excellent support for the arylketene intermediate.[30]

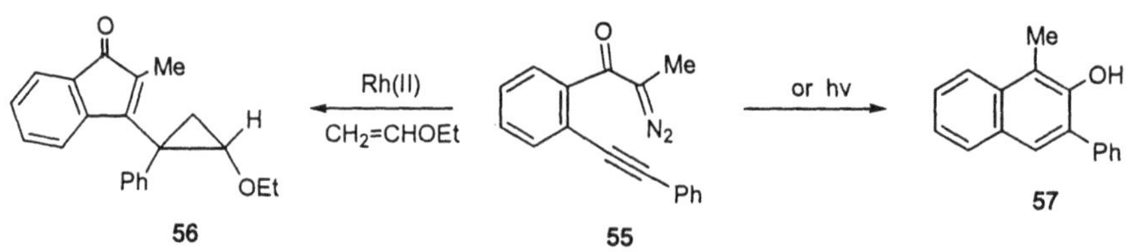

Scheme 16.

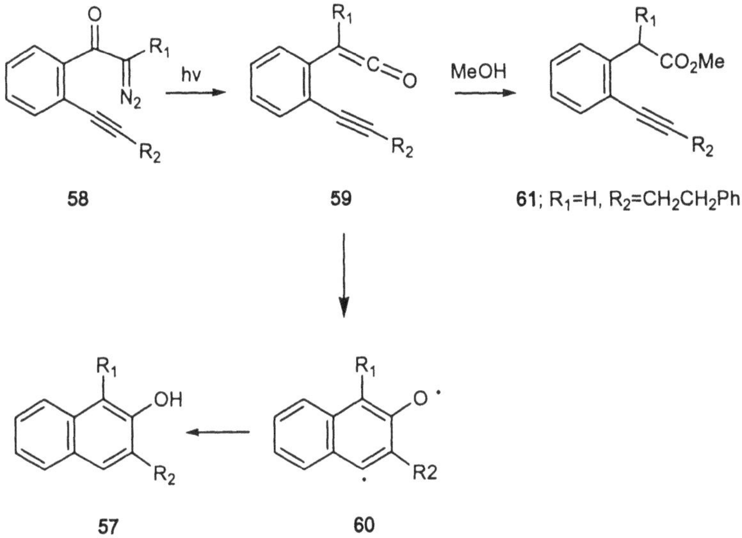

Scheme 17.

IV. REACTIONS OF CONFORMATIONALLY CONSTRAINED SYSTEMS

Regardless of the precise mechanistic pathway followed by these cyclizations, vinyl carbenoid intermediates such as **63** are seemingly formed from the Rh(II)-catalyzed reaction (Scheme 18). This cyclization represents a novel approach for creating substituted cyclopentenones from acyclic precursors. Added to the arsenal of other synthetic methods such as the Nazarov cyclization and the Pauson–Khand $Co_2(CO)_8$-mediated cyclization, this reaction expands the availability of these biologically important ring systems.[33] Our group has exploited intermediates such as **63** to obtain a broad spectrum of products ranging from simple cyclopropanation and insertion adducts, to polycyclic heterocycles resulting from ylide formation and

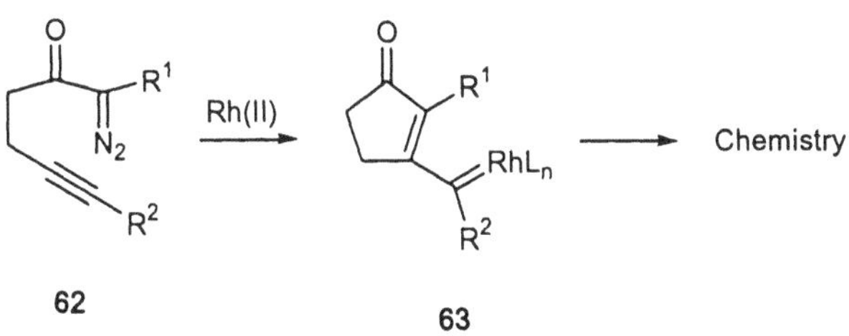

Scheme 18.

subsequent 1,3-dipolar cycloadditions. Some of the results obtained are outlined below.

A. Cyclopropanation

Many of the earlier systems studied involved trapping the vinyl carbenoid intermediate as a cyclopropane via reaction with external or tethered alkenes.[10,13,14,34,35] What is most interesting about this reaction is the formation of three new rings in one step from an acyclic precursor. A typical example is outlined in Scheme 19. Treatment of **55** with a catalytic quantity of rhodium(II) acetate in the presence of 2 equiv of vinyl ether afforded cyclopropane **56** in 91% yield.[34]

In the case of intramolecular trapping with tethered alkenes, two basic structural variations are possible. These will depend on the point of attachment of the alkenyl group and each variation will lead to very different cyclization products. In Type I systems, the alkenyl group is tethered onto the alkynyl carbon atom and this is illustrated with α-diazo ketone **64** (Scheme 20). Treatment of **64** with a catalytic quantity of rhodium(II) acetate gave indenone **65** in 60% yield.[14]

On the other hand, for Type II systems, the alkenyl group is tethered onto the diazo carbon as in diazo ketone **66** (or **67**) (Scheme 21). In these cases, bicycloalkanes **69** or **70** were formed in 67 and 81% yields, respectively.[35] Attempts to trap the cyclized carbenoid intermediate **68** by carrying out the reaction in the presence of a large excess of ethyl vinyl ether failed to give bimolecular adducts such as **71**.

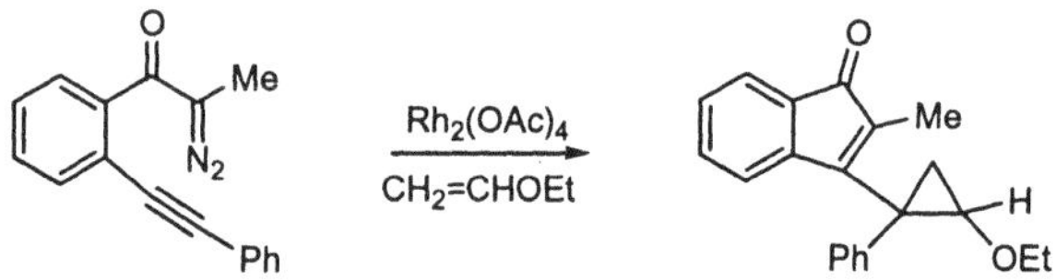

55 56

Scheme 19.

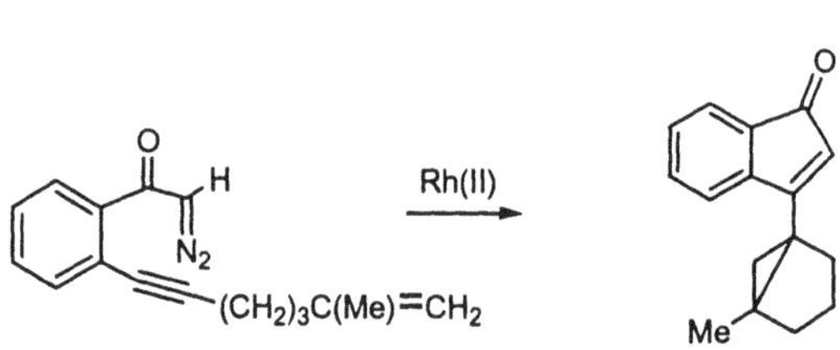

64 65

Scheme 20.

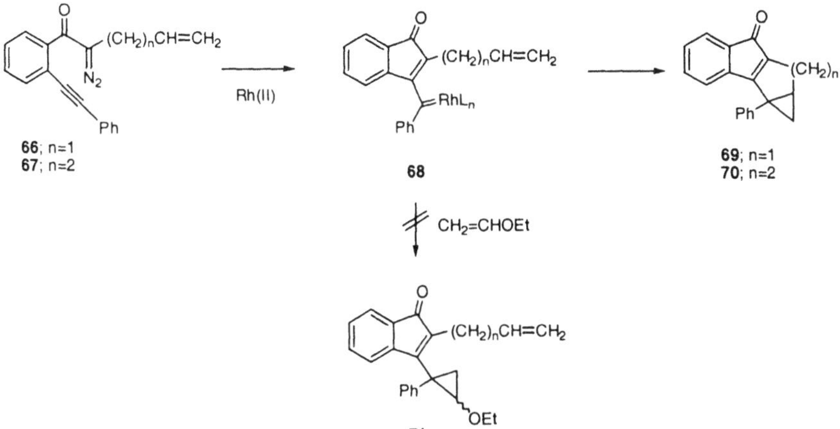

66; n=1
67; n=2
68
69; n=1
70; n=2
71

Scheme 21.

This observation indicates that bimolecular trapping of the cyclized rhodium carbenoid is significantly slower than intramolecular cyclopropanation.[35]

A number of experiments designed to probe the scope and generality of the intramolecular alkyne cyclopropanation reaction were carried out in an effort to exploit this tandem sequence as a synthetic method. Initial efforts focused on the rhodium(II)-catalyzed reaction of *o*-(6,8-nonadien-1-ynyl)-α-diazoacetophenone (**72**). Treatment of **72** with a catalytic quantity of rhodium(II) mandelate gave cycloheptadiene **74** in 58% yield. In a similar manner, treating the closely related diazo ketone **73** (R = CH_3) with rhodium(II) mandelate also gave cyclopent[*g*]azulenone **75** (Scheme 22).[11] The formation of the fused cycloheptadienes **74** and **75** can be rationalized by assuming that the reaction proceeds through the divinylcyclopropane intermediate **77**. When the internal double bond of the diene possesses the *E* geometry, intramolecular cyclopropanation gives rise to a *cis*-divinyl cyclopropane, which rapidly undergoes a Cope rearrangement[36] under the conditions used.[11] It should be noted that intramolecular cyclopropanation of dienes by simple carbenoids followed by rearrangement of the resulting vinylcyclopropane has been effectively used in several elegant syntheses.[37] The overall process is also closely related to work by Davies who developed a synthesis of fused seven membered carbocycles based on a formal intramolecular [3+4] cycloaddition of vinyl carbenoids with dienes.[38]

B. Migration and Insertions

Carbenes are known to insert into C–H bonds and rearrange by hydrogen or alkyl group migration.[39] Often these processes are in competition with each other and therefore mixtures of products result. A typical example of this competition was

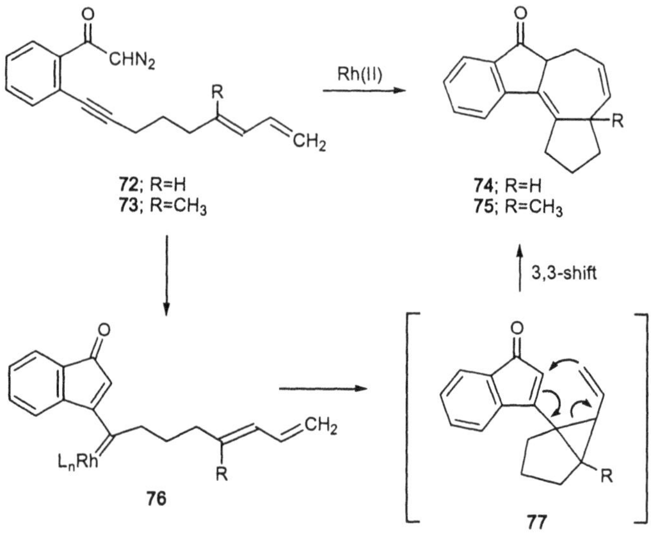

Scheme 22.

encountered on treatment of **45** with a catalytic quantity of rhodium(II)octanoate at 25 °C in benzene (Scheme 23). Under these conditions, a 1:1-mixture of indenones **46** and **47** was formed in 94% yield. The observed products are derived from vinyl carbenoid **78** which reacts by way of either a 1,2-hydrogen shift to give **47** or insertion into the δ-hydrogen bond to produce **46**.[19]

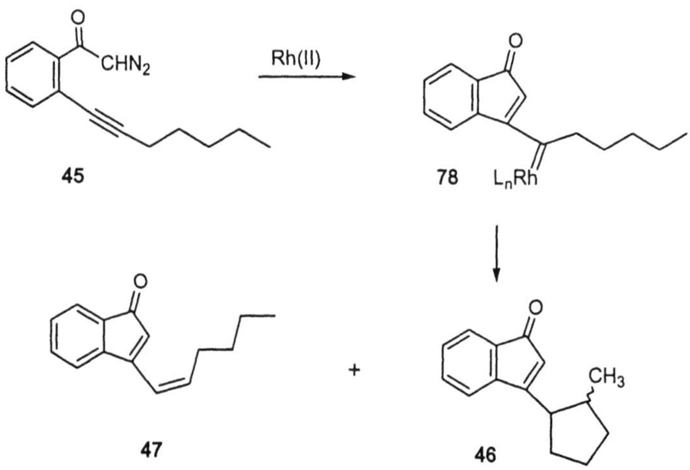

Scheme 23.

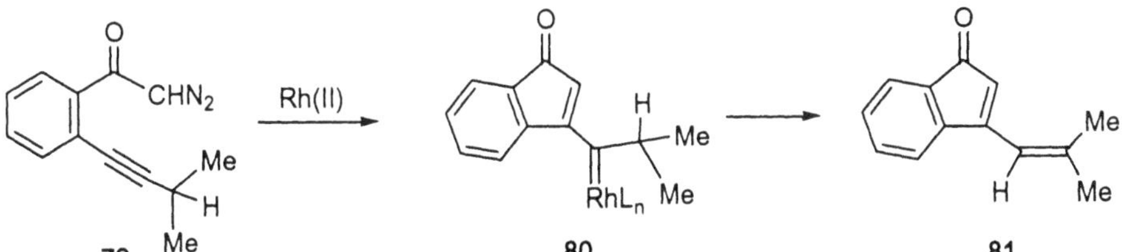

79 80 81

Scheme 24.

The formation of indenone **47** from this reaction implies that other 1,2-shifts to a carbenoid center could also occur. The 1,2-shift of a singlet carbene to form an alkene formally involves the migration of hydrogen to a vacant π-orbital.[40] It is generally assumed that the hydrogen atom carries its electron pair into the vacant orbital of the singlet carbene. The ease of migration is generally related to the alignment of the vacant π- and σ-CH orbitals.[40,41] The ability of groups to migrate to a divalent carbon generally correspond to the order: hydrogen > aryl > alkyl.[42] In the course of our own studies, we found that the treatment of **79** with rhodium(II) mandelate gave indenone **81** in 97% yield (Scheme 24). This product was expected, since migration of a hydrogen should be faster than a methyl group.

A similar reaction using α-diazo ketone **82** afforded 82% of a 5:3-mixture of the *E*- and *Z*-enol ethers **83** in 82% yield (Scheme 25). No signs of a methoxy migrated

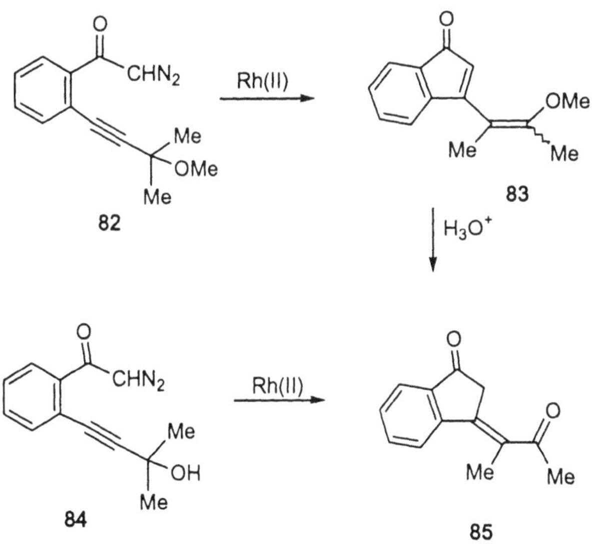

82 83 84 85

Scheme 25.

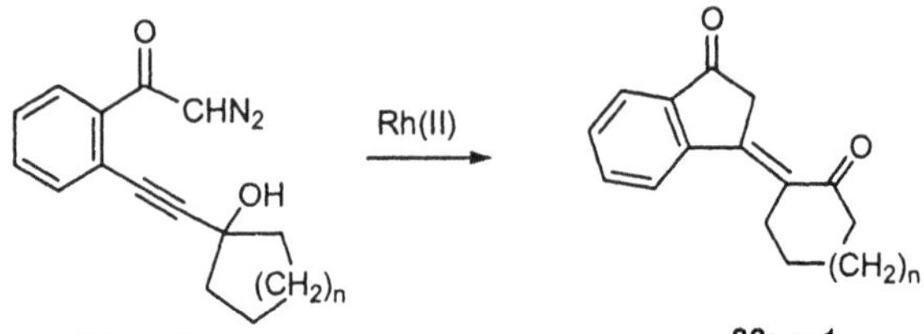

Scheme 26.

product was evident in the crude reaction mixture. The mixture of stereoisomers was cleanly hydrolyzed to indanone **85**. This same indanone could also be obtained on treatment of the related alcohol **84** with a rhodium(II) carboxylate catalyst. The preferential migration of the methyl group with both of these diazo ketones seems related to the ability of the methoxy (or hydroxy) substituent to stabilize the developing positive charge in the transition state for the rearrangement.[23]

The success achieved with the Rh(II)-catalyzed rearrangement of **84** was also extended to a series of cyclic acetylenic alcohols. Thus, treatment of (1-oxycycloalkyl)ethynyl-substituted α-diazoacetophenone **86** (or **87**) with a catalytic amount of rhodium(II) octanoate gave the ring expanded indanone **88** (or **89**) as a 1:1-*cis/trans* mixture in 82% (or 86%) yield (Scheme 26).[23]

As was mentioned in an earlier section, changing the point of tether attachment can often result in the formation of very different cyclization products (vide supra).

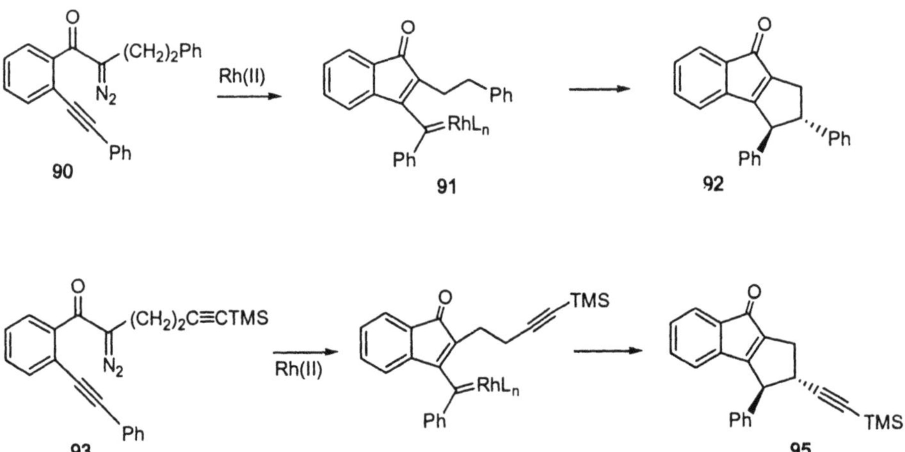

Scheme 27.

Scheme 28.

This is demonstrated in the cyclization of α-diazo ketones **90** and **93** which have also been studied in the author's laboratory.[35] The reaction of **90** and **93** with rhodium(II) acetate in benzene gave bicyclic indenones **92** and **95** in 70 and 83% yields, respectively (Scheme 27). It is interesting to note that in both of these cases, C–H insertion by the carbenoid into the aliphatic side chain competes effectively with alternative pathways such as C–H insertion into the aryl ring of **90** or cyclization onto the acetylenic π-bond of **93**.[35] It is also interesting to note that 1,2-H migration to the carbenoid center to give alkenes such as **97** are not observed in any of these Type II examples (Scheme 28).

C. Cyclopropenation

Cyclization of vinyl carbenoids to produce cyclopropenes is another common reaction that is often encountered with these systems.[43] For example, treatment of α-diazo ketone **98** with a catalytic quantity of rhodium(II) acetate afforded cyclopropene **99** in 95% yield (Scheme 29). Upon standing in the presence of oxygen, this highly strained system underwent ring opening to produce indenone **100**. Cyclopropene **99** was also found to undergo ready Diels–Alder cycloaddition with added diphenyl isobenzofuran (DPIBF) to give a 2:1 mixture of *exo*/*endo* adducts **101** in 85% yield.[29]

Interestingly, if the methyl group adjacent to the diazo center in **98** (R = Me) is replaced with a hydrogen atom as in **102** (R = H), it was not possible to isolate nor detect the suspected cyclopropene. Instead, the transient cyclopropene **103** readily underwent [2+2] dimerization to afford the novel dimer **104** in 54% yield. Bimolecular cycloaddition across the double bond in cyclopropene is known to proceed quite readily since ring strain is reduced by 26 kcal/mol.[44] The transition state associated with this [2+2] cycloaddition reaction is very sensitive to both steric and electronic factors.[45] FMO theory predicts that the preferred [2+2] cycloaddition path of **103** will involve reaction of the HOMO of the cyclopropene with the LUMO of the indenone π-bond (larger coefficients) to give the crossed dimer **104**. Introduction of a substituent group on either the indenone or cyclopropene ring results in increased steric interactions which apparently retard this mode of cycloaddition. Cyclopropene **103** was also trapped by Diels–Alder cycloaddition with DPIBF to give cycloadduct **105**.[29]

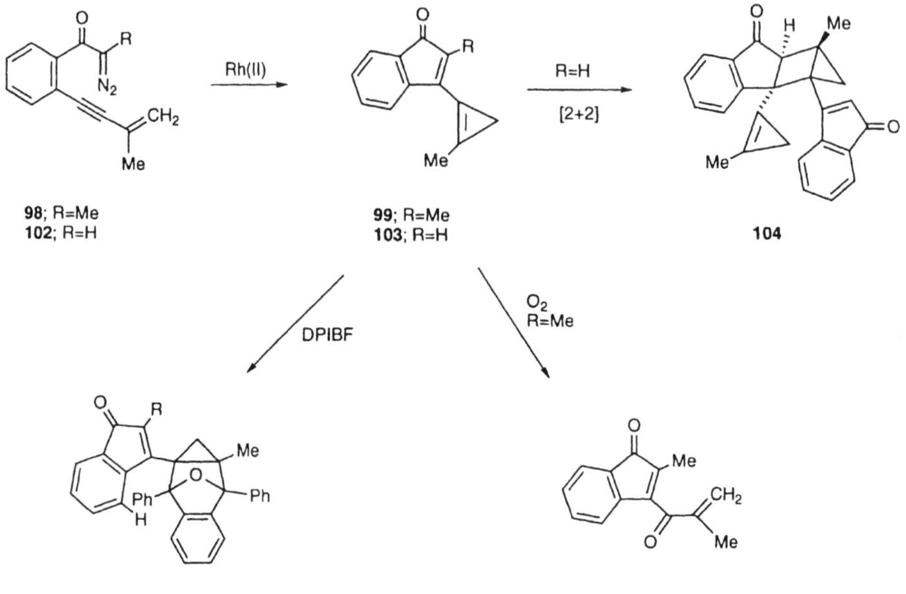

Scheme 29.

As was discussed in the first section of this review, the reaction of rhodium carbenoids with tethered alkynes has been examined in some detail so as to probe certain mechanistic features of the reaction (Schemes 5 and 6).[20,21] A double *internal/external alkyne* cyclization of acetylenic α-diazo ketone **106** with 1-hexyne was studied in our laboratory. Stirring this mixture in the presence of rhodium(II) acetate at 25 °C for 1 h afforded the novel cyclopentadiene derivative **108** in 81% yield (Scheme 30). Control experiments established that the initial product that was first formed was indenone **107**. This product is the result of the vinyl carbenoid adding across the acetylenic π-bond of 1-hexyne. When the reaction

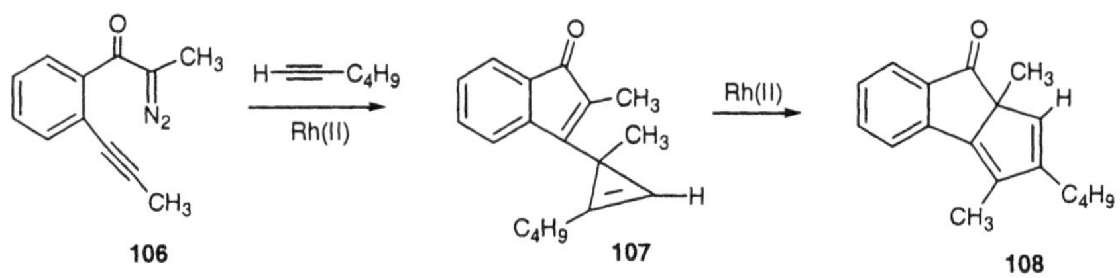

Scheme 30.

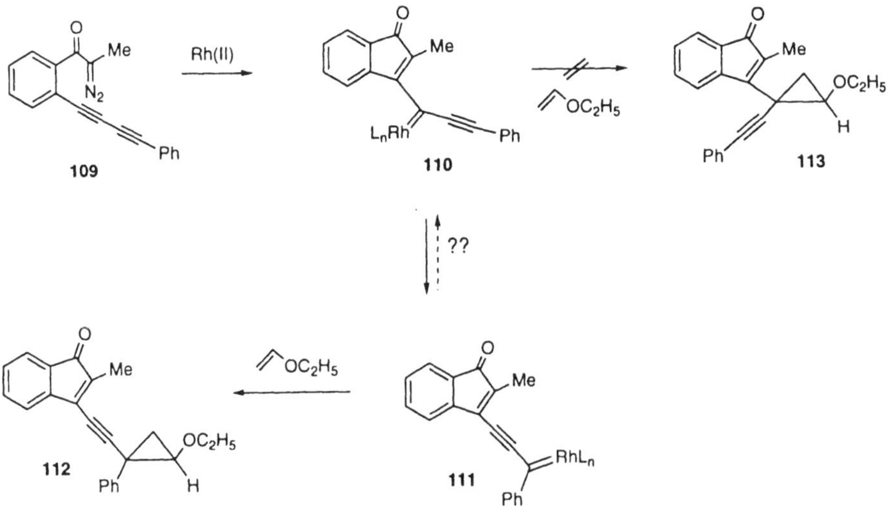

Scheme 31.

was carried out for only 10 min at 25 °C, indenone **107** could be isolated in 85% yield. Further stirring of **107** with rhodium(II) acetate induced a subsequent rearrangement and ultimately produced **108** in 92% yield.[46]

The ease with which these systems undergo the rhodium(II)-catalyzed cyclization to give cyclopropenyl-substituted indenones suggested that a similar transformation might occur with diacetylenic systems.[47] Such a study was carried out using diazo ketone **109** (Scheme 31). A critical issue is whether the cyclization will occur to give products derived from the fully rearranged carbenoid **111** or from the initially formed carbenoid **110**. In fact, treatment of **109** with a catalytic quantity of rhodium(II) acetate at 25 °C in the presence of ethyl vinyl ether afforded cyclopropane **112** with notable efficiency (90% chemical yield) and selectivity (> 95% isomeric purity). No signs of the isomeric cyclopropane **113** could be detected in the crude reaction mixture.[48] The exclusive formation of cyclopropane **112** was attributed to a slower rate of trapping of vinyl carbenoid **110** by ethyl vinyl ether, perhaps as a consequence of a more congested transition state. Another possibility is that the equilibrium between the two carbenoids lies completely in favor of the more stable phenyl-substituted isomer **111**.[47]

D. Ylide Formation and Subsequent Rearrangements

Over the past several years, our group has studied the Rh(II)-induced α-diazo ketone cyclization onto a neighboring carbonyl group followed by dipolar cycloaddition of the resulting carbonyl ylide dipole as a method for generating oxapolycy-

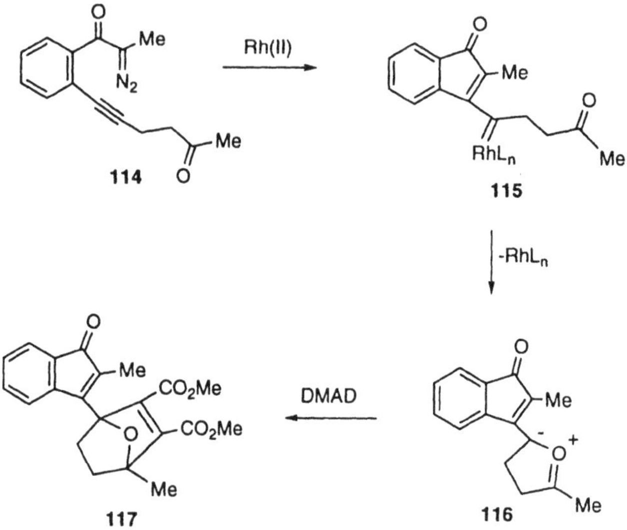

Scheme 32.

clic ring systems.[49] The ease with which α-diazo ketones containing tethered carbonyl groups undergo this tandem process suggested that a similar sequence could also occur with a vinylogous keto carbenoid. In order to test this possibility, the Rh(II)-catalyzed behavior of diazo ketone **114** was studied (Scheme 32). Treatment of **114** with a catalytic amount of rhodium(II) octanoate in the presence of 1 equiv of dimethyl acetylenedicarboxylate afforded cycloadduct **117** in 97% yield. This result can be accounted for in terms of the intermediacy of vinyl carbenoid **115** which cyclizes onto the oxygen atom of the neighboring carbonyl group to give the resonance-stabilized dipole **116**. Dipolar cycloaddition of **116** across the activated π-bond of DMAD affords cycloadduct **117**.[50]

The above domino transformation can also be performed intramolecularly by attaching the trapping agent directly to the carbonyl group. Thus, treatment of diazo ketone **118** with rhodium(II) acetate produced cycloadduct **119** in excellent yield (Scheme 33).[50]

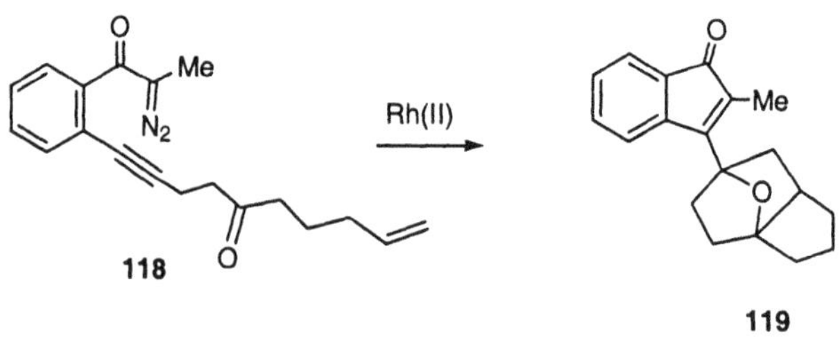

Scheme 33.

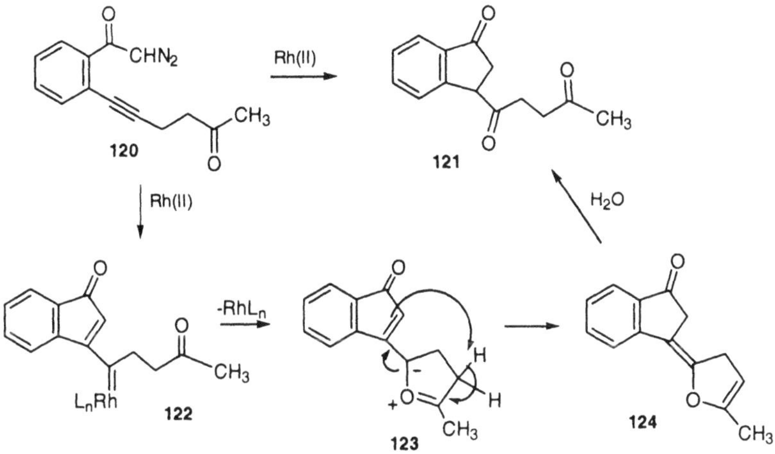

Scheme 34.

Replacing the methyl substituent of α-diazo ketone **114** with a hydrogen atom retarded the bimolecular trapping of the intermediate dipole, and instead a competing process occurred. Thus, the rhodium(II)-catalyzed reaction of diazo ketone **120** with a variety of trapping dipolarophiles did not produce a cycloadduct derived from a carbonyl ylide intermediate. Instead, the only product formed (60%) corresponded to 3-(1,4-dioxo-1-pentyl)-1*H*-indanone (**121**, Scheme 34).[11] It would appear that the rhodium-catalyzed reaction of **120** proceeds by the initial formation of vinyl carbenoid **122** and this is followed by cyclization to give carbonyl ylide **123**. This dipole rapidly undergoes charge dispersal to produce enol ether **124**[51] which is eventually hydrolyzed to give **121**. With this system, internal proton transfer from the carbonyl ylide dipole **123** to produce dione **124** is faster than bimolecular dipolar cycloaddition with an external dipolarophile.[11]

In contrast to the rhodium(II) catalyzed reaction of **120**, treatment of the homologous diazo ketone **125** with a catalytic amount of rhodium(II) mandelate in the presence of *N*-phenylmaleimide afforded a 1:1-mixture of cycloadducts **126** and

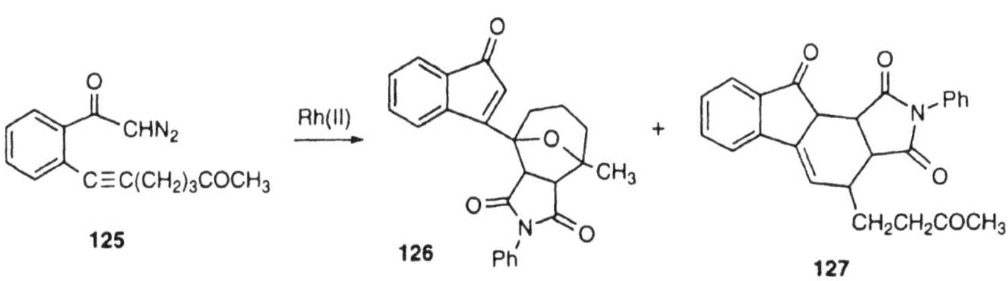

Scheme 35.

127 (Scheme 35).[11] The formation of **126** can be rationalized in a fashion analogous to that invoked to explain the formation of **117**. Lengthening the distance between the diazo ketone and the carbonyl group sufficiently retards the internal hydrogen transfer process, thereby allowing the carbonyl ylide dipole to undergo bimolecular cycloaddition to give **126**. The mechanism by which α-diazo ketone **125** undergoes reaction with *N*-phenylmaleimide to produce trioxoindeno[2,1-*e*]isoindole (**127**) is of considerable interest. Two fundamentally different mechanistic paths seem possible and these are presented in Scheme 36. Path A involves nucleophilic addition of the vinyl carbenoid **128** onto the activated π-bond of *N*-phenylmaleimide to produce zwitterion **129**. A subsequent 1,2-hydrogen shift to generate the more stable allylic cation **130** would have to proceed at a faster rate than bond closure in order to rationalize the formation of **127**. The alternative path B first involves a 1,2-hydrogen shift of vinyl carbenoid **128** to produce diene **131** as a transient species which then undergoes a subsequent Diels–Alder reaction with *N*-phenylmaleimide.[11]

Formation of sulfonium ylides derived from the interaction of a vinyl carbenoid with a sulfur lone pair of electrons has also been examined. The reaction of

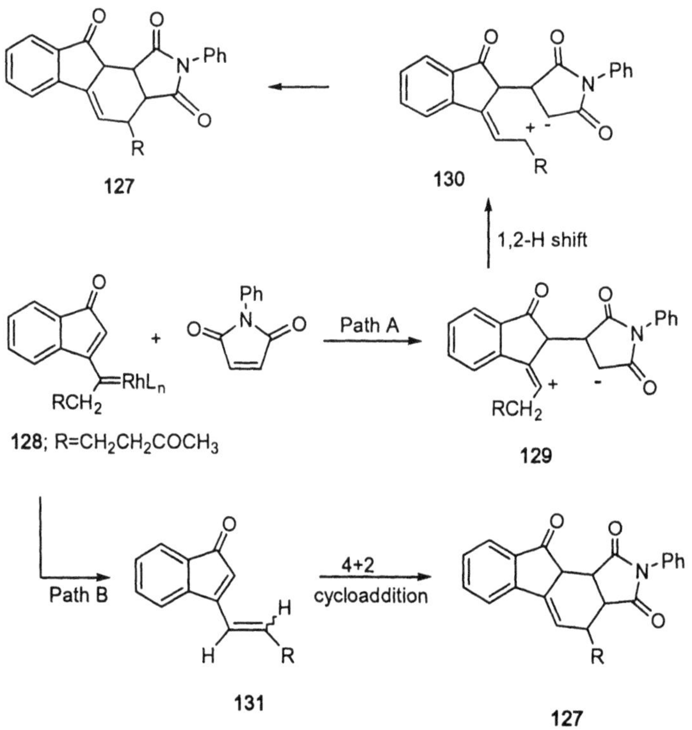

Scheme 36.

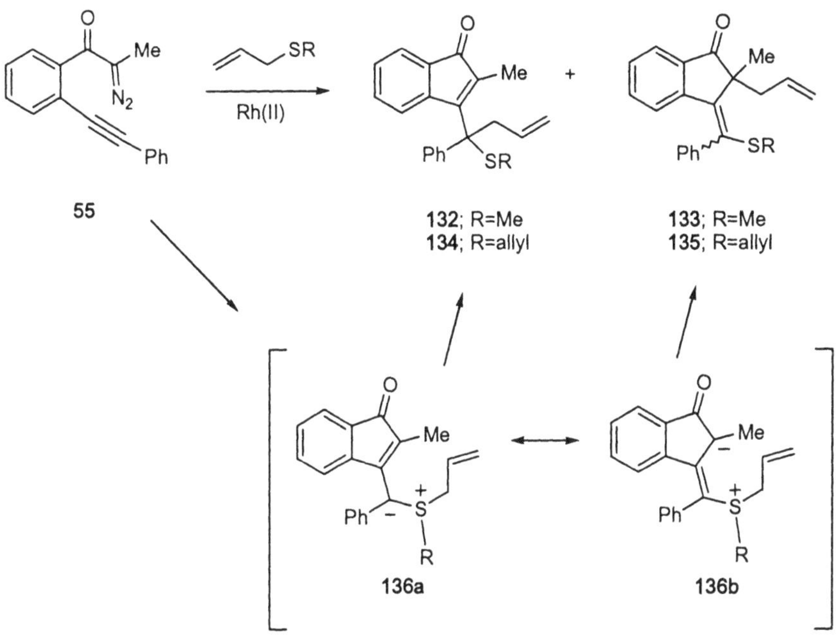

Scheme 37.

electrophilic carbenes and carbenoids with unsaturated divalent sulfur compounds to give sulfonium ylides which then undergo a [2,3] sigmatropic rearrangement is a well-described process in the literature.[52] It is believed that the lone pair of electrons on the sulfur atom adds to the electrophilic carbenoid intermediate and this is followed by a subsequent dissociation of the catalyst to produce the sulfonium ylide.[52a] The symmetry-allowed [2,3] sigmatropic rearrangement is widely recognized as a facile bond reorganization process, especially for allylic sulfides.[53] The reaction of diazo ketone **55** with rhodium(II) acetate in the presence of methyl allyl sulfide behaved similarly, producing indenone **132** (86%) along with a 1:1 -*E/Z* mixture of the isomeric vinyl sulfide **133** in 10% yield (Scheme 37). A related reaction occurred using diallyl sulfide which resulted in the formation of a 9:1 mixture of **134** and **135**. The possibility that the formation of **133** (or **135**) was the result of a Cope rearrangement of **132** (or **134**) was excluded by the finding that the thermolysis of **132** (or **134**) did not produce any detectable quantities of **133** (or **135**). The formation of **132** and **134** occurs by reaction of the sulfur lone pair of electrons with the carbenoid center followed by a subsequent ylide rearrangement via **136a**. Presumably, compounds **133** and **135** arise via a novel antarafacial [3,4] sigmatropic rearrangement of sulfonium ylide **136b**.[33]

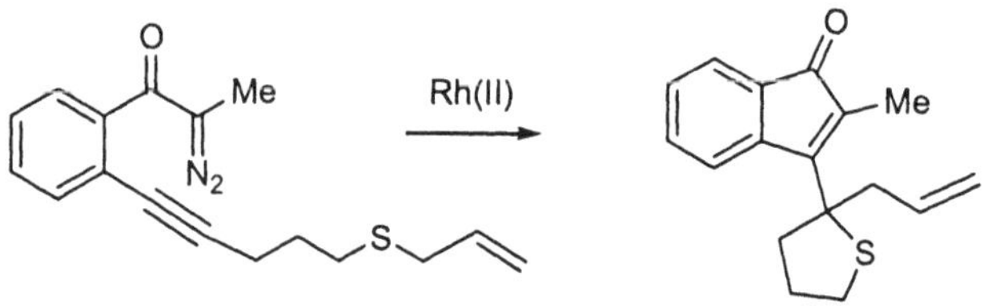

Scheme 38.

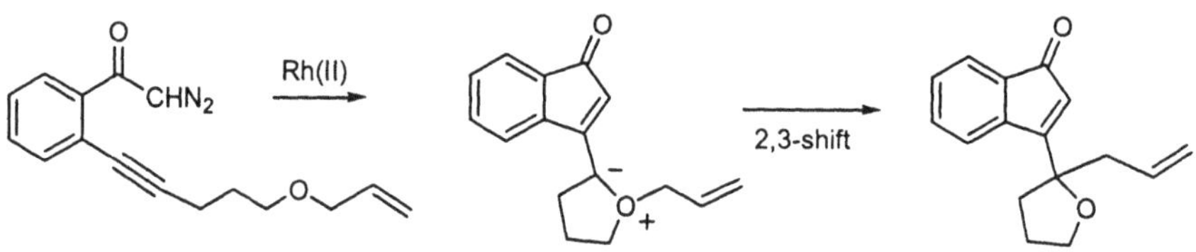

Scheme 39.

The ease with which the intermolecular *sulfonium ylide* [2,3] rearrangement sequence occurred suggests that a similar process might take place intramolecularly by incorporating the allyl sulfide functionality onto the alkyne unit. Indeed, we found that stirring a sample of diazo ketone **137** with rhodium(II) acetate furnished tetrahydrothiophene (**138**) as the sole product (Scheme 38). A number of related intramolecular sulfonium ylide rearrangement reactions were also studied.[33]

There have been several reports in the literature where cyclic oxonium ylides are formed by the intramolecular rhodium carbenoid addition to an ether oxygen followed by either a [1,2] or [2,3] sigmatropic shift.[54] Work in our laboratory demonstrated that the *intramolecular tandem generation*/[2,3]-*sigmatropic rearrangement* of an oxonium ylide derived from a diazo ketone also took place. The overall process corresponds to a formal insertion of a vinyl carbenoid into a C–O bond with concomitant generation of a cyclic ether. This is nicely illustrated by the catalytic decomposition of diazo ketone **139** which furnished the rearranged ether **141** in 81% yield (Scheme 39).[33]

V. CONFORMATIONALLY UNCONSTRAINED SYSTEMS

So far, this chapter has focused on systems in which the α-diazo keto and alkynyl groups have been situated *ortho* to each other on a phenyl ring. Removing this constraint introduces many variations for this reaction. Conformational effects present in the acyclic system may very well dictate what pathway the reaction follows. It is rather easy to prepare systems which possess a heteratom α to the diazo group and this structural modification will not only influence the conformation of the reacting system, but it will also affect the electronic behavior of the rhodium carbenoid intermediate. Outlined below are some of the results that we and others have encountered with these systems.

A. α-Diazo Keto Systems

Early work by Hoye and Dinsmore involved a study of the metal-catalyzed reactions of monostabilized α-diazo ketones such as **142** bearing *gem*-dimethyl substituents α to the carbonyl group (Scheme 40).[10,13,20,27] This study not only helped define the mechanism of these reactions, but it also demonstrated the synthetic utility of the methodology. External alkynes were used to trap the proposed zwitterionic intermediate and the resulting cyclopropene **143** was found to rearrange to give dihydropentalenone (**144**).[20] These authors envisaged product formation to occur by attack of the rhodium metal on the cyclopropene to furnish the vinyl carbenoid **145**. This is followed by cyclization onto the cyclopentenone ring and the resulting intermediate **146** undergoes a [1,5] H shift to give the observed dihydropentalenone.[20] Products derived from dimerization of the initially formed α-keto rhodium carbenoid or from reaction of the carbenoid with the 2-butyne were not detected.

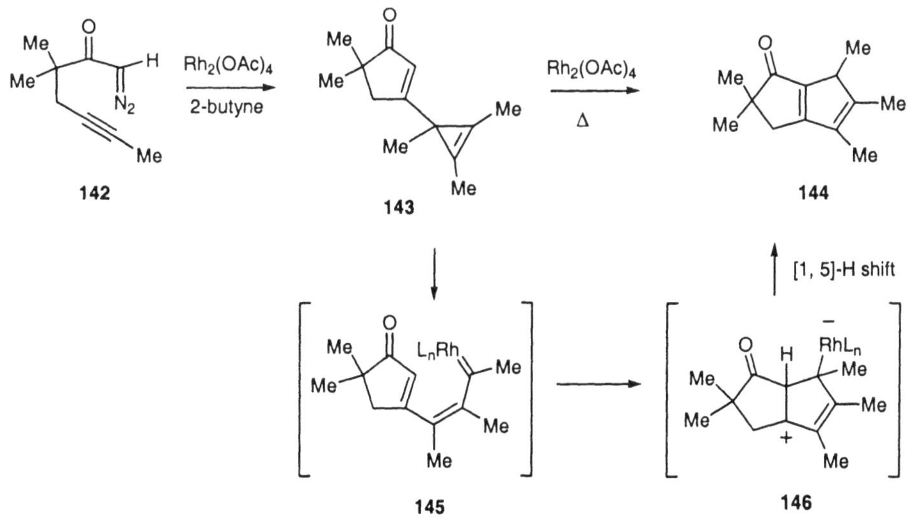

Scheme 40.

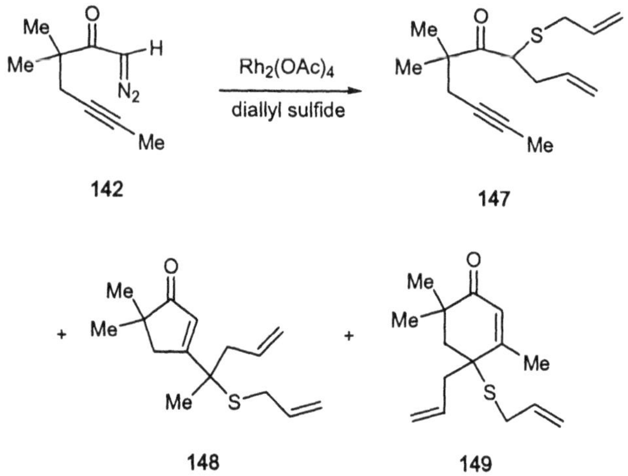

Scheme 41.

Sulfonium ylides were also generated by trapping the initially formed carbenoid intermediates with diallyl sulfide. Thus, treatment of **142** with rhodium(II) acetate in the presence of 1.1 equiv of diallyl sulfide gave 10% of **147** along with 33% of compound **148** (Scheme 41). Ketone **147** results from sulfur trapping of the initially formed α-keto carbenoid, while compound **148** is derived by sulfur trapping of the

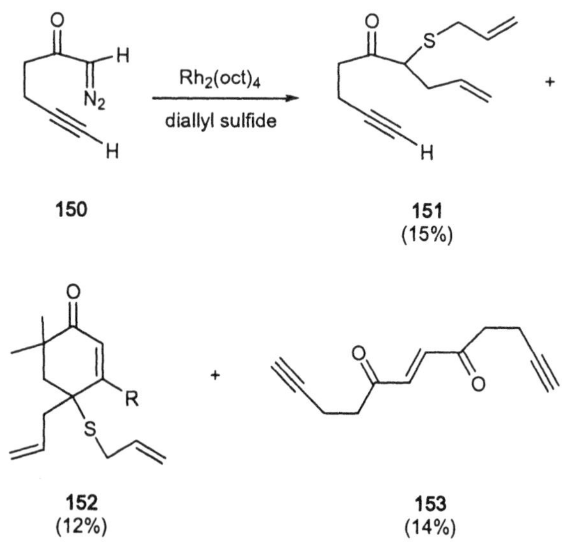

Scheme 42.

cyclized vinyl carbenoid. Both products are the result of a [2,3] sigmatropic rearrangement of the initially formed sulfonium ylide.[27] The fact that diallyl sulfide can trap the vinyl carbenoid intermediate while 2-butyne cannot, seemingly reflects the greater nucleophilicity of the sulfide group. Similar results have been reported by Nakano.[55] This reaction also produced a significant amount of compound **149** (31%), the product of 6-*endo*-vinyl carbenoid formation followed by trapping with the sulfide.[27]

Hoye also demonstrated the importance of having the *gem*-dimethyl substituents on the backbone of the diazo ketone. He found that reaction of the unsubstituted diazo ketone **150** with rhodium(II) octanoate in the presence of diallyl sulfide gave compounds **151–153** (Scheme 42). The change in product distribution when compared with the methylated diazo ketone **142** was attributed to the absence of a Thorpe–Ingold effect.[27] The isolation of enedione **153** indicates that other processes can compete with intermolecular sulfonium ylide formation and cyclization in these systems.

B. α-Diazo Esters

Introduction of a heteroatom α to the diazo carbonyl group may further complicate the chemistry. It is well known that esters exist primarily in the *Z* or *s-trans* (i.e., **154-Z**) conformation about the carbonyl π-bond (Scheme 43). Esters are more stable in this conformation for several reasons, one of which is to minimize the overall dipole effect.[56] In this orientation, intramolecular cyclization of the rhodium carbenoid onto the alkyne π-bond cannot occur. In order for cyclization to take place, there must be rotation about the ester bond to generate the *E* or *s-cis* conformer **154-E**, which can then achieve the necessary geometry to allow the cyclization to proceed.[57]

Another factor that needs to be considered in these systems is that the heteroatom can change the electronic nature of the rhodium carbenoid center. For example, treatment of the monostabilized α-diazo ester **155** with rhodium(II) acetate in refluxing benzene gave rise to a 79% yield of cycloheptatriene **156** (Scheme 44).

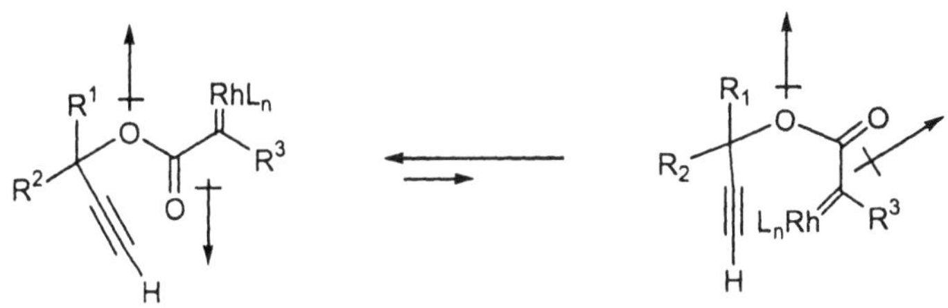

154-Z (*s-trans*) **154-E** (*s-cis*)

Scheme 43.

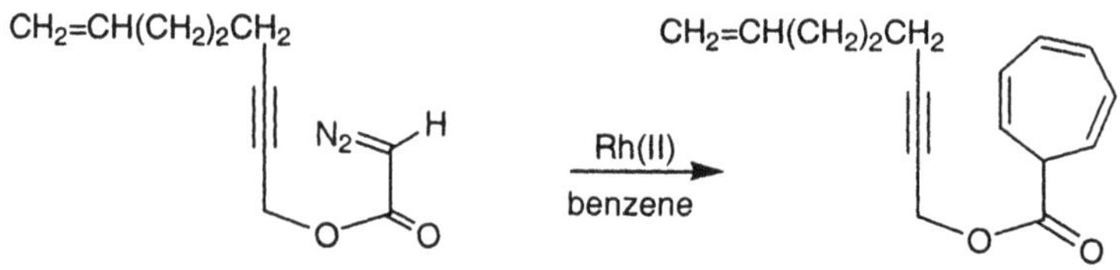

155 **156**

Scheme 44.

This product is derived from insertion of the ketocarbenoid intermediate into the benzene ring.[58]

On the other hand, cyclization of the distabilized diazo ketoester **157** with rhodium(II) octanoate furnished furan **158** in 77% yield (Scheme 45).[58] This transformation proceeds by addition of the rhodium-stabilized carbenoid onto the acetylenic π-bond to produce an electrophilic vinyl carbenoid intermediate (i.e. **160**) which is subsequently attacked by the adjacent carbonyl oxygen bond (Scheme 46). The resulting dipole **161** undergoes subsequent collapse to give furan **162**.[59,60] 6π-Electrocyclization reactions to produce five-membered rings are well-precedented transformations in heterocyclic chemistry.[60,61] Related furan cyclizations have also been observed in *ortho*-constrained systems.[62] The failure of **155** to undergo cyclization is probably related to the fact that the initial carbenoid intermediate is not electrophilic enough to attack the π-system of the alkyne and consequently insertion into the solvent occurs. This has also been observed with other α-diazo ketoester systems.[58]

The conformational aspects of distabilized diazo ketoesters have also been studied. Over the years, numerous investigations dealing with the *gem-dialkyl promoting effect* have been reported in the literature.[63] Consequently, one might expect to encounter a similar rate acceleration in the acetylenic cyclization reaction as the degree of substitution is increased about the α-position of the 2-diazo-3-oxobutanoate system. Surprisingly, the exact opposite was found. Thus, the *gem*-dimethyl propargyl ester **163c** ($R_1 = R_2 =$ Me) afforded the furo[3,4-c]furan ring **164c** at a slower rate (ca. 50%) than the monomethyl propargyl ester **163b** (R_1 =

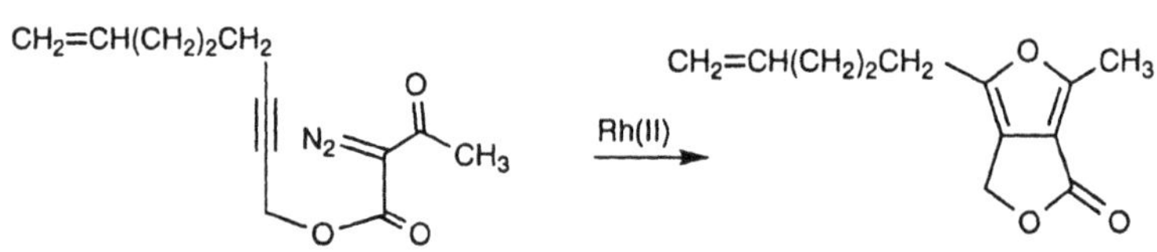

157 **158**

Scheme 45.

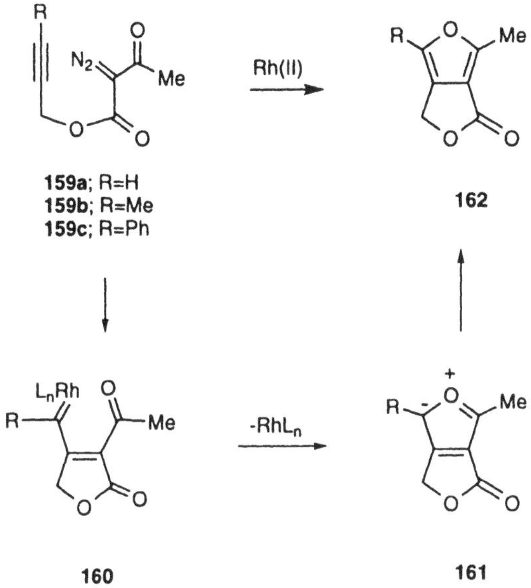

Scheme 46.

Me; $R_2 = H$), which in turn, cyclized at a slower rate (ca. 50%) than the unsubstituted ester **163a** (Scheme 47).[57] At first glance, these results seemingly contradict the Thorpe–Ingold effect which would predict an opposite trend in rates of cyclization. To account for the results, it was suggested that a lower population of the reactive conformer exists with the bulkier propargyl group and, consequently, a retardation in the rate of cyclization occurs. The rationale to account for the observed cyclization rates with **163** is that the unsubstituted propargyl group can more easily attain the required transition state for cyclization as compared to the methyl-substituted

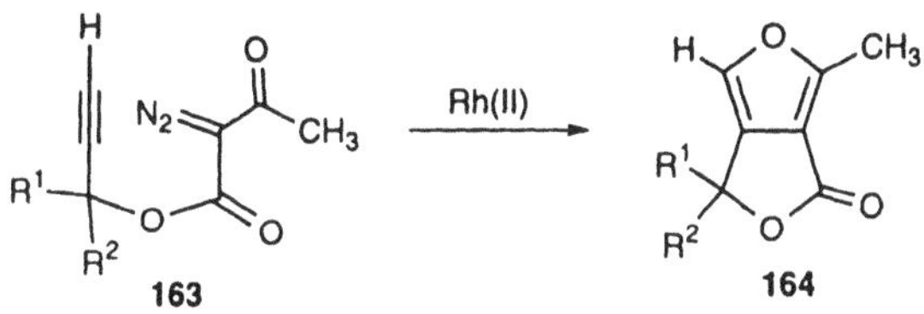

	relative rate
163a; $R^1 = R^2 = H$	1
163b; $R^1 = Me$, $R^2 = H$	0.5
163c; $R^1 = R^2 = Me$	0.25

Scheme 47.

substrate and therefore it cyclizes more rapidly. Energy differences between the two conformations of the diazo ketoesters were calculated using the Still–Steliou Model 2.94 Program. The calculations clearly show a greater energy difference for the conversion of the *s-trans* to *s-cis* ester conformations for the *gem*-dimethyl case (ΔE = 6.5 kcal) than for the monomethyl (ΔE = 4.2 kcal) and unsubstituted cases (ΔE = 4.09 kcal).[57]

The authors' group also examined the competition between C–H insertion and furan formation in systems where both pathways are possible. Insertion of electrophilic rhodium carbene complexes into a C–H bond results in the preferential formation of five-membered rings in acyclic, conformationally mobile systems.[64] The order of insertion reactivity into the C–H bond is generally: methine > methylene >> methyl.[65] There are also several examples in the literature where C–H insertion can lead to four- and six-membered rings.[66,67] The results indicate that site selectivity depends on the nature of the α-diazo carbonyl compound, and also suggests that it is governed by steric, conformational, and electronic factors. We discovered that the C–H insertion reaction can compete with furan formation when an alkyl group is attached to the keto functionality. Thus, a 1:1 mixture of cyclization (**166**) and insertion (**167**) products was observed when diazo ester **165** was treated with rhodium(II) acetate at 80 °C (Scheme 48). This ratio could be slightly altered to favor cyclobutanone formation (i.e. **169:170** = 1:1.5) when a methyl group was placed onto the terminal position of the alkynyl group. In both cases, insertion into the tertiary C–H bond to give a cyclobutanone is favored over insertion into one of the methyl groups. In a related fashion, the unbranched diazo ester **171** preferentially underwent a five-ring insertion reaction over cyclization (**172:173** = 1:3).[57]

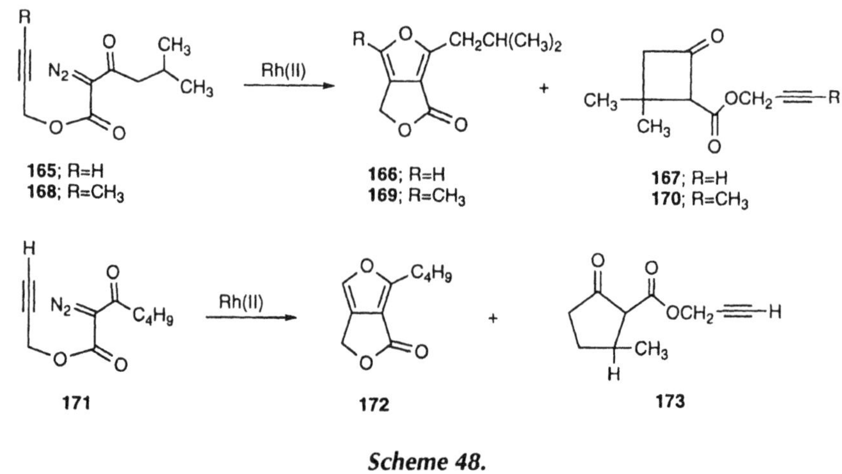

Scheme 48.

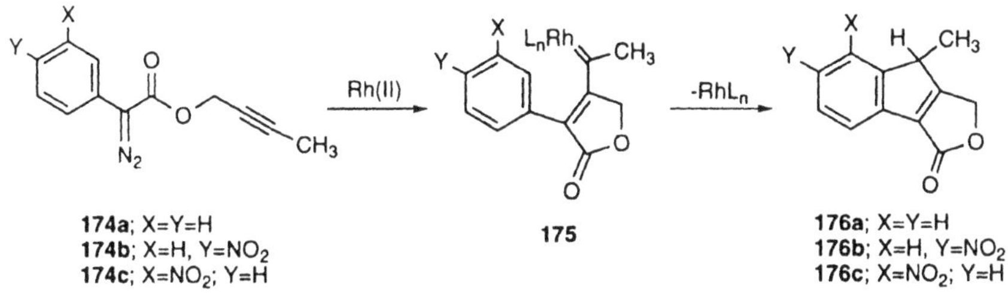

Scheme 49.

The 1,5-electrocyclization process involved in furan formation has also been utilized to produce indeno[1,2-c]furans such as **176a–c** in 45–60% yield (Scheme 49). Treatment of the starting α-diazo esters **174a–c** with rhodium(II) catalysts gave indenes **176a–c** via an electrocyclization of the transient vinyl carbenoid **175**.[57] There seemed to be little effect displayed by the nature of the substituent groups on the aromatic ring as indeno[1,2-c]furans **176b** and **176c** were isolated as the exclusive products. The fact that the insertion reactions occurs *ortho* to the nitro group (i.e. **174c** → **176c**) rather than producing a mixture of *ortho* and *para* isomers suggests that subtle factors play a role in this process as well.[57]

In order to appraise the role of α-sulfonyl stabilization of the carbenoid intermediate in the cyclization process, alkynyl sulfones **177** and **178** were prepared and treated with rhodium(II) acetate at 80 °C (Scheme 50). The major products formed

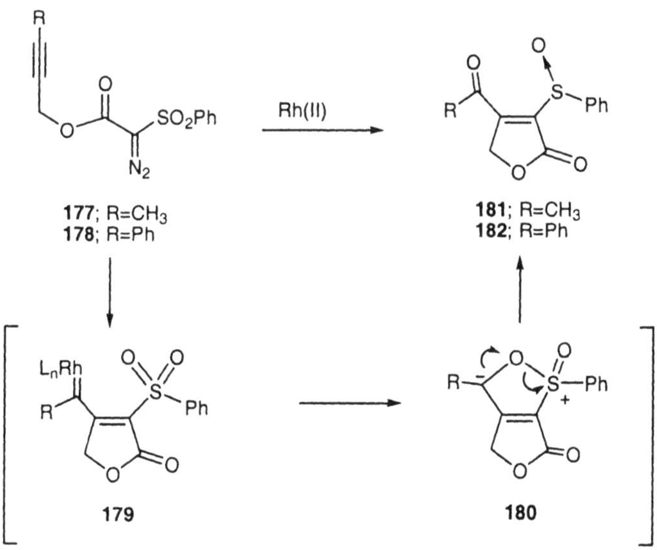

Scheme 50.

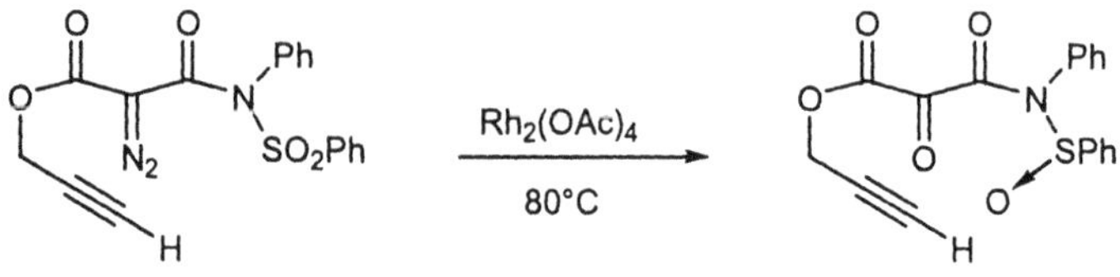

183 184

Scheme 51.

were sulfoxides **181** and **182** in 60 and 90% yields, respectively.[57] This novel oxygen transfer reaction was rationalized by assuming that the oxygen of the sulfonyl group reacts with vinyl carbenoid **179** to produce the dipolar species **180**. Subsequent collapse of **180** gave rise to the ring-opened butenolides.[57] In a similar manner, the vicinal tricarbonyl compound **184** was the sole product obtained from the reaction of the distabilized diazo ketoester **183** with rhodium(II) acetate in refluxing benzene (Scheme 51).[68,69]

The chemistry of bis-(diazocarbonyls) has rarely been examined.[70] This is not surprising when one considers the number of possible complications that may arise from the combination of two reactive carbenoid centers contained within the same molecule. We found that heating the symmetrical bis-(diazo ester) **185** in refluxing benzene with rhodium(II) acetate produced a 90% yield of bis(butenolide) **188** (Scheme 52). The reaction involves one of the diazo ester groups cyclizing on the

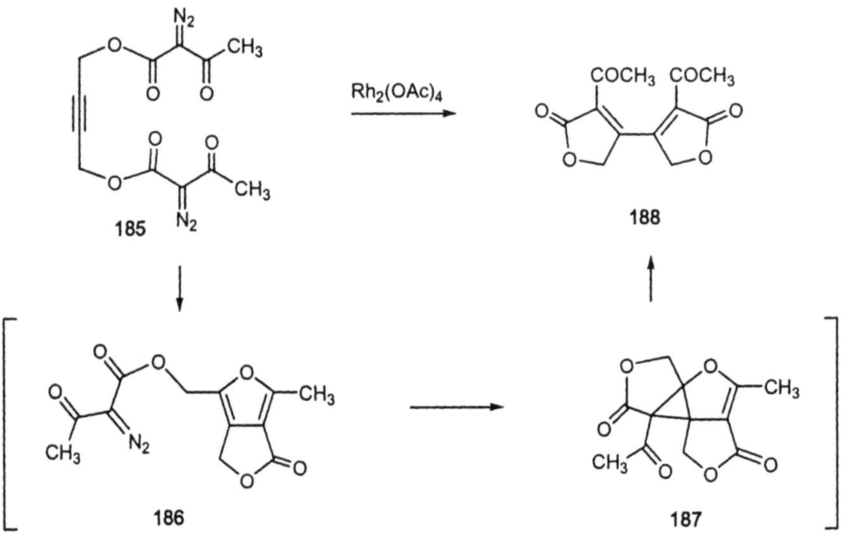

185 188 186 187

Scheme 52.

alkyne functionality to generate a vinyl carbenoid intermediate which subsequently cyclizes again to produce furan **186**. Further reaction of **186** with rhodium(II) acetate generates a second rhodium carbenoid which undergoes intramolecular cyclopropanation onto the furan ring. The resulting cycloadduct **187** undergoes furan fragmentation to produce the symmetrical product **188**.[57]

C. α-Diazo Amides

Rotamer population can play a significant role in determining the chemoselectivity of rhodium(II)-catalyzed reactions of α-diazo amide systems containing tethered alkynes. The reaction of diazo amide **189** with rhodium(II) octanoate was found to undergo attack on the π-system of the acetylenic tether to give a transient vinyl carbenoid (Scheme 53). The next step involved an internal cyclopropanation reaction to produce **190** in 41% yield. Cycloheptatriene **191**, which is derived by insertion of the carbenoid into the *N*-benzyl substituent, was also isolated from this reaction in 33% yield.[58] Rotamer populations nicely account for the behavior of this system. Amide rotamers generally interconvert in solution with lifetimes of 10^{-1}–10^{-2} s.[71] The geometry of a typical amide C–N bond will be fixed during the entire lifetime of the acyl rhodium carbenoid intermediate. Assuming that both amide rotamers are equally reactive toward π-addition, the relative amounts of compounds **190** and **191** that are formed are determined by the equilibrium concentration of the starting rotamers.[58]

The complex nature of these reactions becomes even more apparent in further studies with related amide systems. In order to avoid the aromatic insertion reaction (i.e. **189** → **191**), the methyl-substituted diazo amides **192** and **194** were studied. The rhodium(II)-catalyzed decomposition of **192** did not afford products derived by internal attack on the acetylenic π-bond. Instead, only pyrrolidinone **193** was obtained in 65% yield (Scheme 54).[58] The exclusive formation of **193** is consistent with the rotamer population of the starting amide controlling the course of the reaction. Overlap of the nitrogen nonbonded electrons with the carbonyl π-bond fixes the amide conformation so that the larger allylic substituent is oriented toward the rhodium carbenoid center so as to minimize $A^{1,3}$-strain with the methyl group on the nitrogen atom.[72] This places the allylic hydrogens close to the carbenoid

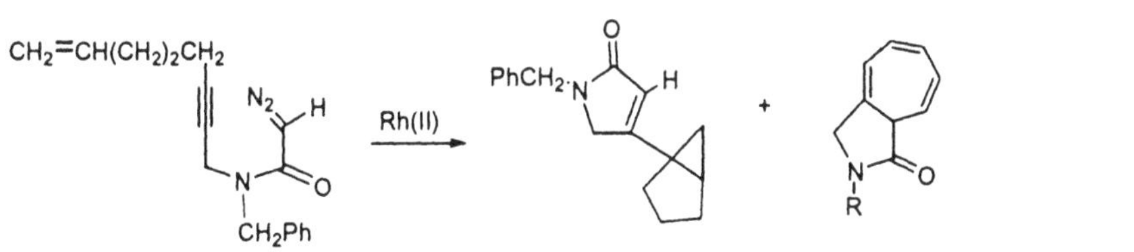

189 **190** **191**; R=$CH_2C{\equiv}C(CH_2)_3CH{=}CH_2$

Scheme 53.

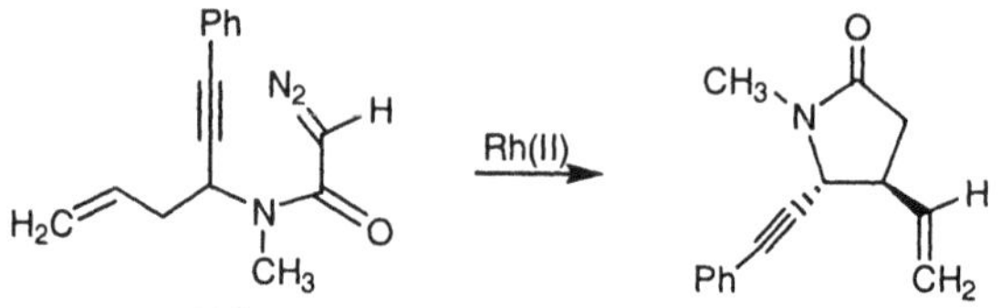

192 193

Scheme 54.

center for an easy C–H insertion. The exclusive production of **193** at 25 °C suggests that at this temperature, the rate of C–H insertion is greater than the rate of conformer interconversion. When C–H insertion is not a viable option, as in **194**, internal cyclopropanation becomes the exclusive process leading to bicyclohexane **195** (Scheme 55). The success of this reaction clearly indicates that internal attack at the alkyne is electronically viable and that conformational factors may dictate the course of the reaction with these acyclic diazo amide systems.[58]

Cyclization reactions of distabilized α-diazo amides have also been studied. Thus, the reaction of **196** and **198** with rhodium(II) octanoate gave rise to bicyclic furans **197** and **199** in 82 and 72% yield, respectively (Scheme 56). Here, electrocyclization of the vinylogous rhodium carbenoid onto the neighboring acetyl group is faster than cyclopropanation with the tethered alkenyl π-system. This result is analogous to that encountered with the distabilized α-diazo ester system (i.e. **157** → **158**).[58]

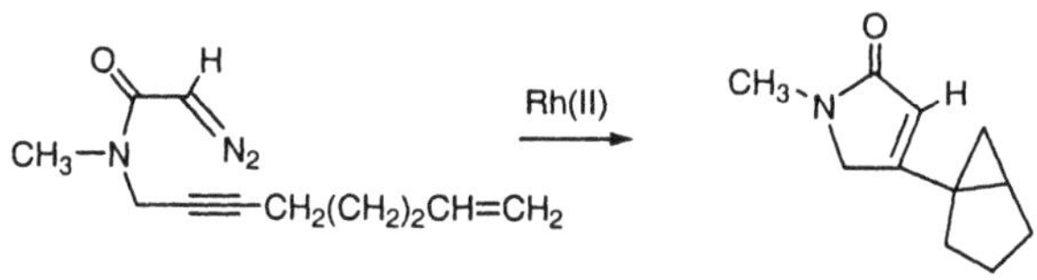

194 195

Scheme 55.

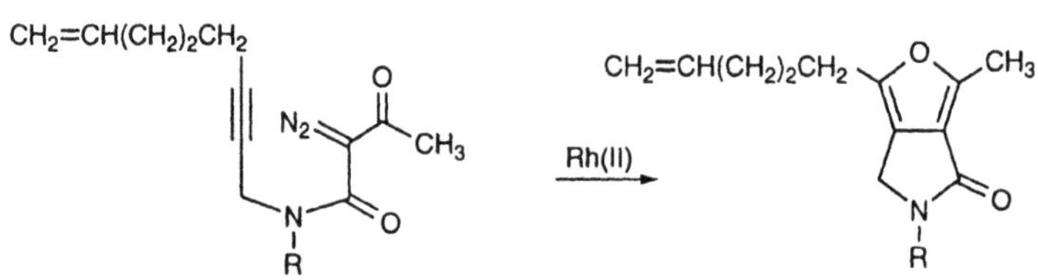

196 R=Bn
198 R=Me

197 R=Bn
199 R=Me

Scheme 56.

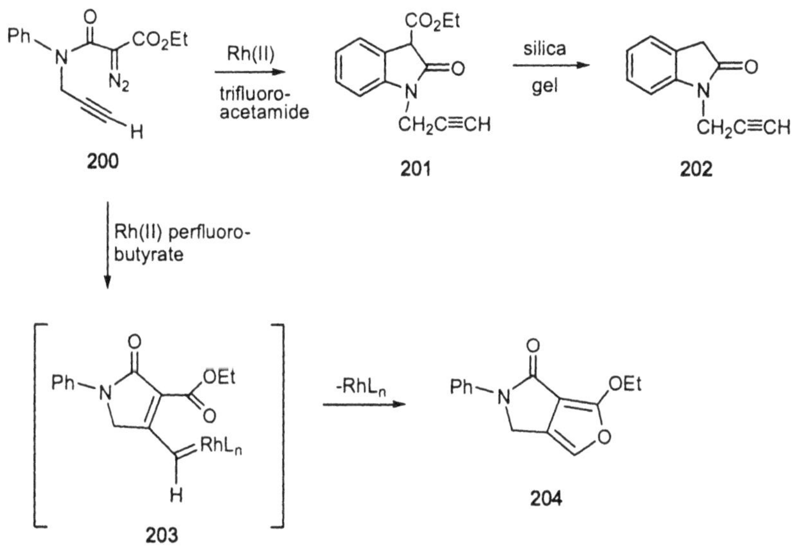

Scheme 57.

In one study, the mode of cyclization of a distabilized α-diazo amide was varied by changing the ligands on the rhodium catalyst. Reaction of **200** with rhodium(II) trifluoroacetamide in benzene at 25 °C provided oxindole **201** in 87% yield (Scheme 57). On the other hand, when rhodium(II) perfluorobutyrate was used as a catalyst, furopyrrolone **204** was formed in 98% yield,[73] in line with previous observations.[74]

VI. CONCLUSION

It is clear from the above discussion that the reaction of α-diazo carbonyl compounds with tethered alkynes is both a mechanistically complex and synthetically useful process. Four major factors dictate the mode of reaction of the initially formed rhodium carbenoid species: (1) the electronics about the carbenoid center is perhaps the most important factor—conformation of the molecule is also quite important; (2) the geometrical orientation can be influenced by both the nature of the carbenoid stabilizing group (amide vs. ester vs. ketone), and by substitution on the carbonyl group; (3) steric factors appear to influence the process in subtle ways; and (4) the polarity of the solvent used in these reactions has also been shown to influence both the mechanism and chemoselectivity of the reaction. These factors can be exploited and manipulated in many ways to generate a wide variety of interesting products. Application of the methodology to the synthesis of natural products is still relatively unexplored.

ACKNOWLEDGMENTS

We thank the National Science Foundation for generous support of this work. We also acknowledge the contributions of the graduate and postdoctoral students who participated in this research area. Their names are given in the literature references.

REFERENCES

1. Collman, J. P.; Hegedus, L. S.; Norton, J. R.; Finke, R. G. *Principles and Applications of Organotransition Metal Chemistry*; Mill Valley, CA, 1987. Doyle, M. P. *Acc. Chem. Res.* **1986**, *19*, 348. Schore, N. E. *Chem. Rev.* **1988**, *88*, 1081. *Tetrahedron Symposia in Print*; Semmelhack, M. F., Ed.; 1985, Vol. 42, pp. 5741–5887. Bishop, K. C. *Chem. Rev.* **1976**, *76*, 461. Deem, M. L. *Synthesis* **1982**, 701. Deem, M. L. *Synthesis* **1972**, 675. Dötz, K. H. *Angew. Chem., Int. Ed. Engl.* **1975**, *14*, 644. Dötz, K. H.; Popall, M. *Tetrahedron* **1985**, *41*, 5797. Dötz, K. H.; Dietz, R. *Chem. Ber.* **1978**, *111*, 2517. Wulff, W. D. In *Comprehensive Organic Synthesis*; Trost, B. M., Fleming, I., Eds.; Pergamon Press: New York, 1990; Vol. 5. Wulff, W. D. In *Advances in Metal-Organic Chemistry*; Liebeskind, L. S., Ed.; JAI: Greenwich, CT, 1989; Vol. 1. McCallum, J. S.; Kunng, F.A.; Gilbertson, S. R.; Wulff, W. D. *Organometallics* **1988**, *7*, 2346. Wulff, W. D.; Xu, Y. C. *Tetrahedron Lett.* **1988**, 415. Boger, D. L.; Jacobson, I. C. *J. Org. Chem.* **1990**, *55*, 1919. Peterson, G. A.; Kunng, F.A.; McCallum, J. S.; Wulff, W. D. *Tetrahedron Lett.* **1987**, 1381. Semmelhack, M. F.; Bozell, Bergman, R. G.; Heathcock, C. H. *J. Am. Chem. Soc.* **1987**, *109*, 2022. Alt, H. G. *J. Organomet. Chem.* **1985**, *288*, 149. Watson, P. L.; Bergman, R. G. *J. Am. Chem. Soc.* **1979**, *101*, 2055. Slough, G. A.; Deshong, P. *J. Am. Chem. Soc.* **1988**, *110*, 2575. Heck, R. F. *J. Am. Chem. Soc.* **1964**, *86*, 1819. Bottrill, M.; Green, M.; O'Brien, E.; Smart, L. E.; Woodward, P. *J. Chem. Soc., Dalton Trans.* **1980**, 292. Corrigan, P. A.; Dickson, R. S. *Aust. J. Chem.* **1979**, *32*, 2147. Huggins, J. M.; Bergman, R. G. *J. Am. Chem. Soc.* **1981**, *103*, 3002. Maitlis, P. M. *Acc. Chem. Res.* **1976**, *9*, 93. O'Connor, J. M.; Pu, L.; Rheingold, A. L. *J. Am. Chem. Soc.* **1990**, *112*, 6232. Harvey, D. F.; Brown, M. F. *J. Am. Chem. Soc.* **1990**, *112*, 7806. Harvey, D. F.; Lund, K. P. *J. Am. Chem. Soc.* **1991**, *113*, 5066. Harvey, D. F.; Brown, M. F. *J. Org. Chem.* **1992**, *57*, 5559.
2. Stork, G.; Ficini, J. *J. Am. Chem. Soc.* **1961**, *83*, 4678.
3. Padwa, A.; Krumpe, K. E. *Tetrahedron* **1992**, *48*, 5385. Burke, S. D.; Grieco, P. A. *Org React.* **1979**, *26*, 361. Hook, J. M.; Mander, L. N.; Urech R. J. *J. Am. Chem. Soc.* **1980**, 102. Hudlicky, T. *J. Org. Chem.* **1982**, *47*, 1522. Padwa, A.; Weingarten, M. D. *Chem. Rev.* **1996**, *96*, 223.
4. Adams, J.; Spero, D. M. *Tetrahedron* **1991**, *47*, 1765. Anciaux, A. J.; Hubert, A. J.; Noels, A. F.; Petiniot, N.; Teyssie, P. *J. Org. Chem.* **1980**, *45*, 695. Taber, D. F.; Ruckle, R. E. *J. Am. Chem. Soc.* **1986**, *108*, 7686.
5. Taber, D. F. In *Comprehensive Organic Synthesis*; Trost, B. M.; Fleming, I., Eds.; Pergamon: New York, 1991, Vol. IV, p. 1046. Taber, D. F.; Raman, K. J. *J. Am. Chem. Soc.* **1983**, *105*, 5935. Taber, D. F.; Petty, E. H.; Raman, K. J. *J. Am. Chem. Soc.* **1985**, *107*, 196. Taber, D. F.; Hoermer, R. S. *J. Org. Chem.* **1992**, *57*, 441.
6. Maas, G. *Top Curr. Chem.* **1987**, *137*, 77.
7. Padwa, A.; Hornbuckle, S. F. *Chem. Rev.* **1991**, *91*, 263.
8. Hook, J. M.; Mander, L. N.; Urech, R. *J. Am. Chem. Soc.* **1980**, *102*, 6628.
9. Hudlicky, T.; Govindan, S. V.; Frazier, J. O. *J. Org. Chem.* **1985**, *50*, 4166. Short, R. P.; Revol, J. M.; Ranu, B. C.; Hudlicky, T. *J. Org. Chem.* **1983**, *48*, 4453. Govindan, S. V.; Hudlicky, T.; Koszyk, F. J. *J. Org. Chem.* **1983**, *48*, 3581. Short, R. P.; Hudlicky, T. *J. Org. Chem.* **1982**, *47*, 1522. Hudlicky, T.; Koszyk, F. J.; Dochwat, D.; Cantrell, G. L. *J. Org. Chem.* **1981**, *46*, 2911. Hudlicky, T.; Kutchan, T. M.; Koszyk, F. J.; Sheth, J. P. *J. Org. Chem.* **1980**, *45*, 5020.
10. Hoye, T. R.; Dinsmore, C. J.; Johnson, D. S.; Korkowski, P. F. *J. Org. Chem.* **1990**, *55*, 4518.
11. Padwa, A.; Krumpe, K. E.; Gareau, Y.; Chiacchio, U. *J. Org. Chem.* **1991**, *56*, 2523.

12. Padwa, A.; Xu, S. L. *J. Am.Chem. Soc.* **1992**, *114*, 5881. Padwa, A.; Kassir, J. M.; Xu, S. L. *J. Org. Chem.* **1991**, *56*, 6971. Baird, M. S.; Buxton, S. R; Whitley, J. S. *Tetrahedron Lett.* **1984**, *25*, 1509. Petiniot, N.; Anciaux, A. J.; Noels, A. F.; Hubert, A. J.; Teyssie, P. *Tetrahedron Lett.* **1978**, *19*, 1239. Padwa, A.; Blacklock, T. J. *J. Am. Chem. Soc.* **1984**, *106*, 4446. Padwa, A.; Carter, S. P.; Nimmesgern, H.; Stull, P. *J. Am. Chem. Soc.* **1988**, *110*, 2894. Padwa, A.; Fryxell, G. E.; Zhi, L. *J. Org. Chem.* **1988**, *53*, 2875.
13. Hoye, T. R.; Dinsmore, C. J. *J. Am Chem. Soc.* **1991**, *113*, 4343.
14. Padwa, A.; Krumpe, K. E.; Zhi, L. *Tetrahedron Lett.* **1989**, *30*, 2633.
15. Mykytka, J. P.; Jones, W. M. *J. Am. Chem. Soc.* **1975**, *97*, 5933.
16. For a review see: Steinmetz, M. G.; Srinivasan, R.; Leigh, W. J. *Rev. Chem. Intermed.* **1984**, *5*, 57. Pincock, J. A.; Boyd, R. J. *Can J. Chem.* **1977**, *55*, 2482.
17. York, E. J.; Dittmar, W.; Stevenson, J.; Bergman, R. G. *J. Am. Chem. Soc.* **1973**, *95*, 5680. Padwa, A. *Acc. Chem. Res.* **1977**, *12*, 310 and references cited therein.
18. Zimmerman, H. E.; Hovey, M. C. *J. Org. Chem.* **1979**, *44*, 2331. Arnold, D. R.; Morchat, R. *J. Am. Chem. Soc.* **1973**, *95*, 7536. Palmer, G. E.; Bolton, J. R.; Arnold, D. R. *J. Am. Chem. Soc.* **1974**, *96*, 3708. Arnold, D. R.; Humphreys, R. W.; Leigh, W. J.; Palmer, G. E. *J. Am. Chem. Soc.* **1976**, *98*, 6225.
19. Padwa, A.; Chiacchio, U.; Garreau, Y.; Kassir, J. M.; Krumpe, K. E.; Schoffstall, A. M. *J. Org. Chem.* **1990**, *55*, 414.
20. Hoye, T. R.; Dinsmore, C. J. *Tetrahedron Lett.* **1991**, *32*, 3755.
21. Padwa, A.; Austin, J. A.; Xu, S. L. *J. Org. Chem.* **1992**, *57*, 1330.
22. Billups, W. E.; Haley, M. M.; Lee, G. A. *Chem. Rev.* **1989**, *89*, 1147. Chenier, P. J.; Southland, D. A. *J. Org. Chem.* **1989**, *54*, 3519. Halton, B.; Bridle, J. H.; Lovett, E. G. *Tetrahedron Lett.* **1990**, 1313.
23. Padwa, A.; Krumpe, K. E.; Kassir, J. M. *J. Org. Chem.* **1992**, *57*, 4940.
24. Hauck, G.; Durr, H. *J. Chem. Res.* **1981**, 180. Gardner, P. D.; Shields, T. C. *J. Am. Chem. Soc.* **1967**, *89*, 5425. Hashem, A.; Weyerstahl, P. *Tetrahedron* **1984**, *40*, 2003. Padwa, A.; Wannamaker, M. W.; Dyszlewski, A. D. *J. Org. Chem.* **1987**, *52*, 4760.
25. Taber, D. F.; Hennessy, M. J.; Louey, J. P. *J. Org. Chem.* **1992**, *57*, 436.
26. Shankar, B. K. R.; Shechter, H. *Tetrahedron Lett.* **1982**, 2277. Ikota, N.; Takamura, N.; Young, S. D.; Ganem, B. *Tetrahedron Lett.* **1981**, 4163. Doyle, M. P.; High, K. G.; Oon, S. M.; Osborn, A. K. *Tetrahedron Lett.* **1989**, 3049.
27. Hoye, T. R.; Dinsmore, C. J. *Tetrahedron Lett.* **1992**, *33*, 169.
28. Bucher, G.; Sander, W. *J. Org. Chem.* **1992**, *57*, 1346.
29. Padwa, A.; Austin, D. J.; Xu, S. L. *Tetrahedron Lett.* **1991**, *32*, 4103.
30. Padwa, A.; Chiacchio, U.; Fairfax, D. J.; Kassir, J. M.; Litrico, A.; Semones, M. A.; Xu, S. L. *J. Org. Chem.* **1993**, *58*, 6429.
31. Danheiser, R. L.; Gee, S. K. *J. Org. Chem.* **1984**, *49*, 1674. Karlsson, J. O.; Nguyen, N. V.; Moore, H. W. *J. Am. Chem. Soc.* **1985**, *107*, 3392. Liebeskind, L. S.; Iyer, S.; Jewell, C. F. *J. Org. Chem.* **1986**, *51*, 3067. Kowalski, C. J.; Lai, G. S. *J. Am. Chem. Soc.* **1988**, *110*, 3693.
32. Wentrup, C. *Adv. Heterocyclic Chem.* **1981**, *28*, 231. Wolff, L. *Liebigs. Ann. Chem.* **1904**, *333*, 1.
33. Padwa, A.; Kassir, J. M.; Semones, M. A.; Weingarten, M. D. *J. Org. Chem.* **1995**, *60*, 53 and references cited therein.
34. Padwa, A.; Austin, D. J.; Chiacchio, U.; Kassir, J. M.; Rescifina, A.; Xu, S. L. *Tetrahedron Lett.* **1991**, *32*, 5923.
35. Mueller, P. H.; Kassir, J. M.; Semones, M. A.; Weingarten, M. D.; Padwa, A. *Tetrahedron Lett.* **1993**, *34*, 4285.
36. Doering, W. V. E.; Roth, W. R. *Tetrahedron* **1963**, *19*, 715.
37. Hudlicky, T.; Natchuz, M. G.; Zingde, G. S. *J. Org. Chem.* **1987**, *52*, 4644 and references cited therein. Corey, E. J.; Wess, G.; Xiang, Y. B.; Singh, A. K. *J. Am. Chem. Soc.* **1987**, *109*, 4717. Corey, E. J.; Xiang, Y. B. *Tetrahedron Lett.* **1987**, 5403. Singh, A. K.; Bakshi, R. K.; Corey, E. J.

J. Am. Chem. Soc. **1987**, *109*, 6187. Wulff, W. D.; Yang, D. C.; Murray, C. K. *J. Am. Chem. Soc.* **1988**, *110*, 2653. Baird, M. S.; Nethercott, W. *Tetrahedron Lett.* **1983**, 605. Marino, J. P.; Kaneko, T. *Tetrahedron Lett.* **1973**, 3975. Marino, J. P.; Browne, L. J. *Tetrahedron Lett.* **1976**, 3245. Piers, E.; Nagakura, *Tetrahedron Lett.* **1976**, 3237. Piers, E.; Ruediger, E. H. *J. Org. Chem.* **1980**, *45*, 1727. Piers, E.; Jung, G. L.; Moss, N. *Tetrahedron Lett.* **1984**, 3959. Wender, P. A.; Filosa, M. P. *J. Org. Chem.* **1976**, *41*, 3940. Wender, P. A.; Eissenstat, M. A.; Filosa, M. P. *J. Am. Chem. Soc.* **1979**, *101*, 2196. Wender, P. A.; Hillemann, C. L.; Szymonifka, M. J. *Tetrahedron Lett.* **1980**, 2205. Wenkert, E.; Greenberg, R. S.; Kim, H. S. *Helv. Chim. Acta.* **1987**, *70*, 2159.

38. Davies, H. M. L.; McAfee, M. J.; Oldenburg, C. E. M. *J. Org. Chem.* **1989**, *54*, 930. Davies, H. M. L; Oldenburg, C. E. M.; McAfee, M. J.; Nordahl, J. G.; Henretta, J. P.; Romines, K. R. *Tetrahedron Lett.* **1988**, 975.
39. Streeper, R. D.; Gardner, P. D. *Tetrahedron Lett.* **1973**, 767. Newmann, M. F.; Buchecker, C. *Tetrahedron Lett.* **1973**, 2875.
40. Jones, W. M. In *Rearrangement in Ground and Excited States*; deMayo, P. Ed.; Academic: New York, 1980, pp. 95–160.
41. Gaspar, P. P.; Hammond, G. S. In *Carbenes*; Moss, R. A.; Jones, M., Jr., Eds; Wiley: New York, 1975, Vol. 2, pp. 207–362.
42. Friedman, L.; Shechter, H. *J. Am. Chem. Soc.* **1961**, *83*, 3159. Phillip, H.; Keating, J. *Tetrahedron Lett.* **1961**, 523.
43. Sevin, A.; Arnaud-Danon, A. *J. Org. Chem.* **1981**, *46*, 2346; Yoshimine, M.; Pacansky, J.; Honjou, N. *J. Am. Chem. Soc.* **1989**, *111*, 2785.
44. Padwa, A.; Ricker, W. F.; Rosenthal, R. J. *J. Am. Chem. Soc.* **1983**, *105*, 4446.
45. Deem, M. L. *Synthesis* **1982**, 701.
46. Padwa, A.; Kassir, J. M.; Xu, S. L. *J. Org. Chem.* **1997**, *62*, 1642.
47. Padwa, A.; Gareau, Y.; Xu, S. L. *Tetrahedron Lett.* **1991**, *32*, 983.
48. Padwa, A.; Austin, D. J.; Gareau, Y.; Kassir, J. M.; Xu, S. L. *J. Am Chem. Soc.* **1993**, *115*, 2637.
49. Padwa, A.; Carter, S. P.; Nimmesgern, H.; Stull, P. *J. Am. Chem. Soc.* **1988**, *110*, 2894. Padwa, A.; Fryxell, G. E.; Zhi, L. *J. Org. Chem.* **1988**, *53*, 2875. Padwa, A.; Chinn, R. L.; Hornbuckle, S. F.; Zhi, L. *Tetrahedron Lett.* **1989**, 301. Padwa, A.; Chinn, R. C.; Zhi, L. *Tetrahedron Lett.* **1989**, 1491. Padwa, A.; Hertzog, D. L.; Chinn, R. C. *Tetrahedron Lett.* **1989**, 4077. Padwa, A.; Dean, D. C.; Zhi, L. *J. Am. Chem. Soc.* **1989**, *111*, 6451. Padwa, A.; Dean, D. C.; Krumpe, K. E. *J. Chem. Soc., Chem. Comm.* **1989**, 921. Hertzog, D. L.; Nadler, W. R.; Zhang, Z. J.; Padwa, A. *Tetrahedron Lett.* **1992**, *33*, 5877.
50. Kassir, J. M.; Semones, M. A.; Weingarten, M. D.; Padwa, A. *Tetrahedron Lett.* **1993**, *34*, 7853.
51. Kharasch, M. S.; Rudy, T.; Nudenberg, W.; Buchi, G. *J. Org. Chem.* **1953**, *18*, 1030. Lottes, A.; Landrebe, J. A.; Larsen, K. *Tetrahedron Lett.* **1989**, 4089. Landgrebe, J. A.; Iranmanesh, H. *J. Org. Chem.* **1978**, *43*, 1244. Gutsche, C. D.; Hillman, M. *J. Am. Chem. Soc.* **1954**, *54*, 2236. Bien, S.; Gillon, A. *Tetrahedron Lett.* **1974**, 3073. Bien, S.; Gillon, A.; Kohen, S. *J. Chem. Soc., Perkin Trans. I* **1976**, 489.
52. Doyle, M. P. *Chem. Rev.* **1986**, *86*, 919. Doyle, M. P.; Tamblyn, W. H.; Bagheri, V. *J. Org. Chem.* **1981**, *46*, 5094. Kondo, K.; Ojima, I. *J. Chem. Soc., Chem. Commun.* **1972**, 860. Tamblyn, W. H.; Hoffmann, S. R.; Doyle, M. P. *J. Organomet. Chem.* **1981**, *216*, C64. Moody, C. J.; Taylor, R. J. *Tetrahedron Lett.* **1988**, *29*, 6005.
53. Hoffmann, R. W. *Angew. Chem., Int. Ed. Engl.* **1979**, *18*, 563. Ando, W. *Acc. Chem. Res.* **1977**, *10*, 179.
54. Pirrung, M. C.; Werner, J. A. *J. Am. Chem. Soc.* **1986**, *108*, 6060. Johnson, C. R.; Roskamp, E. J. *J. Am. Chem. Soc.* **1986**, *108*, 6062. Padwa, A.; Hornbuckle, S. F.; Fryxell, G. E.; Stull, P. D. *J. Org. Chem.* **1989**, *54*, 817. Thijs, L.; Zwanenburg, B. *Tetrahedron* **1980**, *36*, 2145. Eberlein, T. H.; West, F. G.; Tester, R. W. *J. Org. Chem.* **1992**, *57*, 3479. Clark, J. S. *Tetrahedron Lett.* **1992**, *33*, 6193. Clark, J. S.; Krowiak, S. A.; Street, L. J. *Tetrahedron Lett.* **1993**, *34*, 4385.
55. Nakano, H.; Ibata, T. *Bull. Chem. Soc. Jpn.* **1995**, *68*, 1393.

56. Dale, J. *Stereochemistry and Conformational Analysis*; Verlag Chemie: New York, 1978, pp. 83–85. Testa, B. *Principles of Organic Stereochemistry*; Marcel Dekker: New York, 1979, p. 105. Eliel, E. L.; Allinger, N. L.; Angyal, S. J.; Morrison, G. A. *Conformational Analysis*; Interscience: New York, 1965, p. 21.
57. Padwa, A.; Kinder, F. R. *J. Org. Chem.* **1993**, *58*, 21.
58. Padwa, A.; Dean, D. C.; Fairfax, D. J.; Xu, S. L. *J. Org. Chem.* **1993**, *58*, 4646.
59. Padwa, A.; Kinder, F. R. *Tetrahedron Lett.* **1990**, *31*, 6835.
60. Padwa, A.; Kappe, C. O.; Bakulev, V. A. In *Organic Synthesis: Theory and Applications*; Hudlicky, T., Ed.; JAI: Greewich, CT, 1996, Vol. 4, p. 149.
61. Taylor, E. C.; Turchi, I. J. *Chem. Rev.* **1979**, *79*, 181. Huisgen, R. *Angew. Chem., Int. Ed. Engl.* **1980**, *19*, 947.
62. Padwa, A.; Fairfax, D. J.; Austin, D. J.; Xu, S. L. *J. Chem. Soc., Perkin Trans. I* **1992**, 2837.
63. Jung, M. E.; Gervay, J. *Tetrahedron Lett.* **1988**, 2429. For some leading references, see: Jung, M. E.; Gervay, J. *J. Am. Chem. Soc.* **1991**, *113*, 224.
64. Taber, D. F.; Petty, E. H. *J. Org. Chem.* **1982**, *47*, 4808.
65. Taber, D. F.; Ruckle, R. E. Jr. *J. Am. Chem. Soc.* **1986**, *108*, 7686.
66. Cane, D. E.; Thomas, P. J. *J. Am. Chem. Soc.* **1984**, *106*, 5295. Brown, P.; Southgate, R. *Tetrahedron Lett.* **1986**, *27*, 247. Hashimoto, S.; Shinoda, T.; Shimada, Y.; Honda, T.; Ikegami, S. *Tetrahedron Lett.* **1987**, *28*, 637. Sonawane, H. R.; Bellur, N. S.; Ahuja, J. R.; Kulkarni, D. G. *J. Org. Chem.* **1991**, *56*, 1434. Ceccherelli, P.; Curini, M.; Marcotullio, M. C.; Rosati, O. *Tetrahedron* **1991**, *47*, 7403.
67. Doyle, M. P.; Shanklin, M. S.; Oon, S.-M.; Pho, H. Q.; van der Heide, F. R.; Veal, W. R. *J. Org. Chem.* **1988**, *53*, 3384. Adams, J.; Poupart, M. A.; Grenier, L.; Schaller, C.; Ouimet, N.; Frenette, R. *Tetrahedron Lett.* **1989**, *30*, 1749. Doyle, M. P.; Taunton, J.; Pho, H. Q. *Tetrahedron Lett.* **1989**, *30*, 5397. Doyle, M. P.; Bagheri, V.; Pearson, M. M.; Edwards, J. D. *Tetrahedron Lett.* **1989**, *30*, 7001. Lee, E.; Jung, K. W.; Kim, Y. S. *Tetrahedron Lett.* **1990**, *31*, 1023. Box, V. G. S.; Marinovic, N.; Yiannikouros, G. P. *Heterocycles* **1991**, *32*, 245. Doyle, M. P.; Pieters, R. J.; Taunton, J.; Pho, H. Q.; Padwa, A.; Hertzog, D. L.; Precedo, L. *J. Org. Chem.* **1991**, *56*, 820.
68. Unpublished results, C. Straub.
69. Moody, C. J.; Slawin, A. M. Z.; Taylor, R. J.; Williams, D. J. *Tetrahedron Lett.* **1988**, *29*, 6009. Moody, C. J.; Taylor, R. J. *Tetrahedron* **1990**, *46*, 6525.
70. Gillon, A.; Ovadia, D.; Kapon, M.; Bien, S. *Tetrahedron* **1982**, *38*, 1477.
71. Stewart, W. E.; Siddall, T. H. *Chem. Rev.* **1970**, *70*, 517. Rates of rotation of *N,N*-dimethyl benzamides see: Jackman, L. M.; Kavangh, T. E.; Haddon, R. C. *Org. Mag. Reson.* **1969**, *1*, 109. Spaargaren, K.; Korver, P. K.; van der Haak, P. J.; de Boer, Th. *J. Org. Mag. Reson.* **1971**, *3*, 605. Mitsonobu, O. *Synthesis* **1981**, 1.
72. Hoffmann, R. W. *Chem. Rev.* **1989**, *89*, 1841.
73. Brown, D. S.; Elliott, M. C.; Moody, C. J.; Mowlem, T. J.; Marino, J. P.; Padwa, A. *J. Org. Chem.* **1994**, *59*, 2447.
74. Wee, A. G. H.; Liu, B.; Zhang, L. *J. Org. Chem.* **1992**, *57*, 4404. Liu, B.; Wee, A. G. H. *Heterocycles* **1993**, *36*, 445.

RECENT APPLICATIONS OF Cr(0)-MEDIATED HIGHER ORDER CYCLOADDITION REACTIONS TO NATURAL PRODUCT SYNTHESIS

James H. Rigby

ABSTRACT

Recent applications of chromium(0)-promoted [6π+4π] and [6π+2π] cycloaddition reactions are presented. A brief overview of the scope and limitations of the metal-mediated higher order cycloaddition process is provided that puts the subsequent synthesis discussions into the proper context. The principal objective of the review is

Advances in Cycloaddition
Volume 6, pages 97–118.

ISBN: 0-7623-0531-2

to illustrate the utility as well as the versatility of the Cr(0)-promoted cycloaddition methodology for the rapid construction of structurally elaborate and stereochemically rich polycyclic intermediates appropriate for subsequent transformation into a variety of complex natural product targets.

I. INTRODUCTION

Cycloaddition is one of the most powerful and versatile methods for the assembly of ring systems used in contemporary organic synthesis,[1,2] and numerous tactics are currently available for the efficient construction of adducts possessing a range of ring sizes and substitution patterns. Six-membered carbocycles, for example, can be easily made by employing the well-known Diels–Alder reaction, which involves a highly stereoselective combination of a 4π partner (the diene) and a 2π partner (the dienophile). A wide variety of five-membered ring targets can be prepared using one of the many versions of the 1,3-dipolar cycloaddition process that have been developed over the years, and four-membered systems are available, among other ways, from the addition of a pair of 2π reaction partners. The ring-forming event becomes somewhat more challenging in the seven-membered case. However, the advent of reliable [4+3] cycloaddition methods,[3] and, more recently, the corresponding [5+2] cycloaddition processes,[4] have made cycloadditive entry into cycloheptane systems a reasonable synthetic method in a number of contexts (Figure 1). More difficult still is the efficient construction of 8–10-membered ring systems via cycloaddition, and relatively few methods for achieving these cyclizations currently exist.

Figure 2 depicts a series of generic examples of cycloaddition reactions that appear somewhat unusual in that they are characterized by the combination of more extensively conjugated π-systems than are typically seen in the additions used for smaller ring formation. These reactions are commonly referred to as higher order cycloadditions, and recent advances in metal-facilitated versions of these processes promise to make this otherwise obscure family of transformations important members of the modern synthetic repertoire. This review will discuss the most

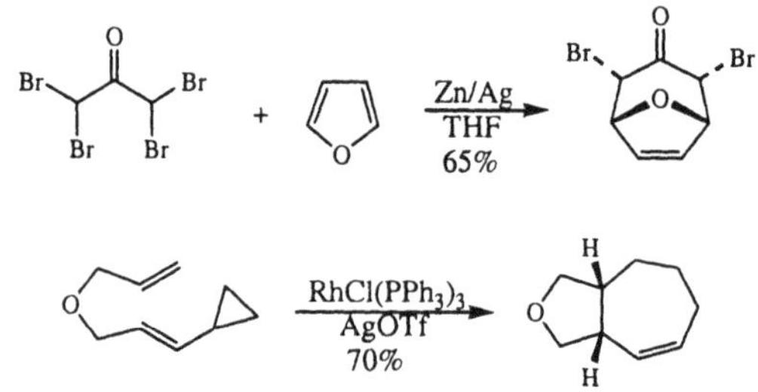

Figure 1. Typical [4+3] and [5+2] cycloaddition reactions.

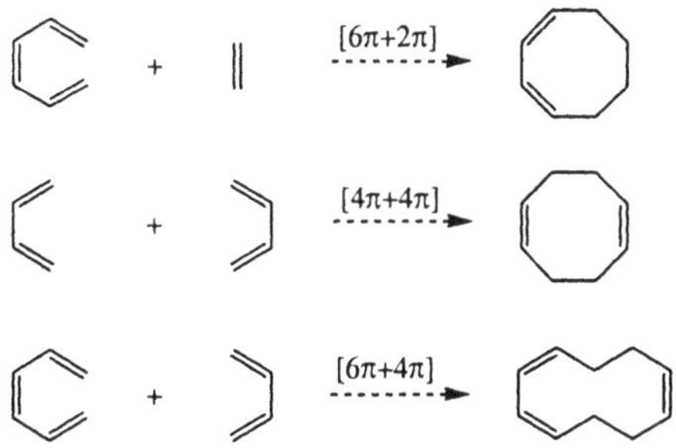

Figure 2. Generic examples of common higher order cycloaddition reactions.

synthetically useful of these recent advances with a particular emphasis on transformations that are known to be facilitated by the presence of a chromium(0) metal center.

The most frequently encountered higher order cycloadditions typically involve [4π+4π], [6π+2π] and [6π+4π] combinations,[5] which can, in principle, provide rapid access to 8- and 10-membered rings, respectively. As a class, these transformations exhibit many of the features that have made other cycloadditions, such as the Diels–Alder reaction, so central to the practice of modern organic synthesis. For example, each process is highly convergent, can accommodate substantial functionalization in both reaction partners, and proceeds with a high degree of predictable stereoselectivity. Unfortunately, higher order reactions often provide only low chemical yields of adducts due to low periselectivity stemming from the extended π-systems involved. These arrays can, and frequently do, participate in multiple competitive pericyclic events, thus diminishing the quantities of higher order adducts produced. Classic examples of several [6π+4π] cycloaddition combinations that illustrate this point are depicted in Eq. 1a,b.[6] Although the higher order pathway that yields [6+4] adduct **A** is thermally allowed, the reaction actually affords numerous other products derived from various competing pericyclic pathways along with a minor quantity of **A**.

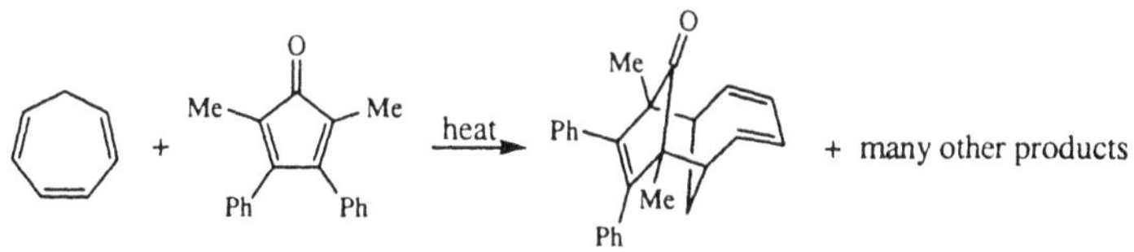

A (1a)

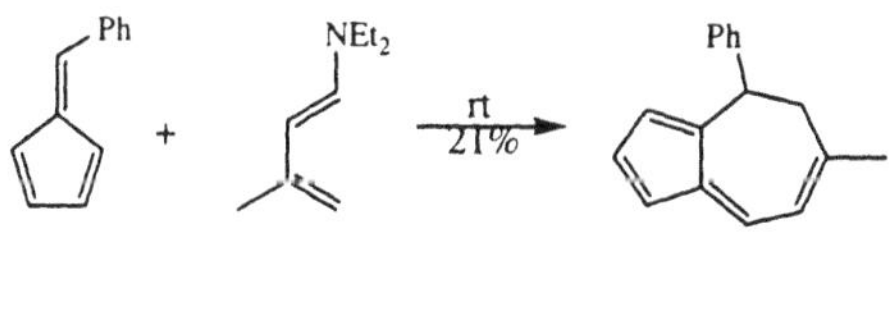

(1b)

Until quite recently the low chemical yields associated with most higher order cycloadditions relegated these reactions to the status of mere laboratory curiosities.[7] However, [6π+4π] cycloaddition appeared to offer a uniquely efficient entry into the intricate molecular structures of a number of natural product targets, including the tumor-promoting diterpene, ingenol (**1**). Thus an important impetus for creating new and more efficient methods for effecting higher order cycloadditions came to the fore. Our initial efforts directed toward bringing this strategy to fruition focused on the well-known thermally allowed tropone–diene [6+4] cycloaddition process for assembly of the core ring system of the ingenane diterpenes.[7a,b] However, the inefficiencies encountered early on with this crucial higher order cycloaddition step prompted us to consider methods for intervening in the reaction with the objective of improving periselectivity, and hence chemical yield, without compromising the other attractive attributes of the process. An intriguing idea for achieving this goal would be to employ an appropriate transition metal as a template that would precomplex the two π-partners prior to the ring-forming event, rendering the reaction temporarily intramolecular in nature (Scheme 1). While there was not a large body of literature in this area at the outset of our investigations, several critical antecedents pointed to the viability of the concept as applied to higher order cycloadditions (Figure 3). For example, Pettit and his coworkers were early proponents of the concept of metal-facilitated cycloaddition,[8] and more recently Wender's laboratory has very nicely developed the intramolecular version of the well-known Ni(0)-butadiene cyclodimerization process into a powerful methodology for construction of eight-membered carbocyclic systems with both stereo- and regiocontrol.[9]

While relatively little relevant precedent was available for bringing this notion to practice in the context of [6+4] cycloaddition, a series of intriguing reports appearing from the Kreiter laboratory suggested that certain chromium(0) complexes could participate with modest efficiency in this type of process under photochemical activation conditions.[10] Recognizing the great preparative potential that higher order cycloaddition could have if reaction efficiency could be improved, we

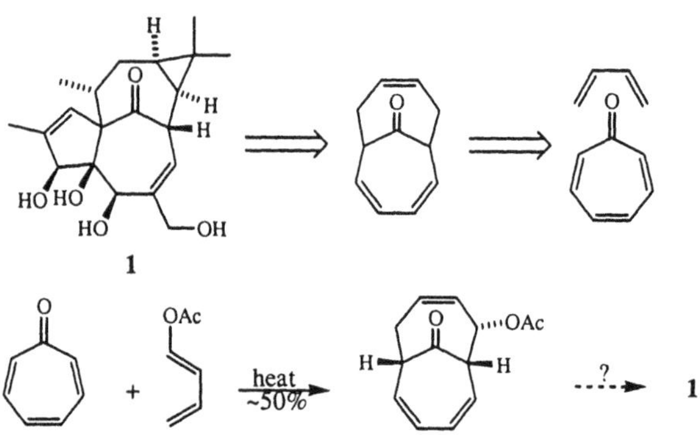

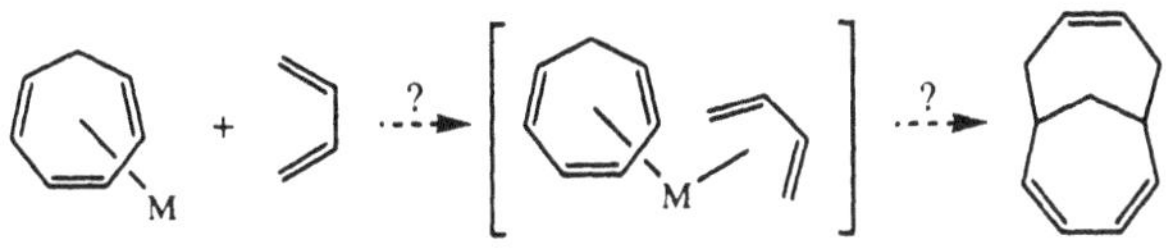

Scheme 1.

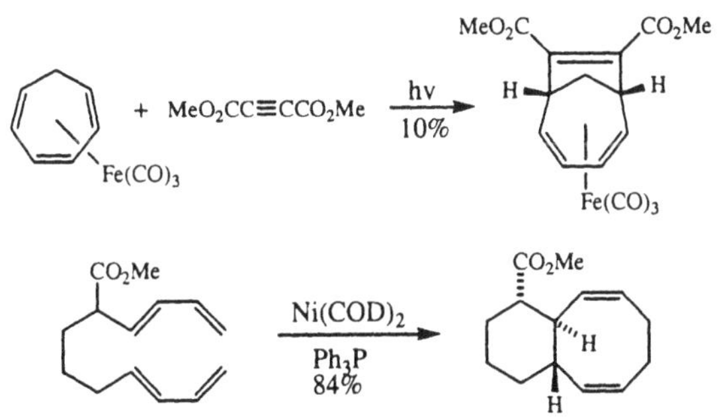

Figure 3. Other examples of metal-mediated higher order cycloadditions.

embarked on a systematic study of transition metal-mediated cycloaddition chemistry with the goal of developing reactions that could be useful in complex natural product synthesis.

II. THE CHROMIUM(0)-PROMOTED [6π+4π] CYCLOADDITION REACTION

This section of the review will describe the most significant developments in the study of the Cr(0)-mediated [6π+4π] cycloaddition process that have occurred in our laboratory since 1995. In addition, there will be particular emphasis placed on the utility of the method for natural product synthesis throughout the following discussions. Thorough accounts of the developmental phases of these investigations, as well as mechanistic treatments, have appeared elsewhere, and the interested reader should consult these sources for further information in this regard.[11]

Equations 2–4 present the salient characteristics of the photochemical Cr(0)-promoted [6+4] cycloaddition process as it is currently practiced. Chemical yields are uniformly high, and, in contrast to the Diels–Alder reaction, wherein diene/dienophile electronics must be carefully matched, reaction efficiencies are independent of the electronic nature of the participants. The reactions feature a high level of stereoselectivity in which the isomer derived from an *endo* transition state prevails in each case. This is particularly noteworthy since the thermal, metal-free [6+4] process is known to proceed via an *exo* transition state, rendering the two reaction pathways stereocomplementary.[5] An additional stereochemical feature of the process is revealed in the conversion of complex **5** to adduct **6**, in which the diene partner reacts with the triene complex exclusively on the face bearing the metal center. Thus, as many as five contiguous stereogenic centers can be reliably produced in one operation using this chemistry. Furthermore, the intrinsic facial bias of the bicyclo[4.4.1]undecane system ensures that additional substituents can be installed with complete control of stereogenicity.

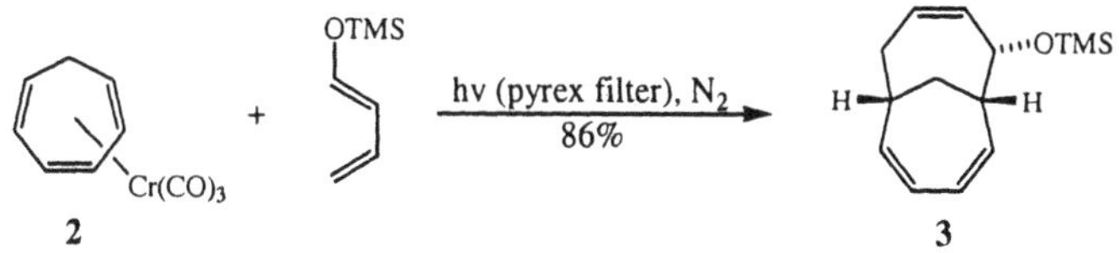

(2)

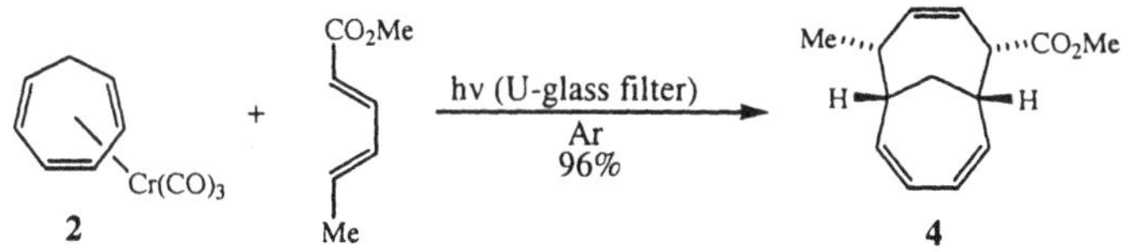

(3)

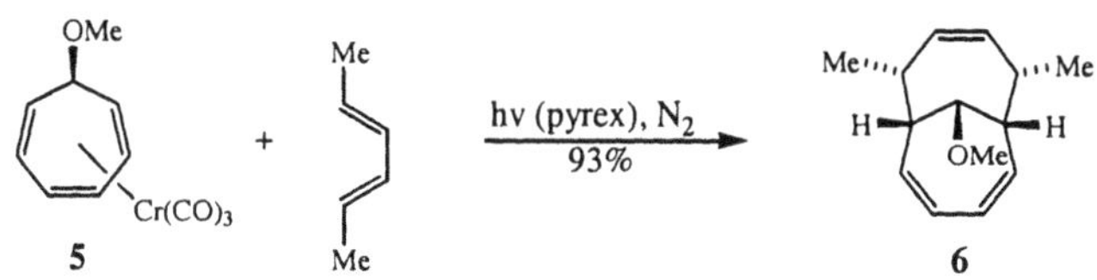

(4)

A critical advance in the development of metal-promoted [6+4] cycloaddition as a synthetically useful tool occurred with the implementation of a thermally activated process employing only substoichiometric quantities of metal.[12] A typical example of this "catalytic" cycloaddition is depicted in Eq. 5. Other sources of "$Cr(CO)_3$," such as (η^6-naphthalene)tricarbonylchromium(0),[13] are also effective precatalysts in this reaction. A critical feature of these reaction conditions is the presence of magnesium powder, which serves to reduce oxidized chromium species that accumulate during reaction back to the catalytically active Cr(0) oxidation state. It is noteworthy that little cycloaddition occurs in the absence of this additive (Figure 4).

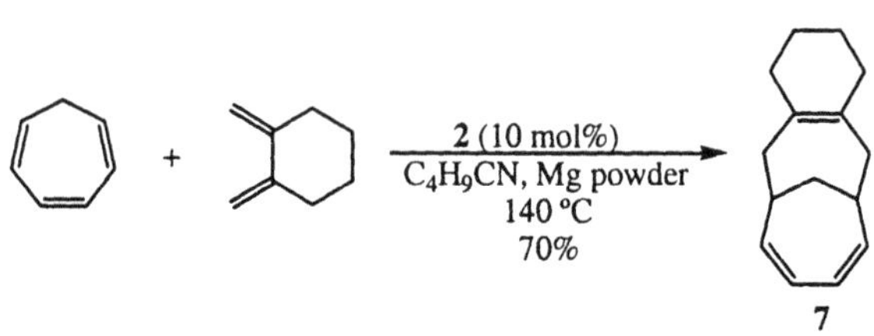

(5)

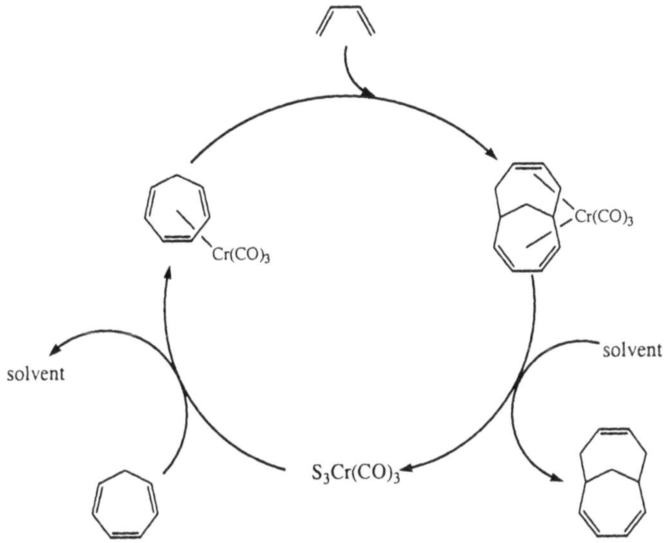

Figure 4. Possible catalytic cycle for the Cr(0)-mediated [6π+4π] cycloaddition process.

The ability, in general, to effect cycloadditions with high levels of asymmetric induction is an issue of contemporary importance, and various auxiliary-controlled methods have been found to provide higher order adducts exhibiting excellent enantiomeric purities. Cycloaddition of the readily available, enantiomerically pure complex **8** (X_c = (+)-camphorsultam), for example, afforded the [6+4] adduct **9** with extremely high levels of diastereoselection. Hydrolysis allowed for recovery of the auxiliary and provided the bicyclo[4.4.1]undecenone product in enantiomerically pure form.[14] Auxiliaries located on the 4π partner can also be effective for inducing asymmetry in the corresponding cycloadducts.[15]

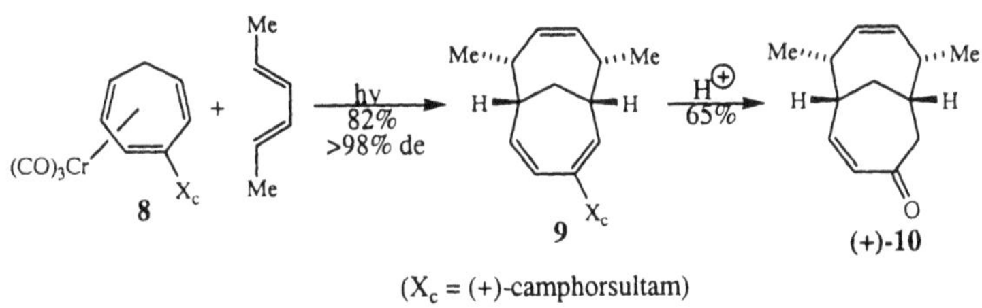

(6)

A fascinating method for accessing enantiomerically pure triene complexes has been identified that exploits the surprising stability of the enol function within the cycloheptatriene ligand framework. Thus, complex **11** can be desilylated to afford the racemic enol complex **12**. Derivatization of this stable material with (−)-α-methoxyphenylacetic acid, followed by diastereomer separation and auxiliary removal furnished both enantiomers of **12** in optically pure form.[16] These species can then be further utilized in cycloadditions of considerable synthetic importance.

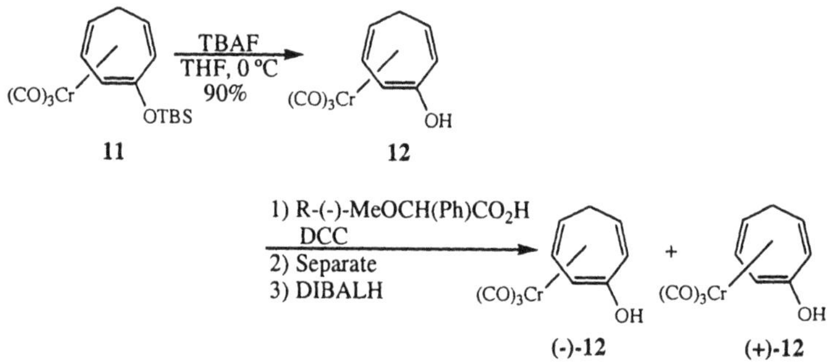

(7)

Another effective route into enantiomerically enriched cycloadducts is via enzymatic resolution of appropriately functionalized bicyclo[4.4.1]undecane intermediates. Various lipases have proven useful for delivering systems with quite good enantiomeric excesses (Eq. 8).[17] It is noteworthy that prior to these investigations,

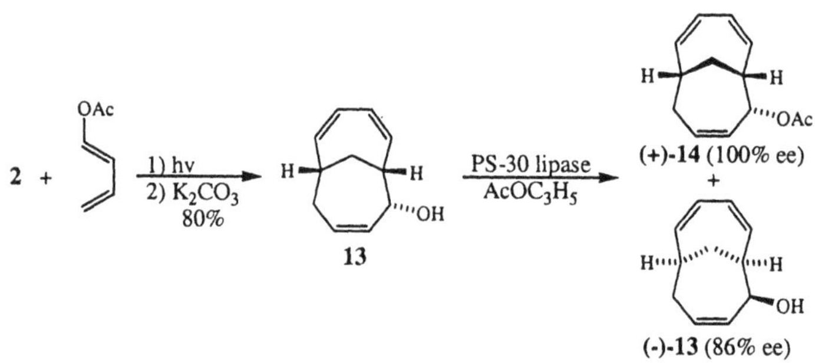

(8)

it was unclear as to whether enzymes would accept these bicyclic systems as substrates. The resultant enantiomerically enriched adducts have been successfully carried forward in an efficient synthesis of enantiomerically enriched, substituted 1,6-methano[10]annulene products as outlined in Eq. 9.[17]

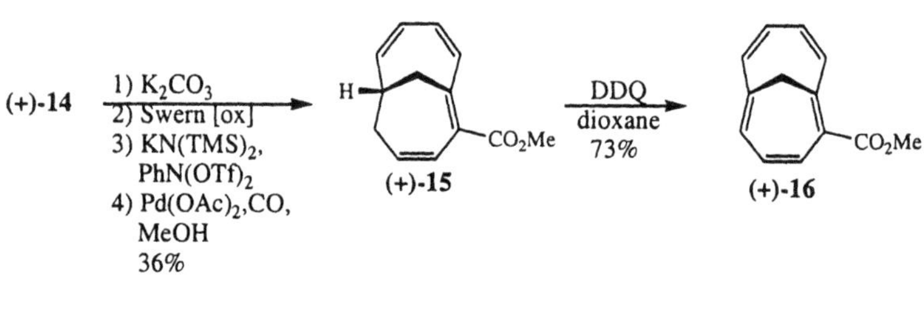

(9)

It was noted earlier that placing an appropriate chiral auxiliary on the diene partner can also afford cycloadducts with useful levels of enantiomeric enrichment. Equation 10 shows a sequence of transformations leading from enantio-enriched bicycle **18** to compound **21**. This latter material represents the C_5 through C_{11} segment of the ansa bridge of the intriguing antibiotic streptovaricin D,[19] and this set of operations nicely illustrates that the products emerging from these higher-order cycloadditions are well-suited to a range of synthetically advantageous post-cycloaddition manipulations. Other stereochemically rich building blocks can be

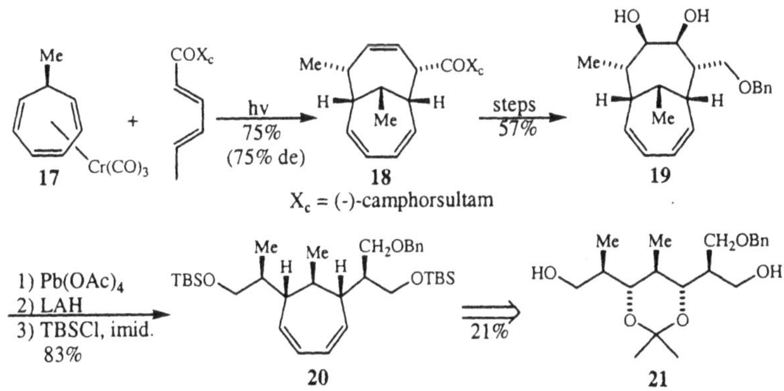

(10)

accessed by processing the 1,3-butadiene function that is produced during the cycloaddition event. Thus, cycloadduct **22** (from 2,4-hexadiene and 7-*exo*-methoxycycloheptatriene) can be conveniently transformed via electrocyclization into tricycle **23**, which can, in turn, be cleaved oxidatively to afford a bicyclo[4.2.1]nonane derivative **24**. It is significant that every ring carbon of this compound possesses a stereogenic center.[20]

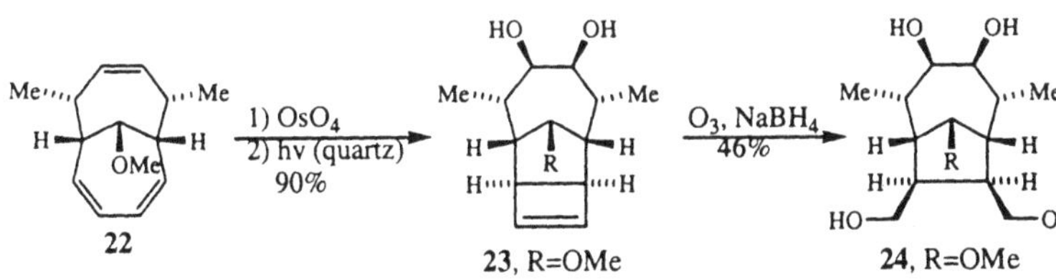

(11)

Another very appealing post-cycloaddition manipulation that is potentially available to certain functionally modified bicyclo[4.4.1]undecane systems would feature a heteroatom extrusion step that would afford various carbocyclic products that can be difficult to make in other ways. As depicted in Scheme 2, an appropriate heterocycle-based complex could undergo conventional [6+4] cycloaddition to afford a bicyclic product with the heteroatom strategically located for convenient excision. Depending on the method of extrusion, either 10-membered carbocycles

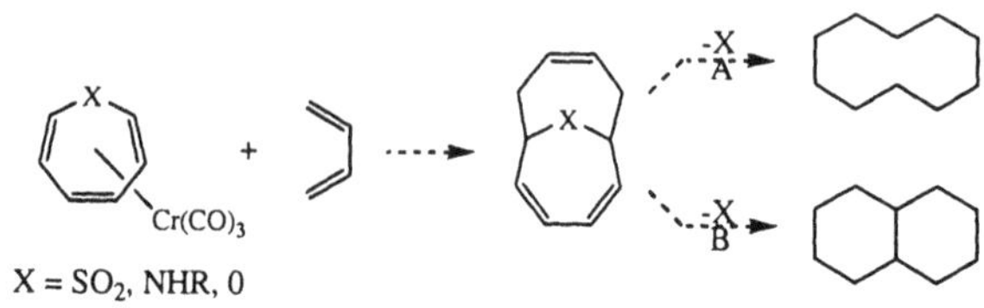

Scheme 2.

or fused bicycles could result. During the formative stages of this investigation it was reasoned that some of the unique chemistry of the carbon–sulfur bond offered a number of possible avenues for bringing this concept to practice.

Photocyclization (uranium glass filter) between the novel thiepin dioxide complex **25**[21] and 1-acetoxybutadiene afforded the bicycle **26** as a single diastereomer, again derived from an *endo* transition state. Subsequent photochemical cheletropic extrusion (quartz filter) of sulfur dioxide afforded the all (*Z*)-cyclodecatetraene **27**. This type of intermediate is currently being used to assemble germacranolide **28**.

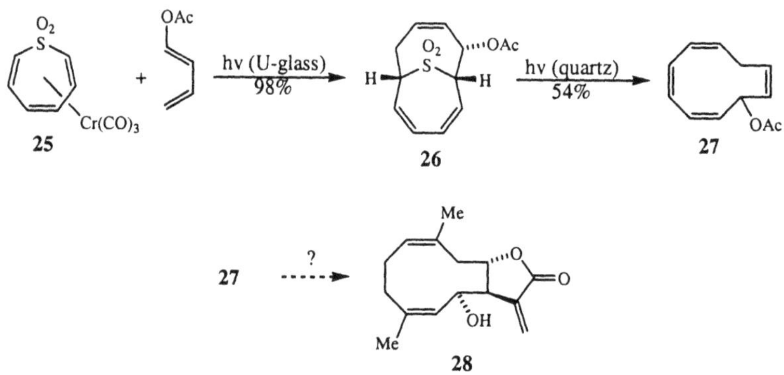

(12)

An alternative extrusion protocol can be envisioned in which the thiepin dioxide cycloadduct could be subjected to Ramberg–Bäcklund conditions to effect SO_2 excision to afford a benzo-fused adduct.

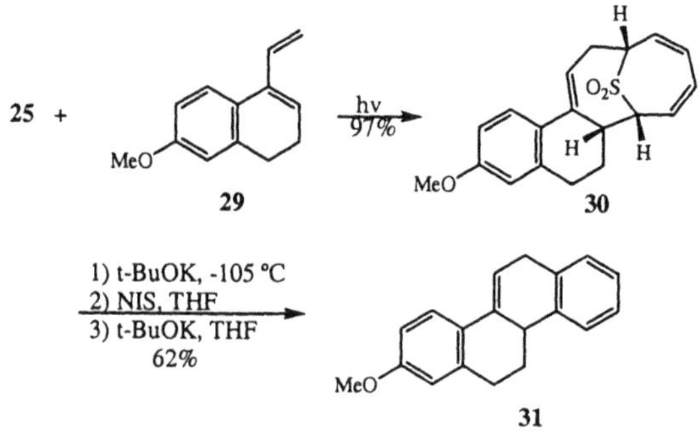

(13)

In a typical example of the concept being brought to practice, complex **25** underwent smooth photocycloaddition with the structurally elaborate diene **29** to afford tetracycle **30** as a single diastereomer in virtually quantitative yield.[22] Subsequent exposure to slightly modified Ramberg–Bäcklund conditions[23] provided the chrysene derivative in good yield. This sequence is quite general and can be successfully applied to a number of complex target molecules. Noteworthy features of this protocol include the simultaneous elaboration of two rings rather than one and the fact that all six of the carbons comprising the incipient arene ring come directly from the thiepin dioxide ligand.

The capability of producing structurally elaborate and stereochemically rich bicyclo[4.4.1]undecane systems through metal-promoted higher order cycloaddition has clearly afforded many new synthetic opportunities that were essentially inaccessible previously due to the general inefficiencies and limited scope of the corresponding thermal, metal-free versions of these reactions. The notion that the bicyclo[4.4.1]undecane core could provide a versatile synthetic building block that could be profitably transformed into a range of target systems not necessarily structurally related to the initial cycloadduct was a direct consequence of the power and efficiency of the metal-promoted higher-order cycloaddition process. Along these lines, a "unified" entry into four distinct diterpene families was devised by considering post-cycloaddition rearrangements that could be carried out on the basic bicyclo[4.4.1]undecane core system. The salient features of this program are delineated in Scheme 3. Direct conversion of the bicyclo[4.4.1]undecane system that emerges from the Cr(0)-[$6\pi+4\pi$] cycloaddition into ingenol is obvious. On the other hand the bond reorganization labeled "a" leading to the isomeric bicyclo[5.4.0]undecane that comprises the BC ring substructure of the related diterpene phorbol is perhaps a less obvious relationship. A range of other conversions are also possible starting from the bicyclo[4.4.1]undecane intermediate that could lead to the taxane system and to substituted nine-membered carbocycles. We have recently brought each of these "post-cycloaddition" manipulations to practice in relevant model studies.

The most obvious application of metal-mediated cycloaddition chemistry, of course, is the construction of the potent tumor-promoting diterpene, ingenol, and, indeed, it was this molecule that originally stimulated our entry into these investigations. Scheme 4 depicts the key strategic considerations for attacking this problem employing Cr(0)-mediated [6+4] cycloaddition. In addition to the highly convergent assembly of the entire ABC tricycle via intramolecular Cr(0)-mediated [$6\pi+4\pi$] cycloaddition,[25] the strategy addresses the installation of the crucial, highly strained "inside, outside" or *trans*-intrabridgehead stereochemical relationship.[24] This interconversion constitutes a major challenge to the synthesis of this target molecule.

A key feature of the ingenol strategy focuses on rapid and convergent construction of the key tricyclic array via intramolecular cycloaddition. This process exploits two consecutive Cr(0)-facilitated pericyclic events to afford the final

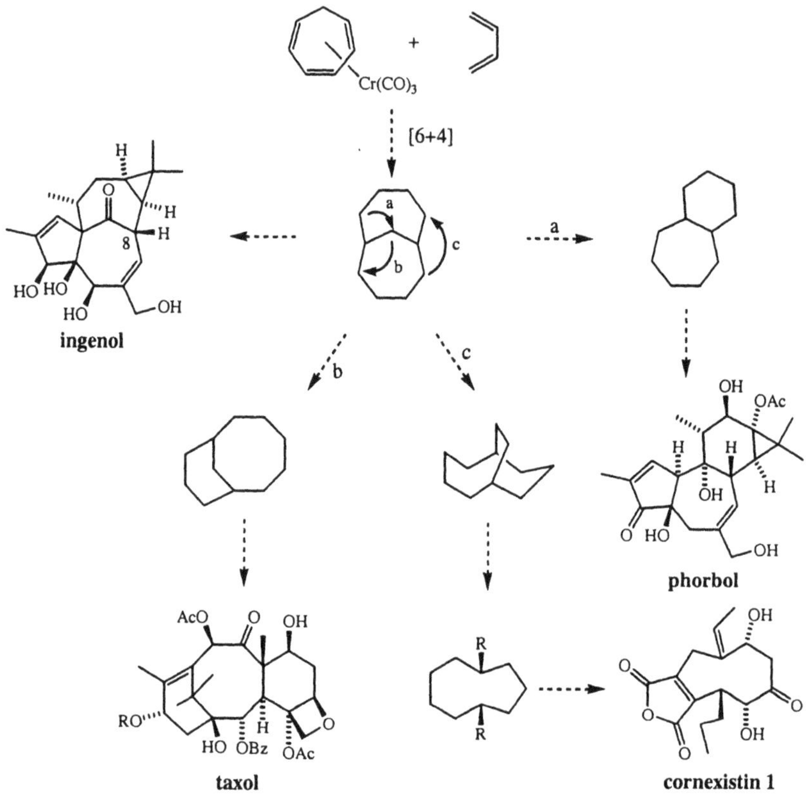

Scheme 3.

product, all carried out in one pot. First, thermal 1,5-hydrogen sigmatropy mediated by the metal center[26] equilibrates the initial triene 7-*exo*-complex into a mixture of all possible positional isomers. Of these various isomers only the 1-substituted complex (**34a**) can undergo [6+4] cycloaddition, thus removing this material from the equilibrium and eventually driving the entire process to the desired product. This reaction scheme is brought to practice as illustrated in Eq. 14, wherein 7-*exo*-cycloheptatriene complex **34** is heated to set up the equilibrium among all

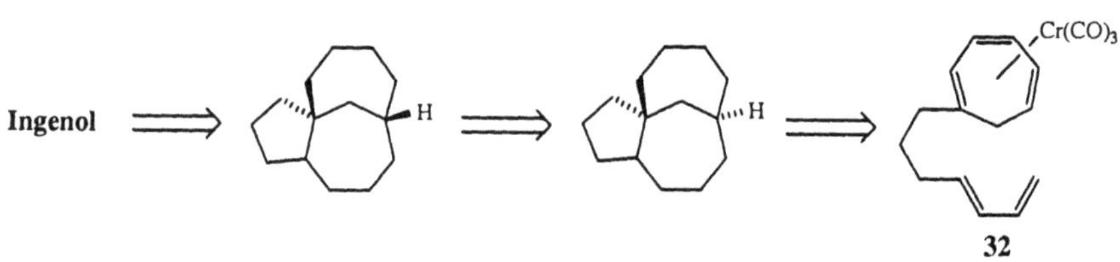

Scheme 4.

possible isomers. Due to geometric constraints only the 1-substituted isomer (**34a**) can undergo effective cycloaddition, ultimately giving product **35** in excellent yield.

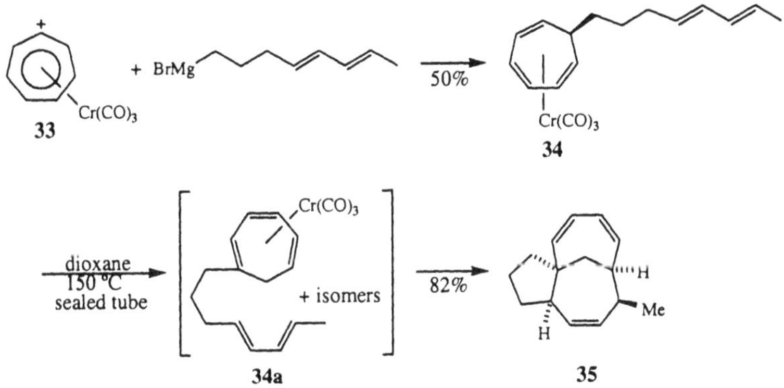

(14)

In a subsequent ingenol model study, complex **36** was most effectively converted to tricycle **37** in high overall yield by employing a two-step process that included a thermal rearrangement followed by a photochemical cycloaddition.[27] Routine functional group manipulation of the adduct afforded epoxide **38** in modest yield. The low yield in this reaction is due to the lack of regioselectivity in the epoxidation step. Treating **38** with lithium diethylamide provided the key dienol **39** required for delivery of the β-oriented hydrogen at C-8 by employing the intriguing, but little used alkoxide accelerated 1,5-H sigmatropy.[28] To our delight, exposing compound **39** to KH/18-crown-6 at 0 °C afforded a good yield of the desired inside–outside-isomer **40**, the structure of which was confirmed by single-crystal X-ray analysis. Thus rapid entry into a highly functionalized and highly strained in–out-ingenol ABC tricycle has been established using Cr(0)-promoted higher-order cycloaddition as the key strategy-level transformation.

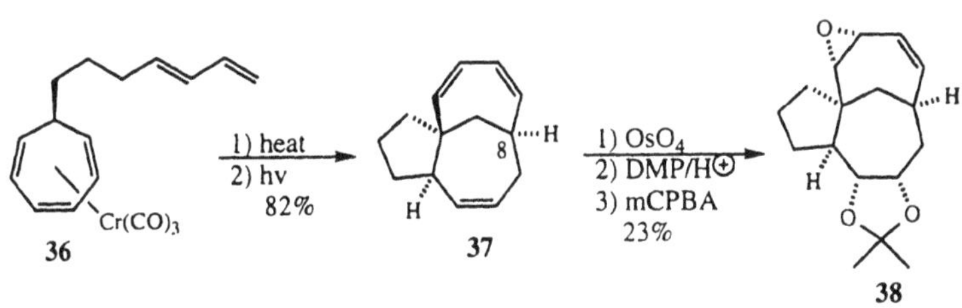

(15)

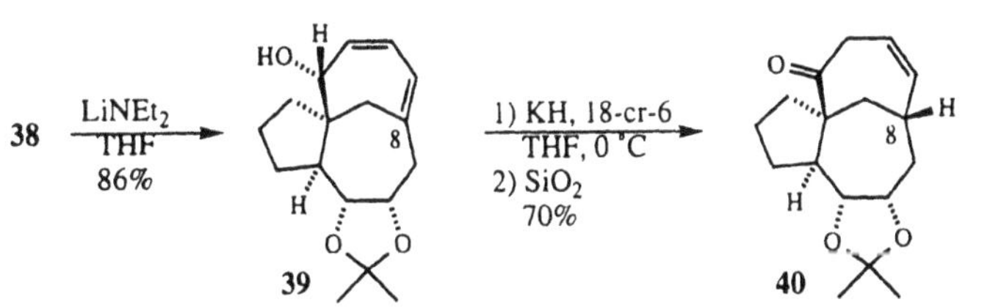

(16)

Phorbol is a diterpene that structurally is closely related to ingenol, and the BC ring moieties of the two compounds can, in principle, be interconverted by simply performing the bond migration labeled (a) in Scheme 3. Ultimately, this projected bond reorganization should be greatly facilitated by the release of the considerable strain associated with the inside–outside topography of the ingenol tricycle; it is noteworthy that there is one report in which ingenol is, in fact, converted into a tigliane (phorbol) ring structure by this process.[29]

In a model study conducted to test the viability of this notion, access to the tigliane (phorbol) BC ring system was envisioned to occur through either a pinacol or α-ketol rearrangement starting from the readily available (out–out)-bicyclo[4.4.1]undecane **42** (Eq. 17).[30] Thus, routine functional group processing from

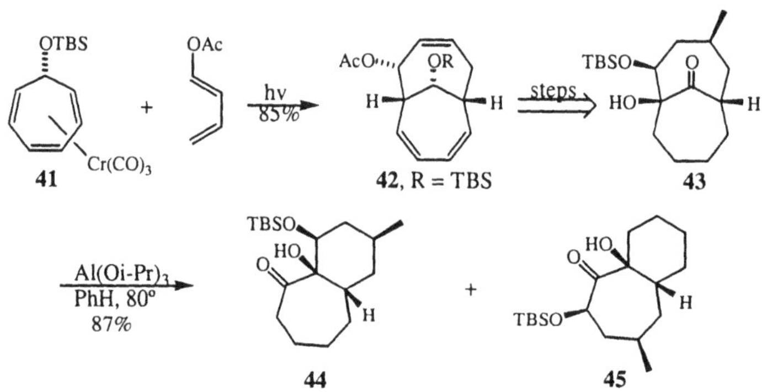

(17)

42 afforded α-ketol **43**, which underwent smooth, but not regioselective, reorganization to the two bicyclo[5.4.0]undecanes **44** and **45**. Despite the production of two products, these results confirm the utility of rearranging appropriately functionalized bicyclo[4.4.1]undecane systems into isomeric bicyclo[n.m.0]-undecane systems as a general route into phorbol as well as related diterpenes targets.

In another application of this strategy, the taxane ABC ring system can also be quickly prepared starting with Cr(0)-mediated [6π+4π] cycloaddition.[30] In the event, efficient reaction of complex **46**, derived from eucarvone, and 1,2-dimethylenecyclohexane, gave adduct **47** in 81% yield. Further processing, in a manner not unlike that in the phorbol model study, furnished α-ketol **48**, which underwent smooth rearrangement to the taxane tricycle **49** in quite high yield when exposed to $Al(OiPr)_3$ in refluxing benzene. An interesting feature of this sequence is that the reversible α-ketol rearrangement is driven exclusively to the desired bicyclo[5.3.1]undecane isomer via base-induced β-elimination of the ring-fusion epoxide in **48**.

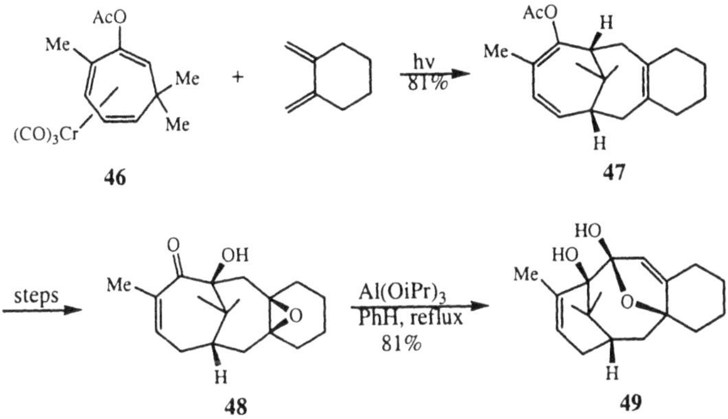

(18)

Finally, nine-membered carbocycles can also be prepared by a third rearrangement pathway. Cycloadduct **42** can be converted into diol **50** using conventional chemistry that was well established in the preceding studies. Subsequent mesylation of the 2° alcohol in this compound and exposure of the resultant mesylate to silica gel precipitated a bond reorganization to afford the isomeric bicyclo[4.3.2]undecane **51**. Oxidative cleavage and reduction yielded the cyclononadiene **52**. This protocol can be used to prepare a number of nine-membered ring diterpenes such as cornexistin I (see Scheme 3).

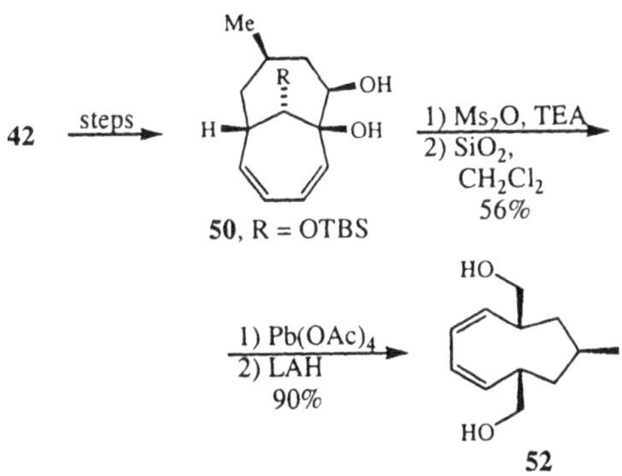

(19)

From the model studies described above, it is clear that functionalized bicyclo[4.4.1]undecane systems are quite versatile and powerful building blocks for complex synthesis. They have proven to be amenable to interconversions into structurally different ring systems that often bear little resemblance to the starting materials. These synthetic opportunities were made possible because of the unique characteristics of Cr(0)-promoted higher-order cycloaddition.

III. THE CHROMIUM(0)-PROMOTED [6π+2π] CYCLOADDITION REACTION

During the investigation of the Cr(0)-mediated [6+4] cycloaddition process, we had occasion to examine the viability of the related [6π+2π] process, which, based on mechanistic reasoning, appeared to be a good candidate for Cr(0) mediation. To our delight, the reaction proceeded with great facility to afford highly functionalized bicyclo[4.2.1]nonane adducts derived exclusively from an *endo* transition state.[32,33] The reaction also proved to be amenable to thermal activation employing a sub-stoichiometric quantity of Cr(0)-precatalyst.[34] Unlike the corresponding "catalytic" [6π+4π] cycloaddition described previously, the [6+2] process was found to be particularly facile with no complications due to slow turnover rates.

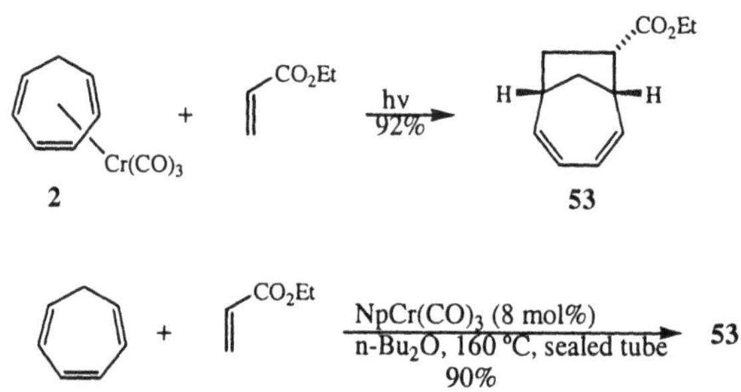

(20)

An interesting feature of the [6+2] reaction was that it could be effected using larger ring triene reaction partners, an attribute not shared by the [6+4] process. Thus, (η^6-1,3,5-cyclooctatriene)tricarbonyl-chromium(0) (**54**) underwent clean photochemical [6+2] cycloaddition with ethyl acrylate to afford compound **55**.[35] The corresponding cyclooctatetraene complex also gave [6+2] adducts in good yields. Furthermore, efforts to effect [6+4] cycloaddition of **54** with a simple diene partner failed, providing only the corresponding [6+2] adduct as a single diastereomer.

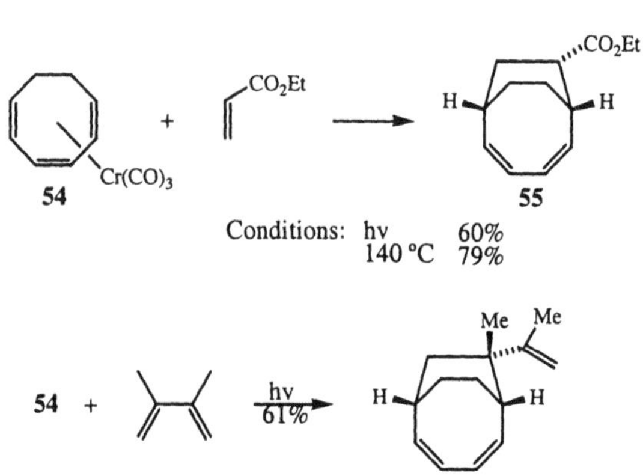

(21)

A particularly noteworthy feature of this process is that the efficiency of the corresponding thermally activated reaction was considerably greater than the photochemical version (Eq. 21).

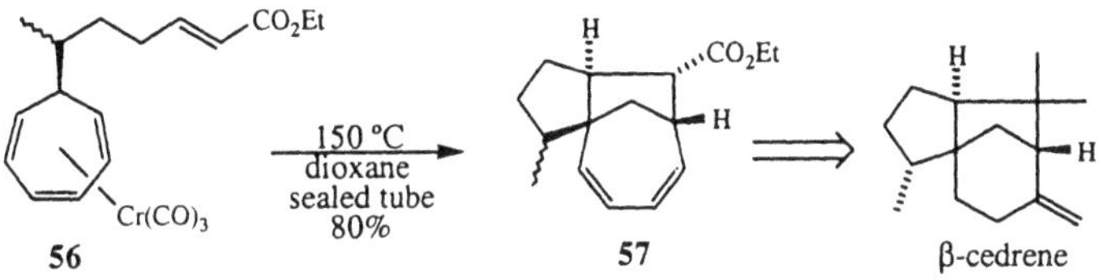

(22)

In a parallel observation, the thermally-induced intramolecular [6+2] process was found to be considerably more effective at delivering adducts than the corresponding [6+4] reaction discussed previously in this document. As before, a tandem Cr(0)-promoted 1,5-H-shift-[6π+2π] cycloaddition process was exploited for this rapid, one-pot assembly of the target tricycle. Thus, readily available **56** afforded **57** in excellent yield, and the latter compound was transformed into β-cedrene in short order using straightforward chemistry.[36]

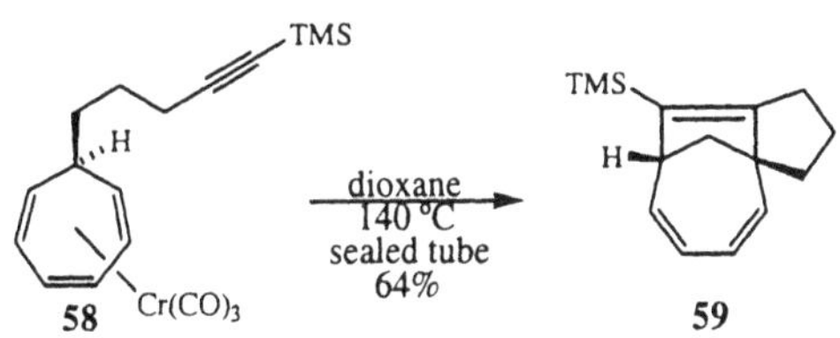

(23)

A closely related thermal cycloaddition of a tethered alkyne also proved effective for providing a range of structurally and functionally elaborate tricyclic products.[37] Indeed, alkyne-based 2π partners have been shown to be particularly versatile participants in various cycloaddition processes in our laboratory.

As an example of this versatility, alkyne cycloaddition with thiepin dioxide complex **25** allowed for rapid access to a variety of substituted cyclooctatetraene products via a [6+2] cycloaddition/photo-SO_2 extrusion sequence.[38] Unusual 2π partners such as cyclooctyne (**60**)[39] were well-behaved addends in this context (Eq. 24).

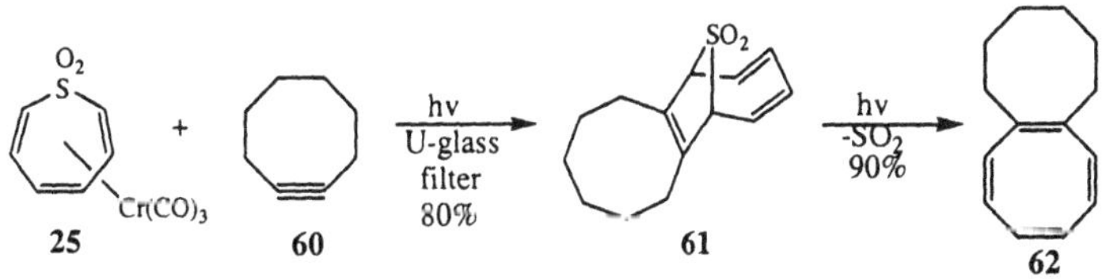

(24)

One of the most remarkable Cr(0)-mediated transformations that has emerged from our studies in this area was an efficient, highly selective three-component triene/alkyne cycloaddition process.[40] Equations 25 and 26 illustrate some of the key features of the reaction. The most notable characteristic of the overall process is the rapid increase in molecular complexity that accompanies the multistep cycloaddition event.[41,42] Fully, four new rings are created when the two alkynes are tethered together as in the example depicted in Eq. 26. Furthermore, the intermolecular examples studied to date have been shown to produce a single regioisomer in each case, thus adding to the potential synthetic utility of the method. Experimental observations made during this study suggest that a stepwise [6π+2π] cycloaddition-homo [6π+2π] cycloaddition sequence is followed in this reaction (Figure 5).

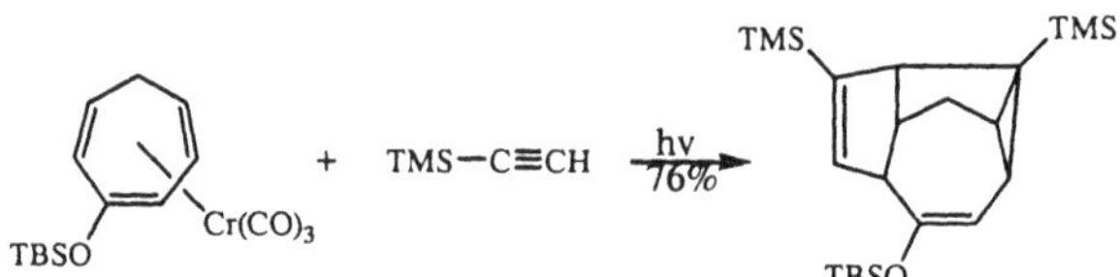

63 64 (25)

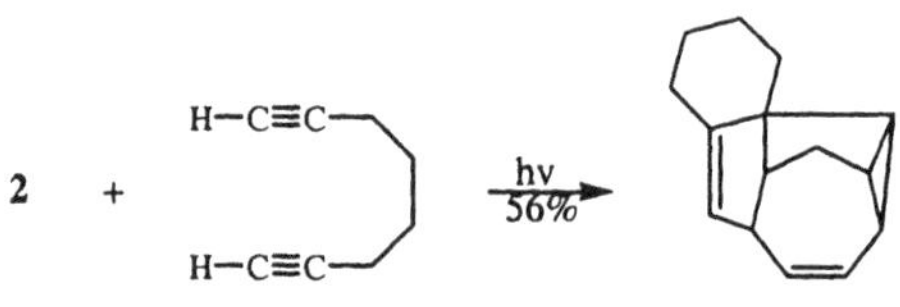

65 (26)

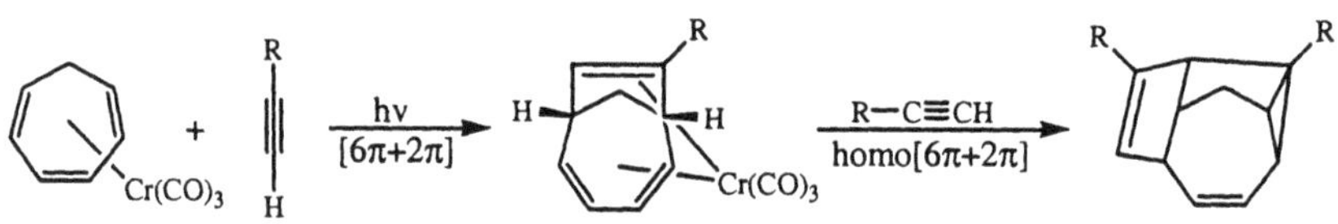

Figure 5. Putative pathway for the [6π+2π+(2π)] cycloaddition.

Nitrogen-based reaction partners have been particularly useful in applications of Cr(0)-promoted [6+2] cycloaddition to natural product synthesis. Isocyanates can be employed as novel 2π partners to afford highly functionalized azabicyclo[4.2.1]nonane products that can be subsequently transformed into the interesting and often difficult to prepare 6-azabicyclo[3.2.1]octane systems (Eq. 27).[43]

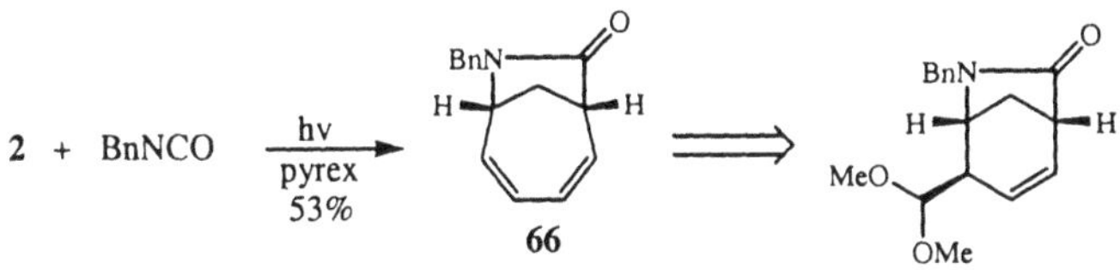

(27)

In closely related chemistry, chromium(0)-promoted [6π+2π] cycloaddition of an azepin-based complex would be expected to provide for a particularly rapid entry into functionalized homotropane ring systems[44] suitable for subsequent conversion into the tropane skeleton via ring contraction.[45] Critical to the success of this strategy is the ability to efficiently contract the homotropane system into the tropane bicycle, and while the 1,3-butadiene moiety that emerges from the cycloaddition event is ostensibly a useful function for effecting this conversion, many otherwise attractive methods failed to deliver significant quantities of the desired product (Scheme 5). In the end, however, a novel application of the well-known Taylor–McKillop reaction succeeded admirably for this purpose.[45]

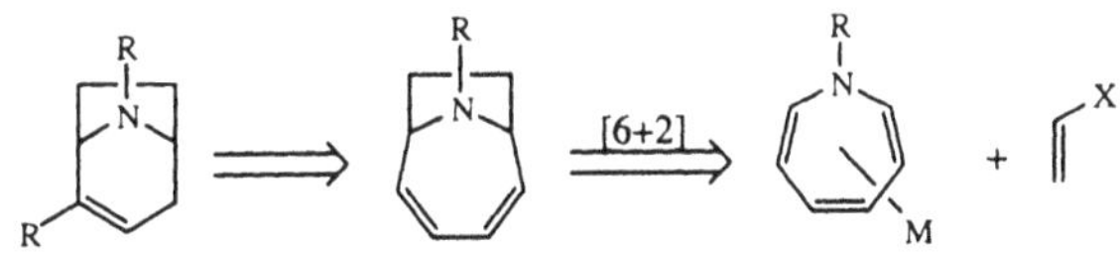

Scheme 5.

In the event, auxiliary-controlled [6+2] photocyclization of azepin complex **67** with (–)-8-phenylmenthyl acrylate afforded a serviceable yield of adduct **68** with high diastereoselectivity.[46] After considerable experimentation, a Tl(III)-mediated oxidative bond reorganization (Taylor–McKillop reaction) afforded tropane **69** in excellent yield and as a single regio- and stereoisomer. The extraordinary regiocontrol exhibited by this reaction may be due to electronic effects of the carboalkoxy group present on the two-carbon bridge. Routine functional group manipulation of **69** afforded enantiomerically pure (+)-ferruginine (**70**) (Eq. 29). This synthesis represented the first total synthesis of this tropane alkaloid in enantiomerically pure form, and the success of this strategy for delivering these species with high stereoselectivity suggests that a range of tropane alkaloids can be assembled in quite similar fashion.

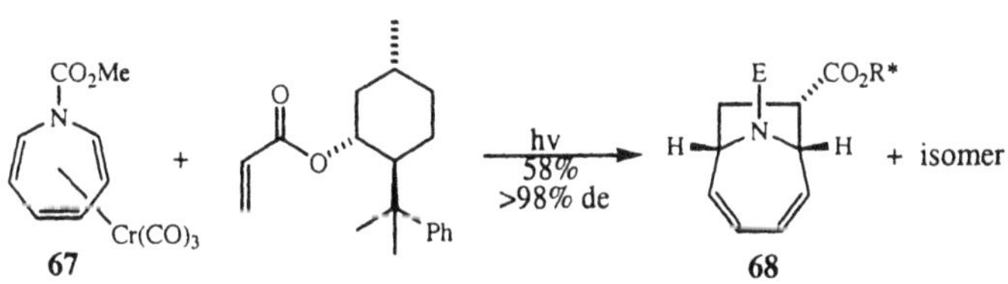

(28)

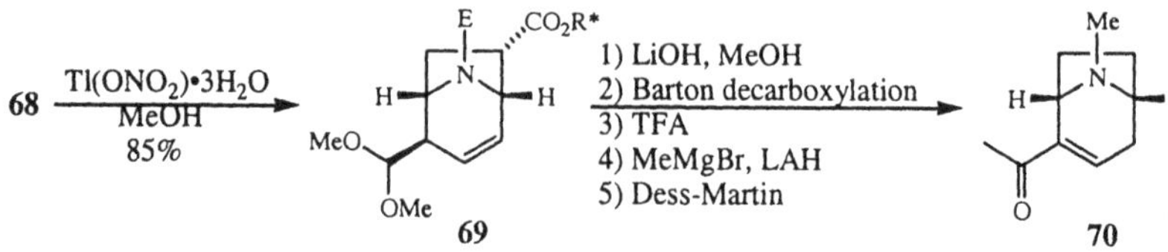

(29)

IV. CONCLUSIONS

Once only a laboratory curiosity, higher order cycloaddition can now be included among the more powerful ring construction methods available in the contemporary synthetic repertoire. With the advent of the chromium(0)-mediated versions of these cycloadditions, many ring systems that were previously either very difficult or even impossible to make using conventional methods are now readily accessible, often in enantiomerically enriched form. The capability of creating polycyclic arrays that are sufficiently functionalized for subsequent conversion into other systems is one of the chief attributes of this new chemistry. Furthermore, since metal-promoted higher order cycloadditions proceed with concomitant high levels of predictable stereoselection, they represent appropriate starting points for complex natural product synthesis. The many examples in this review are a convincing testament to this last point.

Many new applications of this technology to natural product synthesis can be anticipated in the near future, as can the development of a number of other powerful metal-mediated ring construction methods.

ACKNOWLEDGMENTS

I wish to thank the National Institutes of Health for their generous support of this research program over the years. I am also particularly grateful for the hard work and enthusiasm of the many coworkers who have made this research possible.

REFERENCES AND NOTES

1. For a recent overview of commonly used cycloaddition reactions, see: *Comprehensive Organic Synthesis*; Trost, B. M., Fleming, I., Eds.; Pergamon: Oxford, 1991, Vol. 5, various chapters.
2. For a recent review of metal-mediated cycloaddition, consult: Lautens, M.; Klute, W.; Tam, W. *Chem. Rev.* **1996**, *96*, 49.
3. (a) Rigby, J. H.; Pigge, F. C. *Org. Reactions* **1997**, *51*, 351. (b) Harmata, M. *Tetrahedron* **1997**, *53*, 6235.
4. (a) Wender, P. A.; Husfeld, C. O.; Langkopf, E.; Love, J. A. *J. Am. Chem. Soc.* **1998**, *120*, 1940. (b) Etkin, N.; Dzwiniel, T. L.; Schweibert, K. E.; Stryker, J. M. *J. Am. Chem. Soc.* **1998**, *120*, 9702. (c) Wender, P. A.; Takahashi, H.; Witulski, B. *J. Am. Chem. Soc.* **1995**, *117*, 4720. (d) Huffman, M. A.; Liebeskind, L. S. *J. Am. Chem. Soc.* **1993**, *115*, 4895.
5. For a comprehensive review of [6π+4π] cycloadditions, see: Rigby, J. H. *Org. Reactions* **1997**, *49*, 331.

6. (a) Houk, K. N.; Woodward, R. B. *J. Am. Chem. Soc.* **1970**, *92*, 4143. (b) Dunn, L. C.; Houk, K. N. *Tetrahedron Lett.* **1978**, 3411.
7. For a few synthetic uses of thermal higher-order cycloadditions, see: (a) Rigby, J. H.; Cuisiat, S. V. *J. Org. Chem.* **1993**, *58*, 6286. (b) Rigby, J. H.; Moore, T. L.; Rege, S. *J. Org. Chem.* **1986**, *51*, 2398. (c) Funk, R. L.; Bolton, G. L. *J. Am. Chem. Soc.* **1986**, *108*, 4655. (d) Garst, M. E.; Roberts, V. A.; Houk, K. N.; Rondan, N. G. *J. Am. Chem. Soc.* **1984**, *106*, 3882.
8. (a) Davis, R. E.; Dodds, T. A.; Hseu, T. H.; Wagnon, J. C.; Devon, T.; Tancrede, J.; McKennis, J. S.; Pettis, R. *J. Am. Chem. Soc.* **1974**, *96*, 7562. (b) Ward, J. S.; Pettit, R. *J. Am. Chem. Soc.* **1971**, *93*, 262.
9. (a) Wender, P. A.; Ihle, N. C. *J. Am. Chem. Soc.* **1986**, *108*, 4678. (b) Wender, P. A.; Snapper, M. L. *Tetrahedron Lett.* **1987**, *28*, 2221. (c) Wender, P. A.; Ihle, N. C.; Correia, C. R. D. *J. Am. Chem. Soc.* **1988**, *110*, 5904.
10. (a) Özkar, S.; Kurz, H.; Neugebauer, D.; Kreiter, C. G. *J. Organomet. Chem.* **1978**, *160*, 115. (b) Kreiter, C. G.; Kurz, H. *Chem. Ber.* **1983**, *116*, 1494. (c) Michels, E.; Sheldrick, W. S.; Kreiter, C. G. *Chem. Ber.* **1985**, *118*, 964.
11. (a) Rigby, J. H. In *Advances in Metal-Organic Chemistry*; Liebeskind, L. S., Ed.; JAI: Greenwich, CT, 1995, Vol. 4, pp. 89–127. (b) Rigby, J. H.; Krueger, A. C. In *Advances in Detailed Reaction Mechanisms*; Coxon, J. M., Ed.; JAI: Greenwich, CT, 1995, Vol. 4, pp. 1–40. (c) Rigby, J. H. *Acc. Chem. Res.* **1993**, *26*, 579.
12. Rigby, J. H.; Fiedler, C. *J. Org. Chem.* **1997**, *62*, 6106.
13. Desobry, V.; Kündig, E. P. *Helv. Chim. Acta* **1981**, *64*, 1288.
14. Rigby, J. H.; Sugathapala, P.; Heeg, M. J. *J. Am. Chem. Soc.* **1995**, *117*, 8851.
15. Rigby, J. H.; Ateeq, H. J.; Charles, N. R.; Cuisiat, S. V.; Ferguson, M. D.; Henshilwood, J. A.; Krueger, A. C.; Ogbu, C. O.; Short, K. M.; Heeg, M. J. *J. Am. Chem. Soc.* **1993**, *115*, 1382.
16. (a) Rigby, J. H.; Niyaz, N. M.; Sugathapala, P. *J. Am. Chem. Soc.* **1996**, *118*, 8178. (b) Rigby, J. H.; Sugathapala, P. *Tetrahedron Lett.* **1996**, *37*, 5293.
17. Rigby, J. H.; Saha, A.; Heeg, M. J. *J. Org. Chem.* **1997**, *62*, 6448.
18. Rigby, J. H.; Fales, K. R. *Tetrahedron Lett.* **1998**, *39*, 5717.
19. Rinehart, K. L., Jr.; Shield, L. S. *Prog. Chem. Org. Nat. Prod.* **1976**, *33*, 231.
20. Rigby, J. H.; de Sainte Claire, V.; Heeg, M. J. *Tetrahedron Lett.* **1996**, *37*, 2553.
21. Rigby, J. H.; Ateeq, H. S.; Krueger, A. C. *Tetrahedron Lett.* **1992**, *33*, 5873.
22. Rigby, J. H.; Warshakoon, N. C. *J. Org. Chem.* **1996**, *61*, 7644.
23. Paquette, L. A. *Org. React.* **1977**, *25*, 1.
24. Kim, S.; Winkler, J. D. *Chem. Soc. Rev.* **1997**, *26*, 387.
25. Rigby, J. H.; Rege, S. D.; Sandanayaka, V. P.; Kirova, M. J. *Org. Chem.* **1996**, *61*, 842.
26. Roth, W. R.; Grimme, W. *Tetrahedron Lett.* **1966**, 2347.
27. (a) Rigby, J. H.; Hu, J.; Heeg, M. J. *Tetrahedron Lett.* **1998**, *39*, 2265. (b) Rigby, J. H.; de Sainte Claire, V.; Cuisiat, S. V.; Heeg, M. J. *J. Org. Chem.* **1996**, *61*, 7992.
28. Paquette, L. A.; Crouse, G. D.; Sharma, A. K. *J. Am. Chem. Soc.* **1980**, *102*, 3972.
29. Hecker, E. *Pure Appl. Chem.* **1977**, *49*, 1423.
30. Rigby, J. H.; Niyaz, N. M.; Short, K.; Heeg, M. J. *J. Org. Chem.* **1995**, *60*, 7720.
31. Rigby, J. H.; Fales, K. R. *Tetrahedron Lett.* **1998**, *39*, 1525.
32. Rigby, J. H.; Henshilwood, J. A. *J. Am. Chem. Soc.* **1991**, *113*, 5122.
33. Subsequent to our initial disclosure, two reports describing [6+2] cycloaddition of alkynes to (η^6-cycloheptatriene) tricarbonyl chromium(0) appeared: (a) Fischler, I.; Grevels, F.-W.; Leitich, J.; Özkar, S. *Chem. Ber.* **1991**, *124*, 2857. (b) Chaffee, K.; Sheridan, J. B.; Aistars, A. *Organometallics* **1992**, *11*, 18.
34. Rigby, J. H.; Short, K. M.; Ateeq, H. S.; Henshilwood, J. A. *J. Org. Chem.* **1992**, *57*, 5290.
35. Rigby, J. H.; Scribner, S.; Heeg, M. J. *Tetrahedron Lett.* **1995**, *36*, 8569.
36. Rigby, J. H.; Kirova-Snover, M. *Tetrahedron Lett.* **1997**, *38*, 8153.
37. Rigby, J. H.; Kirova, M.; Niyaz, N.; Mohammadi, F. *Synlett.* **1997**, 805.

38. Rigby, J. H.; Warshakoon, N. C. *Tetrahedron Lett.* **1997**, *38*, 2049.
39. Brandsma, L.; Verkruijsse, H. D. *Synthesis* **1978**, 290.
40. Rigby, J. H.; Warshakoon, N. C.; Heeg, M. J. *J. Am. Chem. Soc.* **1996**, *118*, 6094.
41. Subsequent to our initial disclosure, a related transformation appeared: Chen, W.; Chaffee, K.; Chung, H.-J.; Sheridan, J. B. *J. Am. Chem. Soc.* **1996**, *118*, 9980.
42. A related reaction has been mediated by iron: Goddard, R.; Woodward, P. *J. Am. Chem. Soc., Dalton Trans.* **1979**, 711.
43. (a) Rigby, J. H.; Pigge, F. C. *Synlett.* **1996**, 631. (b) Rigby, J. H.; Ahmed, G.; Ferguson, M. D. *Tetrahedron Lett.* **1993**, *34*, 5397.
44. Javier Sardina, F.; Howard, M. H.; Morningstar, M.; Rapoport, H. *J. Org. Chem.* **1990**, *55*, 5025.
45. McKillop, A.; Hunt, J. D.; Kienzle, F.; Bigham, E.; Taylor, E. C. *J. Am. Chem. Soc.* **1973**, *95*, 3635.
46. Rigby, J. H.; Pigge, F. C. *J. Org. Chem.* **1995**, *60*, 7392.

INDOLE AS A DIENOPHILE IN INVERSE ELECTRON DEMAND DIELS–ALDER AND RELATED REACTIONS

Lily Lee and John K. Snyder

Advances in Cycloaddition
Volume 6, pages 119–171.

ISBN: 0-7623-0531-2

ABSTRACT

A review of cycloadditions employing the indole 2,3-double bond as the 2π component in reaction with conjugated diene systems, and with 1,3-dipoles and related mesoionic compounds; 212 literature citations.

I. INTRODUCTION

The development of indole chemistry dates back to the mid-nineteenth century to the production of the highly valued indigo dye, and progresses through the discovery of such biologically important indole-containing molecules as tryptophan, heteroauxin, and serotonin as indole derivatives.[1] In more recent times, the discovery that many medicinally important alkaloids contain the indole skeleton[2] fueled the already mounting efforts to detail the scope of the reactions of indole.[1,3] Among these investigations were numerous studies to annulate a third ring directly onto the C2/C3 enamine substructure of the parent indole skeleton, methodology that would complement and expand upon the widely employed Pictet–Spengler[4] and Bischler–Napieralski[5] chemistry to prepare β-carbolines from tryptamine and tryptophan. Robinson-type annulations using a nucleophilic indole for a Michael addition to α,β-unsaturated carbonyl compounds followed by aldol-type closures (Figure 1, Eq. 1) first appeared in the literature in the 1950s,[6] and new reports detailing ever more effective procedures for the conjugate addition of indole to α,β-unsaturated carbonyl compounds continued to emerge through the present.[7]

In more modern fashion, transition metal-catalyzed cyclizations have also been designed to annulate a third ring onto the indole 2,3-double bond.[8] The cobalt-me-

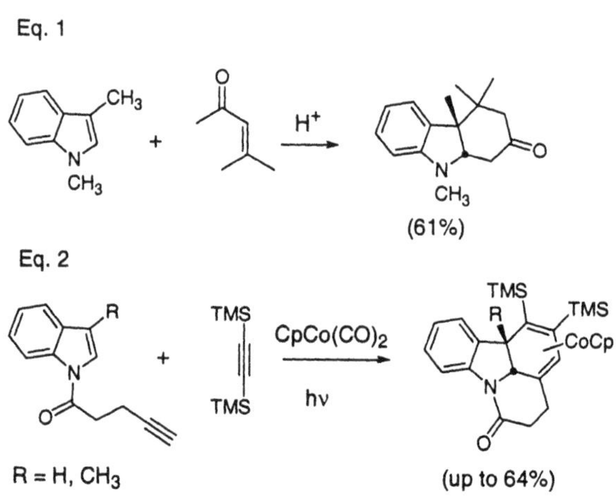

Figure 1. **(Eq. 1)** An example of Robinson-type annulations with indole reported by LeQuesne.[6d] **(Eq. 2)** Vollhardt's cobalt-catalyzed [2+2+2] annulation with alkynes.[9]

diated [2+2+2] cycloaddition of indole and alkynes reported by Vollhardt illustrates such a process involving net cycloaddition chemistry (Figure 1, Eq. 2).[9]

There are also many examples of 2- and 3-vinylindoles as a 4π component in cycloadditions to annulate a third ring onto the indole skeleton, and these subjects have been reviewed.[10] The reactivity of these compounds has been extensively developed by several groups, and has been utilized in the syntheses of various indole alkaloids, exemplified by Kuehne's approach to the Aspidosperma alkaloids via in situ-generated 2-vinylindole derivatives (the secodine approach; Figure 2, Eq. 1).[11] Indole-2,3-quinodimethanes are also well established as valuable intermediates in synthetic strategies for a variety of indole alkaloids[10b,12] illustrated by Magnus' approach to the Aspidosperma alkaloids with the in situ generation of the indole-2,3-quinodimethane transient (Figure 2, Eq. 1),[13] as well as Gribble's[14] ellipticene

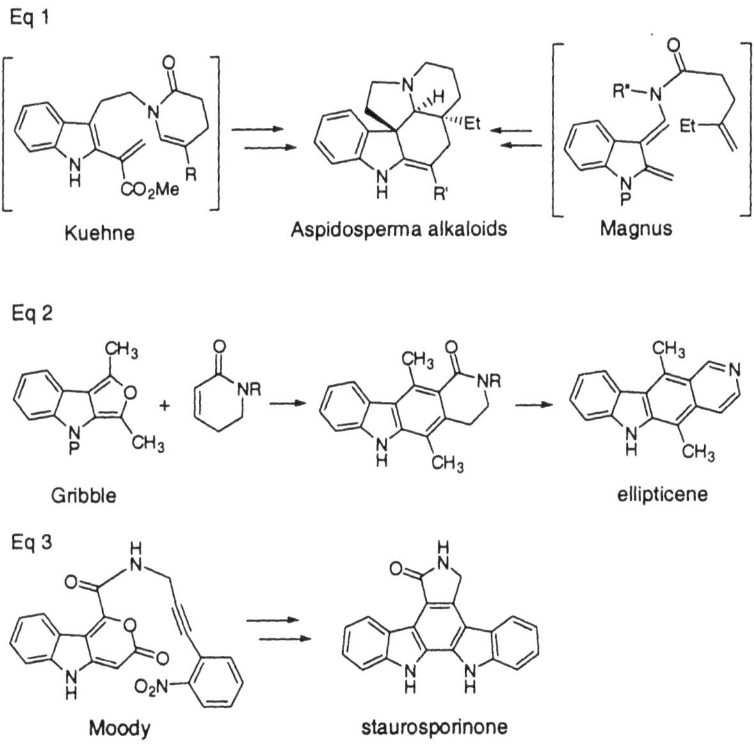

Figure 2. Examples of synthetic strategies using 2-vinylindoles and indole-2,3-quinodimethanes. **(Eq. 1)** Kuehne's 2-vinylindole (secodine),[11] and Magnus' indole-2,3-quinodimethane[13] approaches to the Aspidosperma alkaloids. **(Eq. 2)** Gribble's stable indole-2,3-quinodimethane approach to ellipticene.[14] **(Eq. 3)** Moody's stable indole-2,3-quinodimethane approach to staurosporinone.[15]

preparation with stable indolequinodimethane equivalents (Figure 2, Eq. 2). Moody has also used a stable quinodimethane strategy to prepare several carbazoles including staurosporinone (Figure 2, Eq. 3).[15]

This chapter addresses cycloadditions employing the indole 2,3-double bond as the 2π component in reaction with conjugated diene systems and with 1,3-dipoles. The majority of the reactions reviewed here with dienes are inverse electron demand Diels–Alder reactions, though formal cycloadditions with dienes that do not adhere to the criteria of inverse electron demand reactions are also discussed. Many of these reactions have been shown to be stepwise mechanistically, but no restrictions concerning the concerted nature of the chemistry are placed on the material covered.

II. INVERSE ELECTRON DEMAND DIELS–ALDER REACTIONS OF INDOLE

Inverse electron demand Diels–Alder reactions ($HOMO_{dienophile}$ - $LUMO_{diene}$ controlled cycloadditions) have become an important synthetic tool over the past two decades in organic chemistry.[16] Laudable discoveries and applications of this chemistry to the synthesis of both natural and unnatural heterocycles have emerged from several laboratories, with the work of Boger[17] and Taylor[18] being especially notable. Many of the earliest examples of inverse electron demand cycloadditions employed enamines as electron rich dienophiles in reaction with heteroaromatic azadienes such as 1,2,4,5-tetrazines and 1,2,4-triazines.

Eventually, the latent enamine functionality of the 2,3-double bond of indole (**1a**) drew attention for application in inverse electron demand cycloadditions, though the aromaticity of the 10π electron system was anticipated to reduce the reactivity of indole relative to the simple enamines previously employed in reactions with the tetrazines and triazines. Such chemistry with heteroaromatic azadienes proceeds with rapid loss of the nitrogen gas from the initially formed cycloadduct intermediate **A** (Figure 3). Aromatization of intermediate **B** then leads to carbazoles (X =

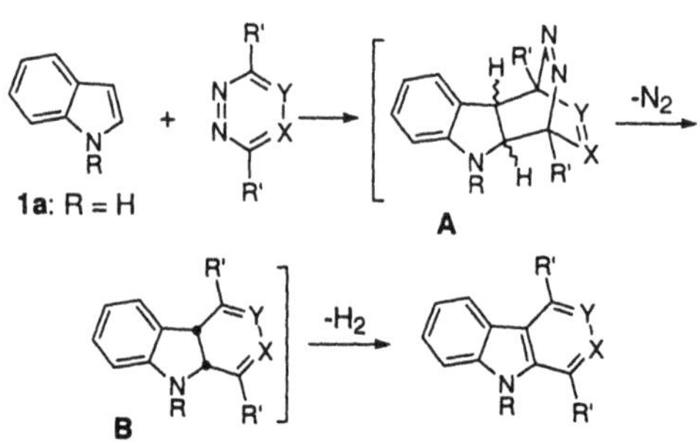

Figure 3. General strategy to annulate a C-ring onto indole using cycloadditions with heteroaromatic azadienes (X, Y = N, CR″).

Y = C–R″), β-carbolines (X = N, Y = C–R″), and other heterocycles. The first reported inverse electron demand cycloadditions of indole appeared in 1976 reported by Seitz[19] and Takahashi,[20] both with 1,2,4,5-tetrazines. Since then, a number of groups have reported success in utilizing indole as the electron-rich component in inverse electron demand cycloadditions with a variety of dienes, and many of these reactions have become key steps in natural product and analogue syntheses.

A. Reactions with Heteroaromatic Azadienes

Reactions with 1,2,4,5-Tetrazines

1,2,4,5-Tetrazines have long been known to serve as excellent electron-deficient dienes in inverse electron demand Diels–Alder reactions with electron-rich dienophiles such as enamines.[16a–c] Indeed, the earliest two examples of indole utilized as a dienophile in inverse electron demand cycloaddition reactions were with 1,2,4,5-tetrazines, and these early reports delineated the two main reaction pathways open to the chemistry following the initial cycloaddition and release of nitrogen gas. Seitz reported the Diels–Alder reaction of *N*-methylindole (**1b**) with dimethyl 1,2,4,5-tetrazine-3,6-dicarboxylate (**2a**) to produce pyridazinoquinolone **4** (45%) via rearrangement of the initially formed nonaromatized cycloadduct **3a**, which could not be detected (Figure 4, Eq. 1).[19] A very similar rearrangement had been observed earlier by Acheson leading to a minor product in his studies of the reaction of indole with dimethyl acetylenedicarboxylate (DMAD).[21] This latter reaction proceeded through initial formation of 3-vinylindole **5** followed by a Diels–Alder reaction with an additional equivalent of DMAD and subsequent rearrangement (Figure 4, Eq. 2). Interestingly, the reaction of 7-azaindole (**1c**) with **2a** yielded only the aromatized adduct **6a** (57%, Figure 4, Eq. 3).[19]

Later, Seitz also reported analogous results in the reaction between indole (**1a**) and 3,6-bis(trifluoromethyl)-1,2,4,5-tetrazine **2b** to give the corresponding cycloadduct **6b** (Figure 5, R = X = H).[22] While the yield of this reaction was only 26%, presumably due to additional consumption of **2b** as the aromatization agent, Haider was later able to obtain the same aromatized cycloadduct **6b** in 58% yield using 3-thiomethylindole (**1d**) with **2b**.[23] The improvement in yield with the 3-thiomethyl substituent was ascribed to the more facile aromatization by methanethiol elimination. Similar to the chemistry reported between tetrazine **2a** and *N*-methylindole (**1b**) discussed previously (see Figure 4, Eq. 1), **1b** underwent a cycloaddition with **2b** followed by pyrrole ring opening to produce **7a**.[22]

Also in 1976, Takahashi obtained cycloadduct **6c** (35%) from the reaction of indole (**1a**) with 3,6-di-(2-pyridyl)-1,2,4,5-tetrazine (**2c**) by heating in toluene (Figure 6).[20] It is interesting to note that Seitz did not report any aromatized cycloadduct from the reactions of *N*-methylindole with **2a**[19] and **2b**[22] in his original communications, while neither Takahashi nor Seitz mentioned any rearrangement

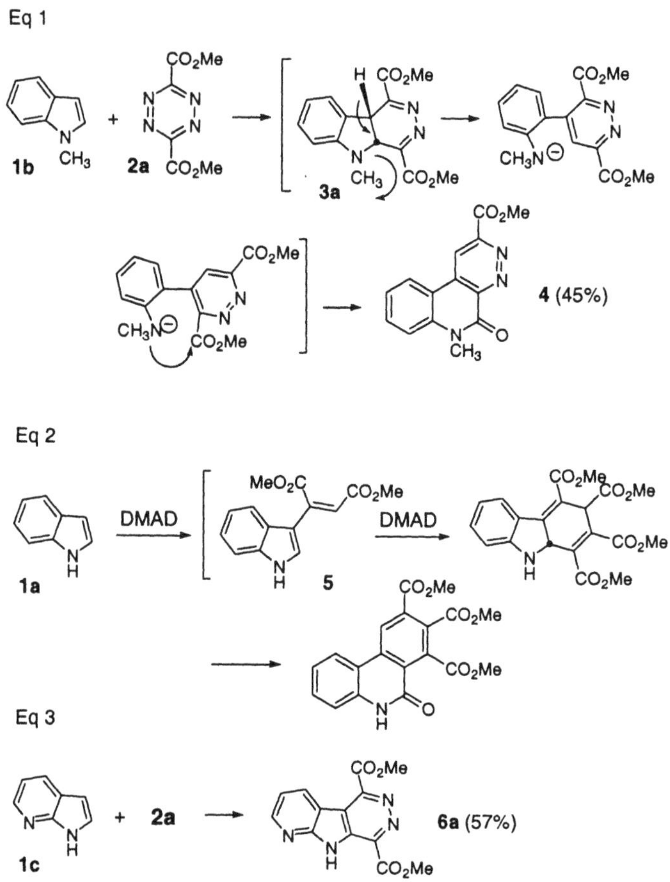

Figure 4. **(Eq. 1)** Reaction of *N*-methylindole (**1b**) with dimethyl 1,2,4,5-tetrazine-3,6-dicarboxylate (**2a**) reported by Seitz.[19] **(Eq. 2)** Initially reported rearrangement of a dihydrocarbazole in the reaction of indole with dimethyl acetylenedicarboxylate (DMAD) reported by Acheson.[21] **(Eq. 3)** Reaction of 7-azaindole (**1c**) with **2a** reported by Seitz.[19]

in reactions of indole or 7-azaindole with the free 1-NH groups. Both pathways from the dihydrocycloadduct intermediates **3**, aromatization and rearrangement with opening of the pyrrole ring, are now known to be common to the cycloaddition of indoles with 1,2,4,5-tetrazines.

Our own work began over 10 years ago with investigations to detail the reactivity of indole in a broad range of inverse electron demand Diels–Alder reactions with heteroaromatic azadienes beginning with a reexamination of the cycloadditions with tetrazine **2a**. Addition of **2a** to a solution of indole (**1a**) produced a vigorous reaction giving adduct **6d** in poor yield (30%), with the major product being **8a**

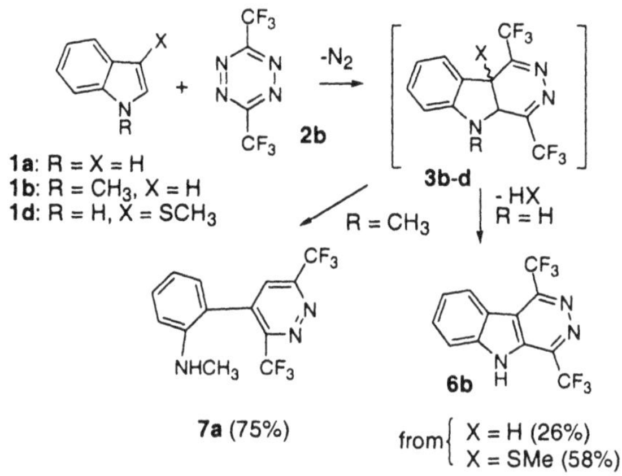

Figure 5. Reactions of 3,6-bis(trifluoromethyl)-1,2,4,5-tetrazine (**2b**) with indole (**1a**), *N*-methylindole (**1b**), and 3-thiomethylindole (**1d**) reported by Seitz[22] and Haider.[23]

(56%, Figure 7).[24] Dihydrotetrazine **9** was also consistently isolated in amounts approximately equimolar to those of **6d**. This production of **9** was an unexpected, but auspicious discovery, indicating that the aromatization of intermediate **3e** was accomplished by the starting tetrazine **2a** as the oxidizing reagent. This process competed with the nucleophilic addition of a second equivalent of indole (**1a**) to the convex face of intermediate **3e** to produce **8a**.

The cycloaddition improved significantly in favor of **6d** (68%) using 1.9 equiv of **2a** with inverse addition (adding indole to **2a**),[25] and this reaction has now been optimized to 95% yield of **6d**.[26] The excess **2a** was consumed in the dehydrogenation of intermediate **3e**, producing dihydrotetrazine **9**, which can be recycled to **2a**,[27] equimolar to the amount of **6d**. Under these conditions, **8a** was formed in only 9% yield. Formation of **8a** was completely avoided using *N*-acylindole derivatives (not shown), which suppressed the nucleophilicity of indole but not its dienophilicity, and the cycloadducts were isolated in much improved 85–89% combined yields

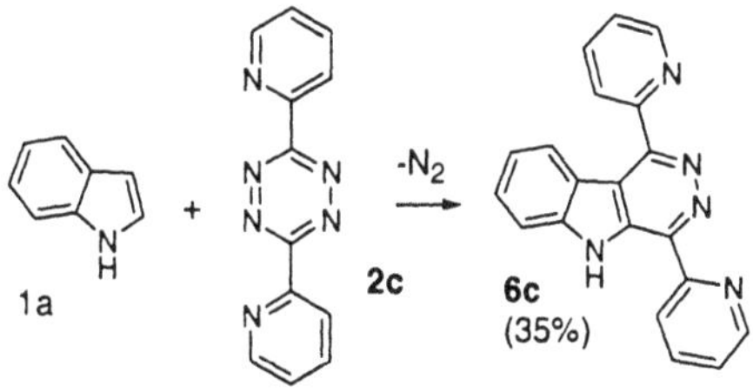

Figure 6. Reaction between indole (**1a**) and 3,6-di-(2-pyridyl)-1,2,4,5-tetrazine (**2c**) reported by Takahashi.[20]

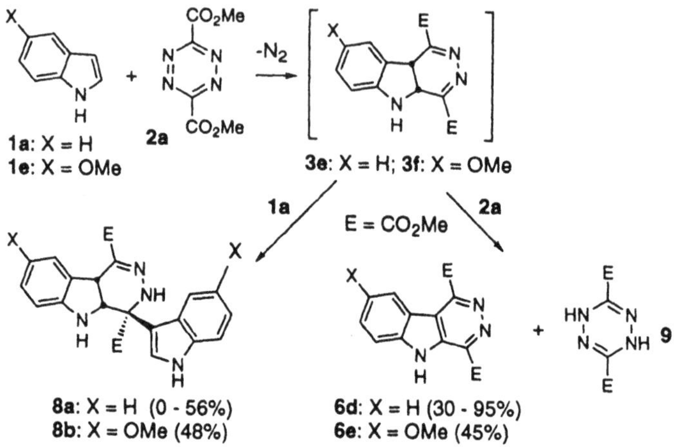

Figure 7. Product analysis from the reactions of indole (**1a**) and 5-methoxyindole (**1e**) with tetrazine **2a**.[24,25]

(acylated + deacylated, 5–17% *N*-deacylation of the product pyridazino[4,5-*b*]indoles occurred during chromatography on silica gel, with the benzoyl group being more stable than the acetyl group to the chromatography).[24,25]

5-Methoxyindole **1e** also reacted with **2a** under the optimized conditions (1.9 equiv **2a**), but a significant amount of **8b** (48%) was still obtained in addition to cycloadduct **6e** (45%) presumably due to the enhanced nucleophilicity of **1e** in comparison to **1a** (Figure 7).[24] *N*-Acetylation to reduce the 5-methoxyindole nucleophilicity led to an excellent yield of cycloadduct N^4-acetyl **6e** (99+%, not shown).[24]

N-Methylindole (**1b**) similarly underwent a cycloaddition with **2a** to produce adduct **6f** (53%) along with rearranged **4** (35%, Figure 8, Eq. 1);[24,25] the latter compound had been reported earlier by Seitz (see Figure 4, Eq. 1).[19] It was presumed that the difference in the results of this work, which produced aromatized cycloadduct **6f** as the major product from the reaction of **2a** with **1b**, from that of Seitz, who observed only the rearranged product **4**, was the number of equivalents of tetrazine **2a** employed. Seitz used an equimolar amount of **1b** and **2a**, while our experiments employed 1.9 equiv of **2a** with the additional equivalent used to promote the aromatization of intermediate **3a**. However, when the reaction of **1b** with **2a** was repeated using equimolar amounts of reagents, **6f** was still obtained in 45% yield.[24]

When the more electron-withdrawing tosyl group was used to protect the indole nitrogen in **1f**, only ring-opened **7b** was isolated (65%, Figure 8, Eq. 2).[24,26] Presumably the cycloaddition proceeded comparably to that of the parent indole, but the ring opening pathway from dihydropyridazinoindole intermediate **3f** was favored over the bimolecular dehydrogenation due to the stability of the sulfona-

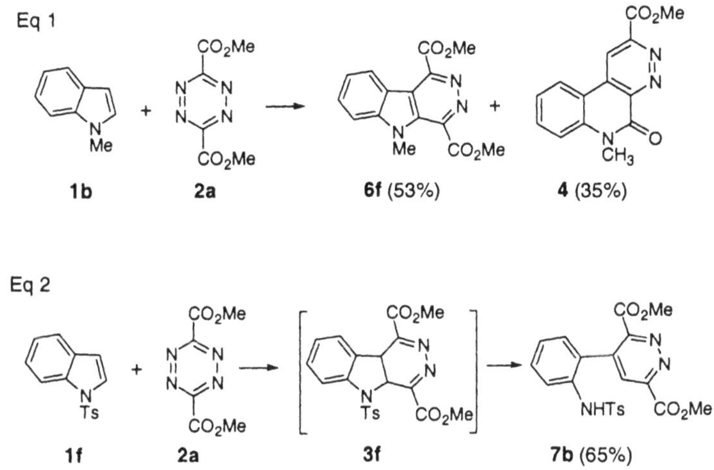

Figure 8. **(Eq. 1)** Product analysis from the reaction between *N*-methylindole (**1b**) and tetrazine **2a**. **(Eq. 2)**[24,25] Formation of ring-opened **7b** in the reaction between *N*-tosylindole (**1f**) and tetrazine **2a.**[24,26]

mide anion which failed to cyclize to the quinolone as observed in previous rearrangements (see Figures 4 and 8, Eq. 1). Haider has since reported similar results in the reaction of *N*-methanesulfonylindole with 3,6-bis(trifluoromethyl)-1,2,4,5-tetrazine (**2b**),[23] and as noted previously (see Figure 5), Seitz had reported that the reaction of *N*-methylindole with the bis(trifluoromethyl)tetrazine **2b** resulted only in the corresponding pyrrole ring-opened adduct **7a.**[22] Thus, such ring-opening pathways which lead to aromatized biaryls from dihydrocycloadducts like **3** are not uncommon in the cycloadditions of indole.

The cycloadditions between indole and the mono- and diketotetrazines **2d** and **2e** gave only aromatized cycloadducts **6g** and **6i**, respectively (Figure 9, Eq. 1).[28] The reaction with **2d**, which is the only example of a cycloaddition between any indole derivative and a nonsymmetrically disubstituted 1,2,4,5-tetrazine, proceeded with excellent regioselectivity to produce **6g** (72%) as the main product with only trace amounts (< 1%) of regioisomer **6h** detected in the ^{1}H NMR spectrum of the crude product mixture. This regioselectivity contrasted with the poor selectivity observed in the reaction between **2d** and ethoxyacetylene which gave a 1:1 mixture of pyridazine regioisomers **10** and **11** (Figure 9, Eq. 2). The greatly enhanced regioselectivity in the reaction with indole was ascribed to a more nonsynchronous or stepwise pathway in comparison to the reaction with ethoxyacetylene. This selectivity for regioisomer **6g** was anticipated from simple resonance considerations: addition of the nucleophilic indole C3 position to C3 of tetrazine **2d** enables resonance stabilization of the developing negative charge into the more electro-

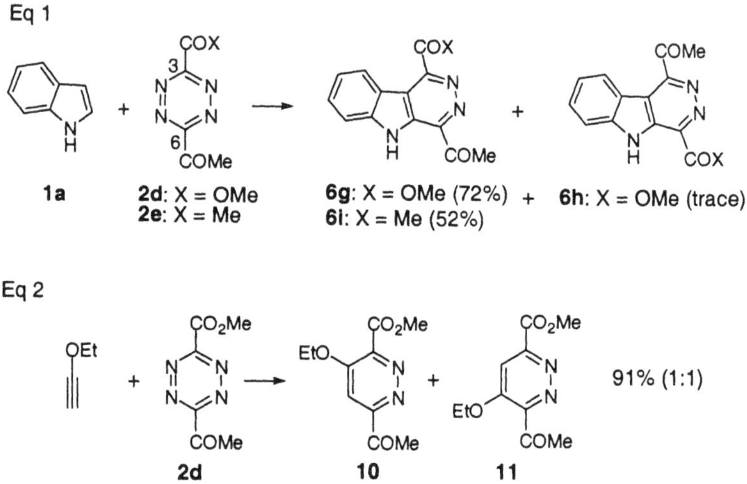

Figure 9. **(Eq. 1)** High regioselectivity in the reaction of indole with nonsymmetrically substituted 1,2,4,5-tetrazine **2d**, and the reaction of indole with diketotetrazine **2e**. **(Eq. 2)** For comparison, nonregioselective cycloaddition of ethoxyacetylene with **2d**.[28]

philic ketone carbonyl. The lower yield in the reaction with tetrazine **2e** reflects the lower stability of this tetrazine.

Current work in our labs is focusing on the use of pyridazino[4,5-*b*]indoles **6** as stable indole-2,3-quinodimethane equivalents in intramolecular cycloadditions to produce carbazoles. To date, regioselective Weinreb amidation[29] with alkynyl amines to either the C1 ester group with N5 protected derivatives of **6d**, or the C4′ ester group with unprotected **6d** followed by thermally promoted cycloadditions have been successful, though results are still preliminary (Figure 10, Eq. 1).[24,30] All attempts to effect intermolecular cycloadditions with **6d** or any of its N5 protected derivatives with a great variety of electron-rich dienophiles including enamines and alkoxyacetylenes have met with failure (Figure 10, Eq. 2).[24,25] In contrast, Haider has shown that the bis(trifluoromethyl)pyridazinoindole **6b** will participate in intermolecular cycloadditions with enamines in respectable yields and with excellent regioselectivity, though often requiring reaction times on the order of days (Figure 10, Eq. 2).[31] This success presumably reflects the greater electron-withdrawing nature of the trifluoromethyl substituents in **6b** in comparison to the ester groups in **6d**, thereby enabling reaction with electron-rich dienophiles due to the lower $LUMO_{diene}$.[32]

In the exploration of the cycloadditions between tetrazine **2a** and 2- and 3-vinylindoles, Pindur has reported that substitution on the remote olefin terminus of the 2-vinylindoles diverts the chemistry from the vinyl substituent to the indole

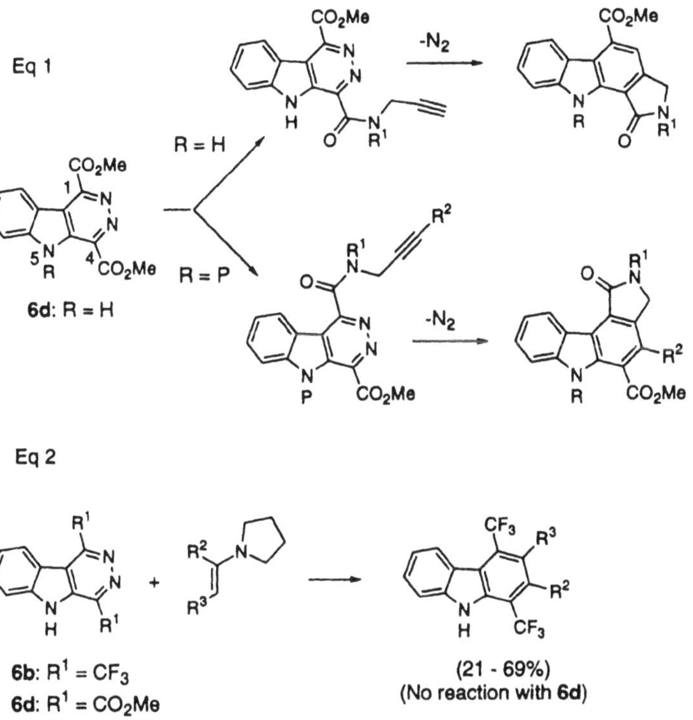

Figure 10. **(Eq. 1)** Intramolecular cycloadditions of pyridazino[4,5-*b*]indoles derived from **6d** with tethered alkynes.[24,30] **(Eq. 2)** Haider's intermolecular cycloadditions of pyridazino[4,5-*b*]indole **6b** with enamines.[31]

2,3-double bond when the indole C3 position is unsubstituted (Figure 11; compare Eqs. 1 and 2), though the products from the latter cycloadditions were isolated in only very low yields.[33] Both dihydropyridazinoindoles (**14**) and rearranged products (**15** and **16**) were isolated depending upon the nature of the substitution on the 2-vinyl group. When the 2-vinylindole is also substituted with a methyl group at the indole 3-position, the cycloaddition returns to the vinyl group presumably for steric reasons, producing **13** as a mixture of diastereomeric atropisomers (Figure 11, Eq. 1). As expected, 3-vinylindoles underwent cycloadditions with **2a** solely on the more electron-rich 3-vinyl group to yield either aromatized (R^2 = H) or tautomeric dihydropyridazines ($R^2 \neq$ H) **17** (Figure 11, Eq. 3).

Other 3-substituted indoles, however, do indeed undergo inter- and intramolecular cycloadditions with 1,2,4,5-tetrazines at the indole 2,3-double bond.[24,34] The thermally promoted cycloaddition between skatole (**1g**, 3-methylindole) and **2a** proceeded smoothly to produce cycloadduct **18** (82%) as a 1:1.5 mixture of rotamers about the C4–CO bond, as established by variable temperature NMR

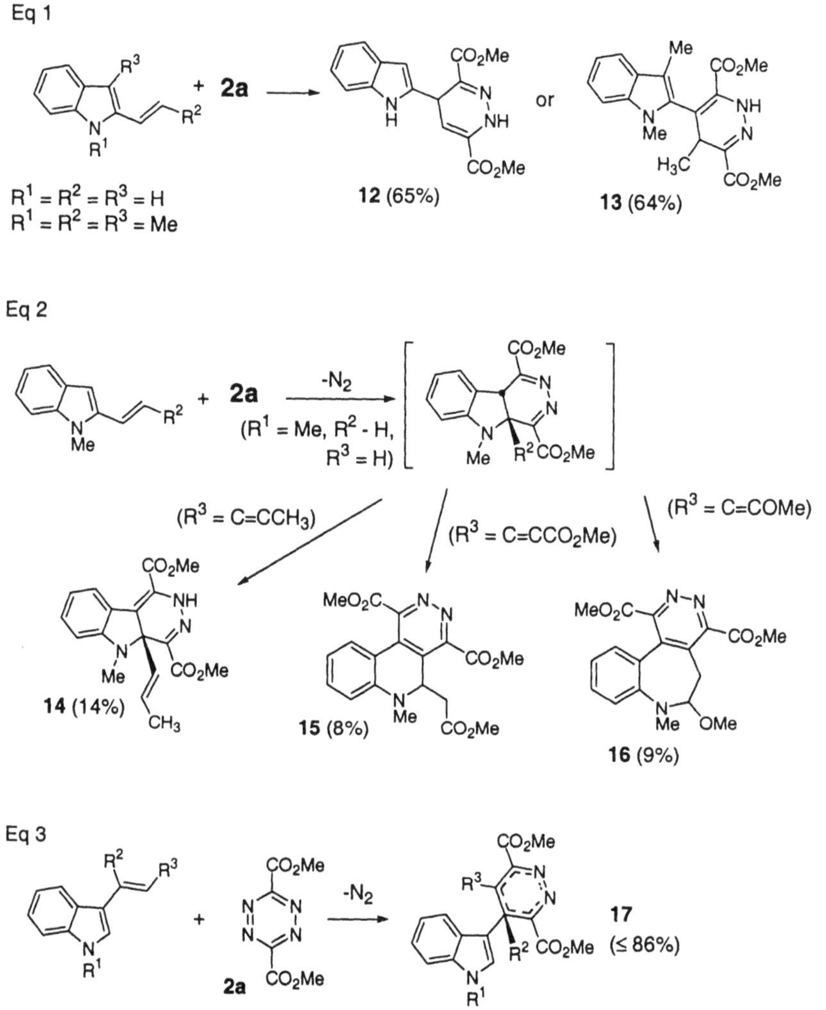

Figure 11. **(Eq. 1)** Cycloadditions of 2-vinylindoles with tetrazine **2a** occur on the vinyl substituent. **(Eq. 2)** Cycloadditions of 2-vinylindoles with a substituent on the vinyl terminus with tetrazine **2a** occur on indole 2,3-double bond. **(Eq. 3)** Cycloadditions of 3-vinylindoles with tetrazine **2a** occur on the vinyl substituent, all reported by Pindur.[33]

(Figure 12).[34b] Methylation following deprotonation led to dimethylated product **19** with the 1,2-dihydropyridazine C-ring rather than the 1,4-dihydro isomer (methylation occurred on C4a rather than N3).[24]

While the reaction of unprotected tryptamine with **2a** produced an intractable mixture of products, ^{t}BOC-protected tryptamine **20a** reacted efficiently with **2a** to

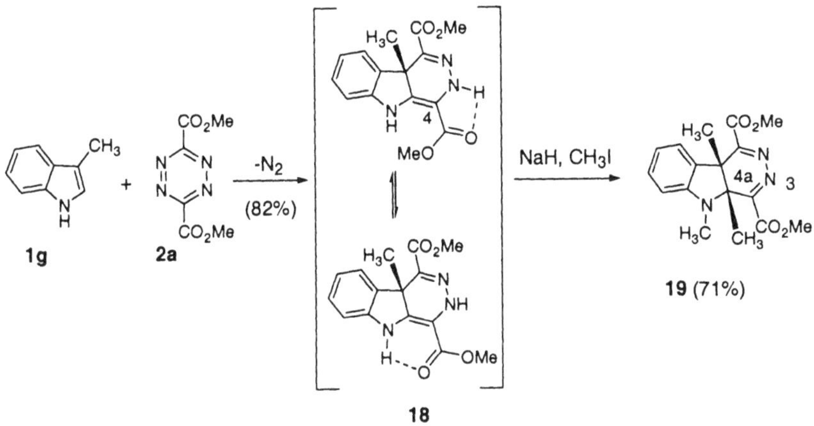

Figure 12. Reaction of skatole (**1g**) with tetrazine **2a** leading to cycloadduct **18** as a mixture of slowly interconverting rotamers, and subsequent methylation of **18** to produce **19**.[24,34]

produce adduct **21a** (Figure 13, Eq. 1); deprotection yielded **21b**.[34] Closure of the fourth ring was confirmed by an HMBC experiment which revealed three-bond heteronuclear coupling between the H11 protons and C1. Interestingly, the cycloaddition of **2a** with ^{t}BOC-protected N^{ω}-methyltryptamine **20b** under the same conditions produced the fully aromatized pyridazino[4,5-*b*]indole **6d** by loss of the

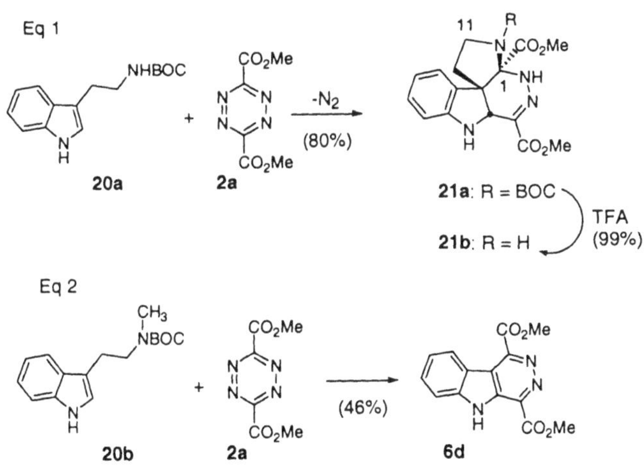

Figure 13. **(Eq. 1)** Reaction of ^{t}BOC-protected tryptamine **20a** with tetrazine **2a** to form adduct **21a**. **(Eq. 2)**[34] Reaction of ^{t}BOC-protected N^{ω}-methyltryptamine **20b** with tetrazine **2a** to form pyridazino[4,5-*b*]indole following loss of the aminoethyl side chain.[34b]

aminoethyl side chain (Figure 13, Eq. 2).[34b] Apparently closure of the pyrrolidine ring following the cycloaddition of **20a**, something not possible with **20b**, prevented aromatization with side-chain loss. Similar loss of this side chain with aromatization[35] had previously been reported in the syntheses of the carbazoles hyellazole[35b] and carazostatin.[35c]

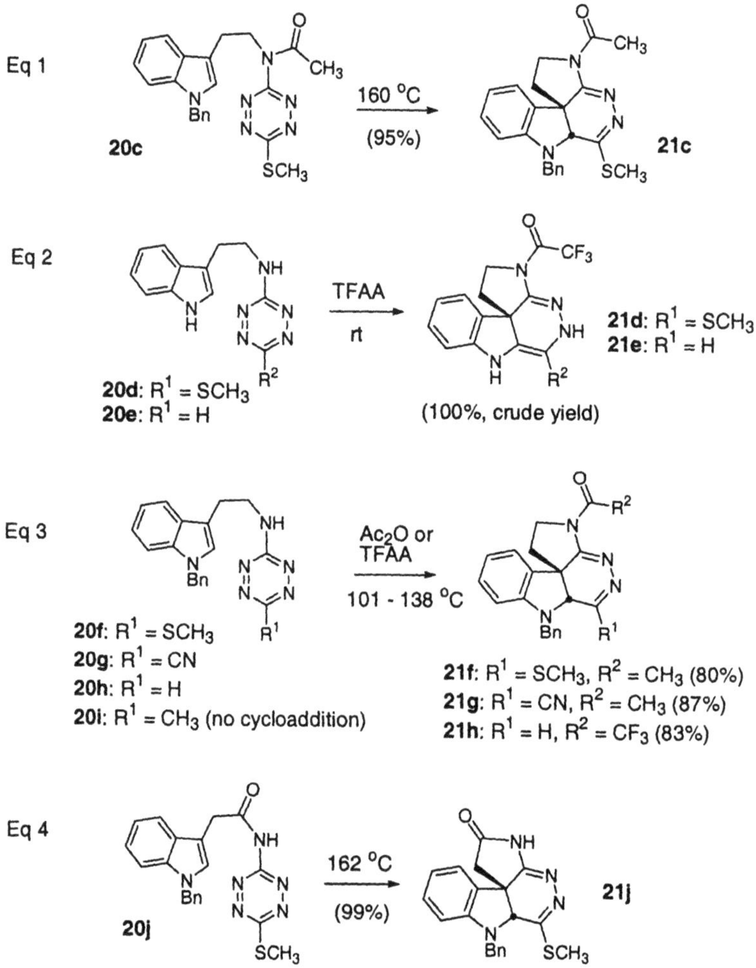

Figure 14. Intramolecular cycloadditions of tryptamine-tethered tetrazines. **(Eq. 1)** Reaction of N-acetylated tryptamine-tethered tetrazine **20c**. **(Eq. 2)** In situ acylations followed by cycloadditions of tryptamine-tethered tetrazines **20d** and **20e**. **(Eq. 3)** In situ acylations followed by cycloadditions of *N*[1]-benzyltryptamine-tethered tetrazines **20f–20h**; tethered tetrazine **20i** didn't react. **(Eq. 4)** Cycloaddition of indolylacetamide-tethered tetrazine **20j**.[34,37]

Intramolecular cycloadditions of tryptamine-tethered 1,2,4,5-tetrazines **20c–20h**, prepared by straightforward S_NAr displacements[36] of methylthiolate or chloride from the tetrazine, also proceeded in good to excellent yields when the tethering tryptamine nitrogen was acylated (Figure 14, Eqs. 1–3).[34,37] Acylation of the tethering nitrogen could also be accomplished in situ using acetic anhydride (reflux) or trifluoroacetic anhydride (TFAA) in dioxane (rt) as the reaction medium (Figure 14, Eq. 3), though use of the former required that the indole nitrogen be protected since acetylation of this nitrogen prevented the cycloaddition from proceeding. Adducts **21d** and **21e** with unprotected indole nitrogens proved to be unstable to chromatography on both silica gel and alumina, though the dominance of the 1,4-dihydropyridazine tautomeric form was apparent from the ^{1}H NMR spectrum. In contrast, the cycloadducts from the N-benzylated tryptamines preferred the 1,2-dihydropyridazine tautomers **21c**, and **21f–21h**. With the tethered methyl tetrazine **20i**, however (Figure 14, Eq. 3), no cycloaddition occurred, only acylation of the tethering nitrogen.[34b,37] The cycloadducts **21c–21h** represent dehydro analogues of the product **21a** obtained from the intermolecular cycloadditions of protected tryptamine **20a** with **2a** described previously (see Figure 13, Eq. 1). Reinforcing the importance of an acylated tethering nitrogen for the cycloadditions to proceed, indolylacetamide-derived tethered tetrazine **20j** produced cycloadduct **21j** in quantitative yield after refluxing in mesitylene for only 10 min (Figure 14, Eq. 4).

Reactions with 1,2,4-Triazines

1,2,4-Triazines, though less reactive than 1,2,4,5-tetrazines, are also very well established as excellent heteroaromatic azadienes for inverse electron demand Diels–Alder reactions with electron-rich dienophiles such as enamines.[16a–c] Similar reactivity with indole and its derivatives has the potential to generate β- and/or γ-carbolines depending upon the regiochemical outcome of the reaction (see Figure 3). The reactions of indole with 1,2,4-triazines required either neat or highly concentrated conditions, yielding a product mixture that was often complicated due to the existence of distinct regiochemical pathways, leading to multiple products that were sensitive to the triazine substituents.[24,38] The three different regiochemical pathways (Figure 15) were cycloadditions with the nucleophilic indole C3 position adding to the triazine C3 (the so-called "C3/C6" pathway, indicating the triazinyl positions in order of relative bond formation in a stepwise route or nonsynchronous transition state), or to the triazine C6 position (the "C6/C3" pathway), and simple nucleophilic addition of indole onto the triazinyl C5 position. This latter reactivity we called the C5/(N2) pathway since it can be viewed as the first step of a nonconcerted cyclocondensation across the triazinyl C5/N2 positions that failed to close in the second step, yielding noncyclized adducts **24**. Cyclocondensations across the triazinyl C5/N2 positions are known with ynamines,[39] and are thought to proceed by a stepwise pathway.[16b,c] The C3/C6 and C6/C3 cycloaddition path-

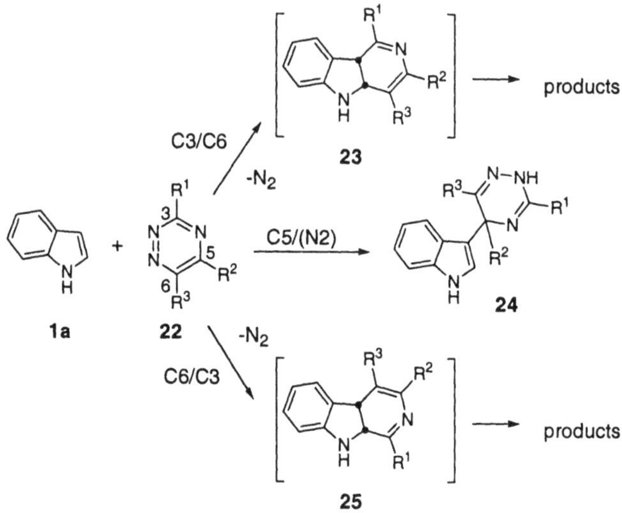

Figure 15. Three regiochemically distinct pathways in the reactions of 1,2,4-triazines with indole: C3/C6 and C6/C3 cycloadditions, and the C5 nucleophilic addition.

ways produce dihydro-γ- and dihydro-β-carboline intermediates **23** and **25**, respectively, following loss of N_2 by the retro Diels–Alder reaction, which subsequently lead to the isolated products.

Triazine **22a** reacted with indole solely by the C3/C6 pathway to produce γ-carboline **26a** as the major adduct, ultimately optimized to 89% yield (Figure 16).[38] Adduct **27a** and benzonaphthyridone **28a** were also isolated in varying, minor amounts depending upon the reaction conditions. Formation of these latter compounds parallels the formation of **8a** and **4**, respectively, in the reactions of indole and derivatives with tetrazine **2a** (see Figures 7 and 8). Consistent with the results of the indole/tetrazine cycloadditions, 2,5-dihydrotriazine **29** was also isolated, presumably resulting from the dehydrogenation of intermediate **23a**, thus accounting for the need to use near 2 equiv of **22a** in order to optimize the yield of **26a**. Other minor products were also isolated and characterized.

Substituted indoles also reacted with triazine **22a** by this C3/C6 pathway to produce mixtures of γ-carbolines and rearranged benzonaphthyridones (Figure 17).[24,38] Increased amounts of rearranged products were obtained with 1-methylindole **(1b)** and 5-methoxyindole **(1e)**: **1b** → **28b** (49%) and **26b** (42%), **1e** → **28e** (20%) and **26e** (72%), again illustrating the enhanced penchant for 2-methylindole to rearrange following the cycloaddition as observed in the chemistry with the tetrazines (see Figures 4, 5, 8, and 11). In contrast, 3-acetoxyindole **(1h)** gave only a poor yield of γ-carboline **26a** (15%). In light of the comparably low yield of cycloadduct in the reaction of **1h** with tetrazine **2a** as well (25%, not shown), the

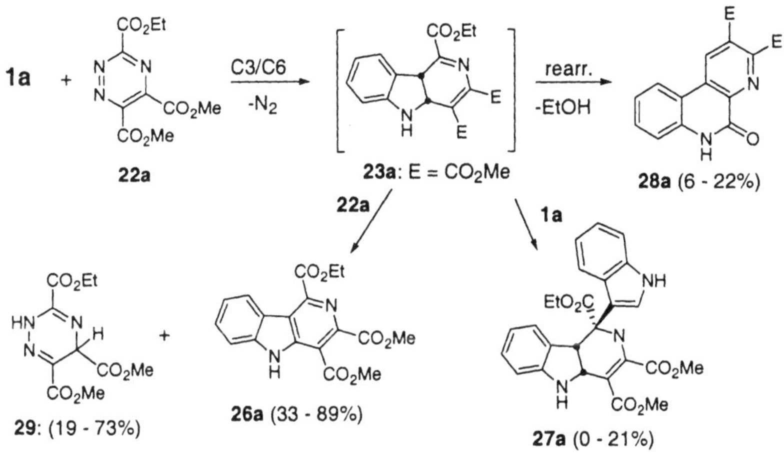

Figure 16. Product analysis from the reaction of indole (**1a**) with trialkyl 1,2,4-triazine-3,5,6-tricarboxylate **22a**.[38]

implication was that the indole dienophile must retain some C3 nucleophilicity to effectively initiate the reaction with electron-deficient heteroaromatic azadienes. Haider, however, was able to obtain good yields of adducts with 3-thiomethylindole with tetrazines, as noted previously (Figure 5).

Other triazines reacted with indole via different regiochemical pathways depending upon their substituents (Figure 18).[24,38] Triazine **22b**, which still bears the C3/C6 diester substituents, also reacted with indole via the C3/C6 pathway, to give γ-carboline **26c** (17%) and benzonaphthyridone **28c** (41%, Figure 18, Eq. 1), but triazine **22c**, which no longer has the C6 ester substituent gave β-carboline **30a** as the major product (50%) via the C6/C3 regiochemical pathway (Figure 18, Eq. 2). Other minor products (not shown) from this C6/C3 pathway accounted for another

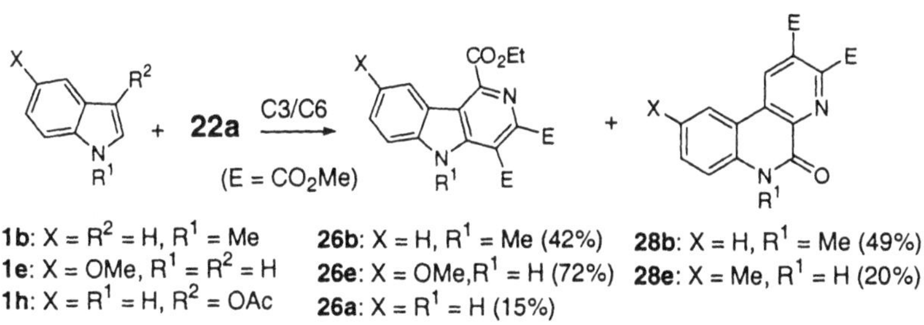

Figure 17. Reactions of 1-methylindole (**1b**), 5-methoxyindole (**1e**), and 3-acetoxyindole (**1h**) with 1,2,4-triazine-3,5,6-tricarboxylate **22a** to produce γ-carbolines **26** and benzonaphthyridones **28**.[38]

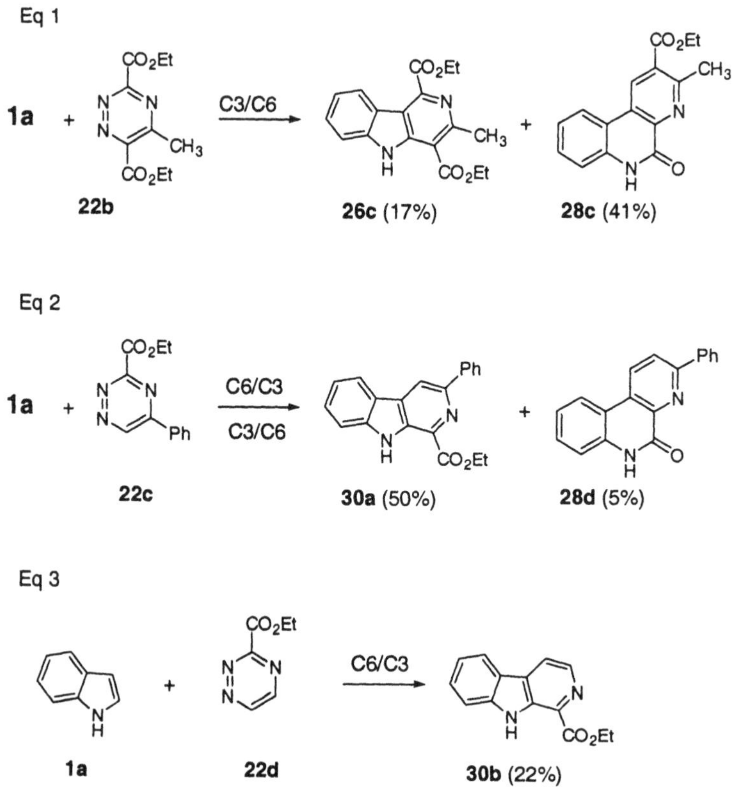

Figure 18. **(Eq. 1)** Cycloaddition of indole (**1a**) with diethyl 5-methyl-1,2,4-triazine-3,6-dicarboxylate (**22b**) by the C3/C6 pathway to produce γ-carboline **26c** and benzonaphthyridine **28c**. **(Eq. 2)** Cycloaddition of indole (**1a**) with ethyl 5-phenyl-1,2,4-triazine-3-carboxylate (**22c**) by the C6/C3 pathway to produce β-carboline **30a**, and by the C3/C6 pathway to produce benzonaphthyridine **28d**. **(Eq. 3)** Cycloaddition of indole (**1a**) with ethyl 1,2,4-triazine-3-carboxylate (**22d**) by the C6/C3 pathway to produce β-carboline **30b**.[24,38]

23% of the product mixture, and a small amount (5%) of the corresponding benzonaphthyridone **28d** originating from the C3/C6 regiochemical pathway was also formed.[38] Thus, **22c** was the only triazine examined which produced cycloadducts from both the C3/C6 and C6/C3 pathways. Unfortunately, this proved to be the best yield of a β-carboline attained to date via an intermolecular reaction between indole and a 1,2,4-triazine. Triazine **22d** also produced a β-carboline, **30b**, by the C6/C3 pathway, but only in 22% yield (Figure 18, Eq. 3).[24,38]

Triazines **22e** and **22f** produced noncyclized adducts **24a** and **24b**, respectively, in reaction with indole (**1a**, Figure 19, Eq. 1).[24,38] These products can be viewed as arising from a C5/N2 stepwise cyclocondensation which "failed to cyclize." As

Eq 1

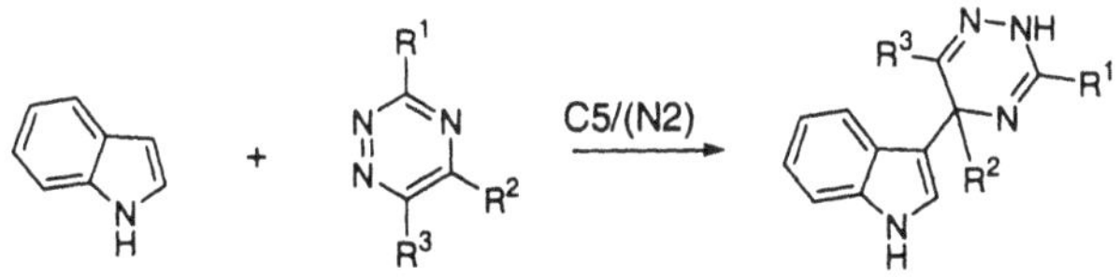

1a

22e: R^1 = SMe, $R^2 = R^3 = CO_2Me$
22f: $R^1 = CO_2Et$, $R^2 = CO_2Et$, $R^3 = CH_3$
22g: $R^1 = CO_2Et$, $R^2 = R^3 = CH_3$ (no reaction)

24a: R^1 = SMe, $R^2 = R^3 = CO_2Me$ (60%)
24b: $R^1 = CO_2Et$, $R^2 = CO_2Et$, $R^3 = CH_3$ (64%)

Eq 2

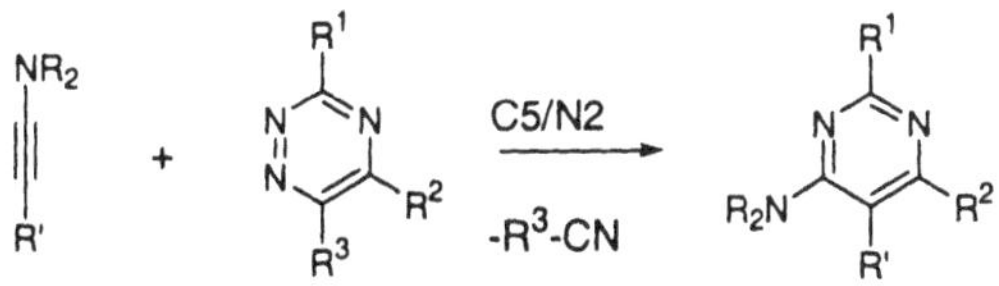

Figure 19. **(Eq. 1)** Reactions of indole (**1a**) with triazines **22e** and **22f** to produce noncyclized adducts **24a** and **24b**, respectively. Diethyl 6-methyl-1,2,4-triazine-3,5-dicarboxylate (**22g**) did not react with indole.[24,38] **(Eq. 2)** General reaction of ynamines with 1,2,4-triazines to produce pyrimidines.[39]

noted previously, such C5/N2 cycloadditions of 1,2,4-triazines have been reported, primarily with ynamines to produce pyrimidines (with loss of RCN, Figure 19, Eq. 2).[39] Dimethyltriazine **22g** did not react with indole under any conditions presumably due to the presence of two electron-donating methyl substituents.

As observed in the reaction of enamines with triazines,[16a–c,40] the regioselectivity of the reactions of indole with 1,2,4-triazines can be rationalized by a stepwise or concerted, highly nonsynchronous cycloaddition mechanism whereby the regiochemistry is controlled by the site of the initial nucleophilic addition of indole to the triazine. In the reaction of indole and triazines **22a** and **22b**, initial nucleophilic addition of indole occurs to C3 on the triazine, which is not only the most electrophilic center on the heterocycle, but also results in delocalization of the developing negative charge into the C6 ester carbonyl. Loss of this resonance-stabilizing C6 carbonyl system while retaining the C3 ester group favors a "flip" in the regioselectivity to the C6/C3 route, though in the two examples **22c** and **22d**, the C6 is also unsubstituted so initial addition of indole to the triazine C6 is also sterically less encumbered.[41] Finally, substitution at the triazine C6 or C3 position by the electron-donating methyl or thioether groups, **22e** and **22f**, respectively, leads to C5 addition and the formation of noncyclized adducts **24**. Nucleophilic addition to C5 of 1,2,4-triazines is well known, and indeed is the preferred site of nucleophilic addition in the parent, unsubstituted triazine.[42]

This study led to the conclusion that some β- and γ-carbolines could be prepared in moderate to excellent yields via this intermolecular cycloaddition strategy, but the regioselectivity of the reaction was too variable, and the yields of desired cycloadduct often too poor to be of broad applicability. An intramolecular cycloaddition between indole and 1,2,4-triazines would provide desired regioselectivity dictated by the location of tether linking the heteroaromatic azadiene and indole dienophile, as well as gain an entropic advantage for otherwise difficult reactions,[18,43] and this approach became the subject of our next investigation (Figure 20). Use of a removable tether would lead to β-carbolines, but retention and further transformation of the new D-ring formed by the tether would provide a very facile route to the canthin-6-one alkaloids. The initial synthetic targets of this strategy thus became the canthin-6-one alkaloids found only in members of the Rutaceae and Simaroubaceae families, which have been reported to have a variety of biological activities.[44]

For the synthesis of the canthin-6-one alkaloids, the strategy was to link indole to the triazinyl C3 position with a trimethylene tether, with methylene tethers of different length also investigated for the preparation of canthine analogues. Di-, tri-, and tetramethylene tethered indolyl triazines **33** were prepared by construction of the triazine ring from either a terminal carboxylate **31** or a terminal nitrile **32** (Figure 21).[45] The intramolecular cycloadditions were easily accomplished by refluxing in 1,3,5-triisopropylbenzene (TIPB, bp 232 °C). The trimethylene tethers produced the desired canthines **35** in excellent yields (73–93%), while the tetramethylene tethers produced cycloadducts **36** with the seven-membered D-ring in modest yields (38–51%). In these and other intramolecular cycloadditions presented later, it is not clear what dehydrogenating agent accomplishes the aromatization of the cycloadduct following release of N_2. The lower yield of cycloadducts **36** was not unex-

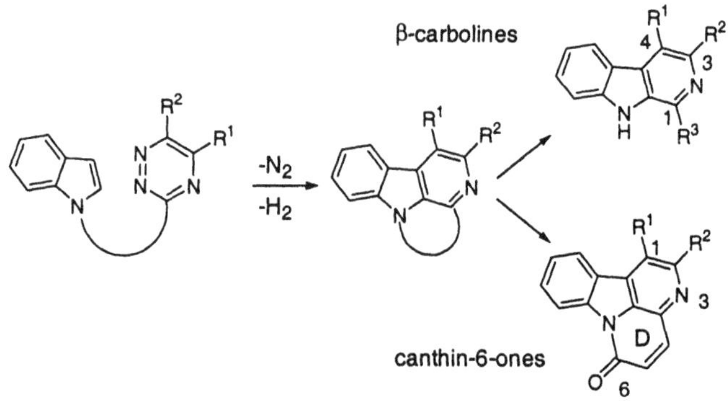

Figure 20. General strategy for the synthesis of β-carbolines and canthin-6-ones from the intramolecular inverse electron demand Diels–Alder reactions of indoles with 1,2,4-triazines.

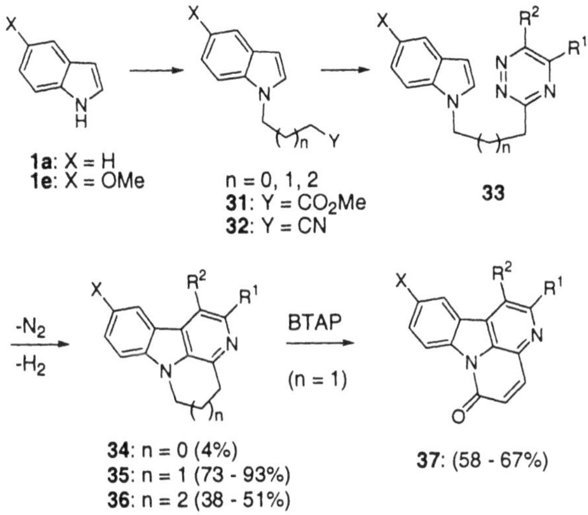

Figure 21. Intramolecular inverse electron demand Diels–Alder reactions of methylene-tethered indoles with 1,2,4-triazines and the synthesis of the canthin-6-one alkaloids **37**.[45,47]

pected since a study by Seitz had shown that the intramolecular cycloadditions of triazines tethered at C3 with acetylenes became dramatically less favorable when the newly formed ring size exceeds six.[46] The cycloaddition of the dimethylene tethered indole/triazine pair **33** (n = 0) produced only 4% of unstable adduct **34** presumably due to ring strain. The canthin-6-one alkaloids **37** were completed by the regiospecific oxidation of the canthines **35** with benzyltriethylammonium permanganate (BTAP) in HOAc/CH_2Cl_2 (58–67%).[45b,47]

Of note in these intramolecular cycloadditions was the production of **35a** and **35b** since these formally represent reactions between indole and trialkylated triazines (Figure 22, Eqs. 1 and 2), thereby illustrating the entropic advantage of the intramolecular chemistry in enabling electronically (and sterically) challenged reactions to proceed.[18,43] In the intermolecular reactions between indole and 1,2,4-triazines, even the dialkylated triazine **22g** with the electron-withdrawing ester substituent failed to react (Figure 22, Eq. 3).

With the successful synthesis of the canthin-6-ones, attention was then turned to other β-carbolines using a similar strategy but with a removable tether (see Figure 20). Two advantages of this cycloaddition approach to β-carbolines over the more commonly employed Pictet–Spengler[4] and Bischler–Napieralski[5] reactions, which provide the tetrahydro-β-carbolines, are: (1) the facile aromatization by rapid loss of N_2 from the initially formed cycloadduct obviates the need for the oxidation of a tetrahydro-β-carboline, which can be capricious, and (2) the ease with which diverse substituents can be built into the β-carboline C1 and C2 positions as these

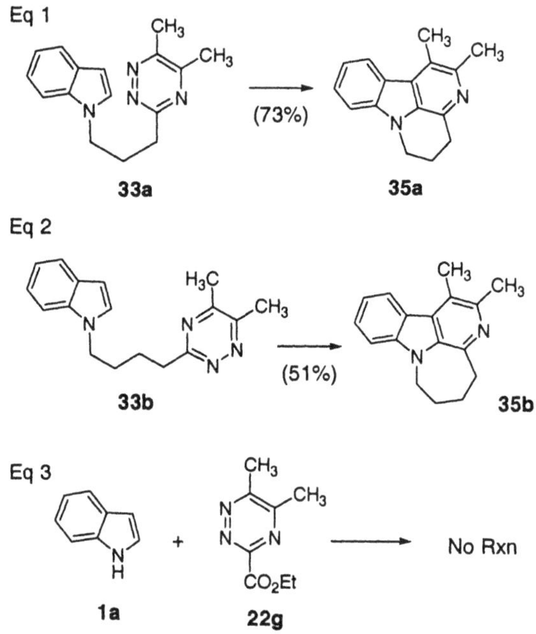

Figure 22. Reactions of indoles with alkylated 1,2,4-triazines. **(Eqs. 1 and 2)** Successful cycloadditions with tri- and tetramethylene tethers.[45a,45b] **(Eq. 3)** Unsuccessful intermolecular reaction between indole (**1a**) and ethyl 5,6-dimethyl-3-carboxylate (**22g**).[24,38]

derive from the readily variable C5 and C6 substituents of the triazine, and hence ultimately from the 1,2-dicarbonyl synthon.[39f] Success in this endeavor required that a removable tether be developed, in contrast to the approach to the canthinones which incorporated the trimethylene tether as the desired canthinone D-ring described previously.

Taking a cue from Kraus' recently reported "urea connection" to tether butadienes with indoles substituted with electron-withdrawing groups at the C3 position[48] (see Section II.B), 5,6-diphenyltriazines were constructed from the carboxylate group of N-protected glycine methyl ester, and subsequently linked to indole via either a urea or thiourea connector, **38** and **39**, respectively (Figure 23).[49] The cycloadditions were accomplished by heating in 1,3,5-triisopropylbenzene (TIPB) in the presence of the antioxidant BHT (1 equiv), with the thiourea linkage giving the best yields of cycloadducts **41** (> 85%). Without the added BHT, the reactions were considerably less clean and yielded lower amounts of cycloadducts. The yields of the cycloadducts **40** using the urea tether were considerably more variable (< 10%–65%). This, along with the greater difficulty in preparing the urea tether in **38** led us to prefer the thiourea linkage.

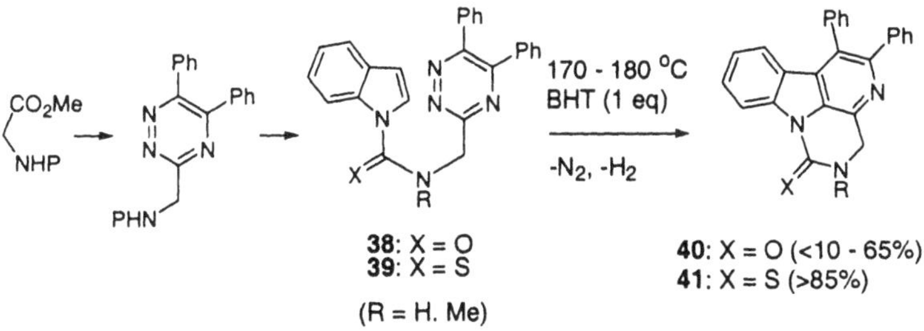

Figure 23. Intramolecular inverse electron demand Diels–Alder reactions between indole and 1,2,4-triazines using urea **38** (X = O) and thiourea **39** (X = S) linkages.[49]

Cleavage of the thiourea linkage was achieved by reduction of **41a** with LAH to give β-carboline **42a** (Figure 24).[49] With $NaBH_4$ in refluxing pyridine, only reduction of the thiocarbonyl occurred, giving **43a** with the tetrahydropyrimidine D-ring as found in the eudistomidins E and F, **44b** and **44c**, respectively, isolated from *Eudistoma glaucus*.[50]

The success of the thiourea linkage as the connective element enabled the production of β-carbolines from the intramolecular cycloadditions between indole and 1,2,4-triazines. The so-formed β-carbolines, however, all carried a vestige of this tether as a C1 substituent (Figure 24). Attention was then turned to the preparation of a fully removable tether that would leave the β-carboline C1 position unsubstituted.

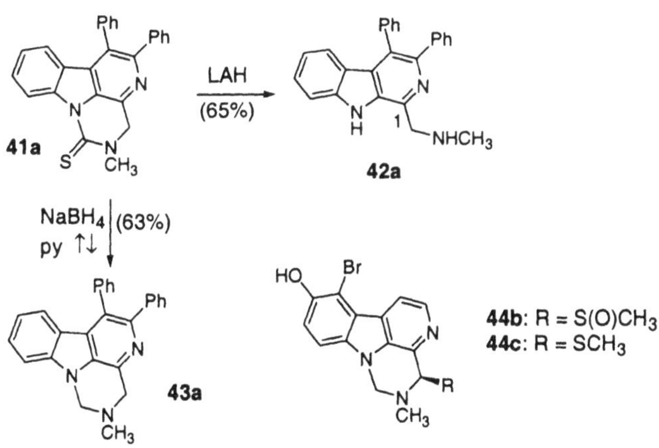

Figure 24. Reductions of thiourea cycloadduct **41a** with LAH to produce β-carboline **42a**, and with $NaBH_4$ in refluxing pyridine to produce **43a** with the tetrahydropyrimidine D-ring.[49] Eudistomidins E and F (**44b** and **44c**, respectively) have the same skeleton as **43a**.[50]

α-Mercaptoacetyl tethers, which linked indoles with 1,2,4-triazines, were prepared by displacement reactions between the N-(bromoacetyl)indoles **45** and the 3-thiotriazinones **46** (Figure 25).[51] Not surprisingly, these triazinyl-tethered indoles **47a** did not participate in intramolecular cycloadditions, presumably due to the poor $HOMO_{dienophile}/LUMO_{diene}$ match as a consequence of the electron-donating sulfide substituent on the triazine and the electron-withdrawing acyl group on the indole nitrogen. Oxidation of the sulfide to the corresponding sulfoxide **47b**, and ultimately the sulfone **47c**, succeeded with MCPBA (and considerable patience; oxidation to the sulfone required 7 days at 0 °C!). The crude sulfones, which were unstable to chromatography on silica gel, underwent cycloadditions in refluxing toluene (bp 110 °C) to provide a mixture of N-acetyl-β-carbolines **42** and cycloadducts **48**. Reduction of this mixture (LAH) without further purification gave the C1 unsubstituted β-carbolines (41–46% from **47a**). Small amounts of sulfones **47c** could be purified, albeit in modest yield (40–55%) due to their instability, and these sulfones underwent cycloadditions to produce the cycloadduct mixtures of **42** and **48** in near quantitative combined yields (≥95%), thereby indicating that the cycload-

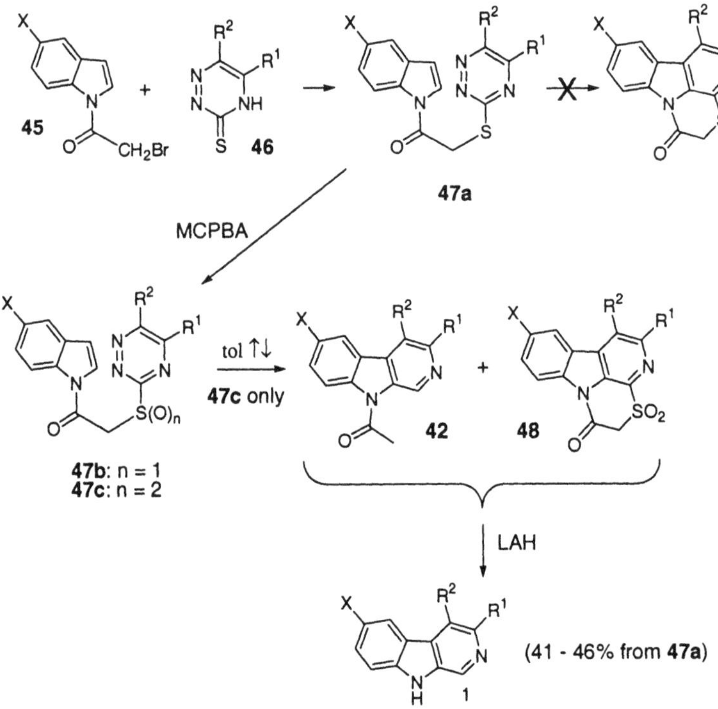

Figure 25. Construction and cycloaddition of β-sulfonoacetyl-linked triazinyl indoles **47c** leading to the preparation of C1 unsubstituted β-carbolines.[51]

dition itself is very favorable. The purified sulfoxides **47b** were inert to cycloaddition chemistry.

The mechanism for the production of **42** was probed using 2- and 3-deuteroindole.[51] Following the cycloaddition and loss of nitrogen, intermediate **49** can either be aromatized to produce **48**, or undergo a [1,5-H] shift to dihydrocarboline intermediate **50**, followed by elimination of SO_2, to give **42** (Figure 26). With 3-deuteroindole, the 1-deutero-*N*-acetyl-β-carboline was produced with the deuterium label entirely transferred to the carboline C1 position, as predicted by the proposed mechanism. In contrast, the deuterium label beginning with 2-deuteroindole was not incorporated into either product, as also predicted by the proposed mechanism.

As with the tetrazines bonded to tryptamine at the tetrazinyl C3 position through the side chain amino group described previously (see Figure 14), triazines could also link to tryptamine by S_NAr displacements of methylthiolate.[24,34,37,52] These triazinyl tryptamines also underwent intramolecular inverse electron demand cycloadditions when the tethering nitrogen was acylated, but only when a acylating medium (TFAA/dioxane or Ac_2O) was employed, and only when the triazinyl C5 and C6 substituents were the electron-withdrawing ester groups as in **51a** (Figure 27, Eq. 1).[24,34,37] As with the tethered tetrazines, this N-acylation could also be accomplished in situ (Figure 27, Eq. 2), and the 1,2-dihydropyridine tautomer dominated in the cycloadducts **52a** and **52b**. The 5,6-unsubstituted, 5,6-diphenyl, and 5,6-dimethyl tethered triazines all failed to react. Furthermore, N^{ω}-acetyl-N^1-benzyltryptamine tethered triazine **51a** did not produce a cycloadduct upon heating up to 232 °C in 1,3,5-triisopropylbenzene, but did give cycloadduct **52a** in refluxing Ac_2O (bp 138 °C), suggesting that a second acylation of a triazine ring nitrogen

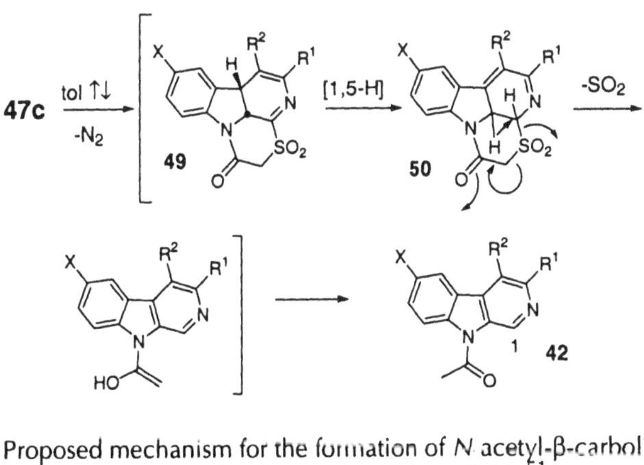

Figure 26. Proposed mechanism for the formation of *N*-acetyl-β-carbolines **42** with the loss of SO_2 from the intramolecular cycloaddition of **47c**.[51]

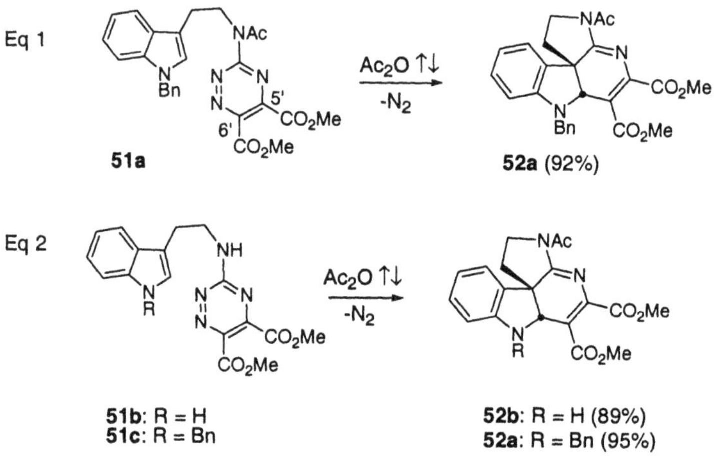

Figure 27. Intramolecular cycloadditions of tryptamine-tethered triazines. **(Eq. 1)** Reaction of *N*-acetylated tryptamine-tethered triazine **51a**. **(Eq. 2)** In situ acetylations followed by cycloadditions of tryptamine-tethered triazines **51b** and **51c**.[24,34,37]

may be essential for the triazine cycloaddition, though this was not the case for the tethered tetrazines.

The similarly derived tryptophan-tethered triazine **51d** also underwent a cycloaddition (TFAA, 8 equiv, refluxing dioxane, bp 101 °C, 1 h) to produce a single adduct **52c** (85%) with no other diastereomer detected (Figure 28).[34b,37] This exclusive

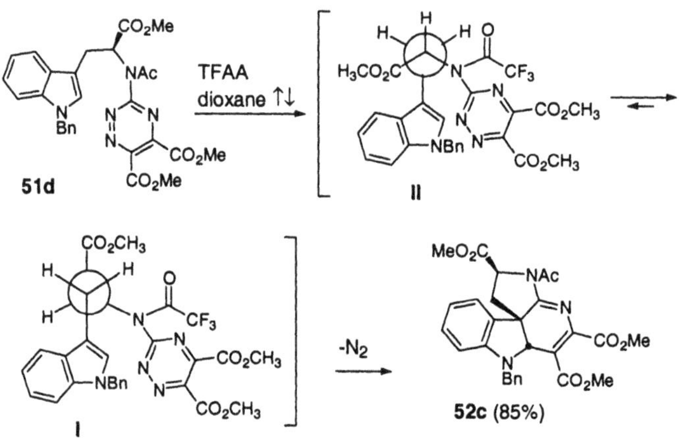

Figure 28. Intramolecular cycloaddition of N^{ω}-acetylated tryptophan methyl ester-tethered triazine **51d** through conformation **I** to produce a single diastereomer **52c**.[34b,37]

diastereoselectivity was thought to arise from the dominance of conformation **I** which leads to a *si* face approach of the triazine to the indole 2,3-double bond. The alternative approach to the *re* face would originate from conformation **II** which is disfavored due to the additional *gauche* interaction.

Reactions with Pyridazines

Pyridazines (1,2-diazines) are well known to be much less reactive than triazines and tetrazines in inverse electron demand Diels–Alder reactions due to their higher LUMO level, but numerous examples of their participation in such reactions are still well documented.[16a–c,53] An attractive feature of successful cycloaddition chemistry of pyridazines with indole derivatives would be a direct entry into the carbazole skeleton (see Figure 3, X = Y = CR). We briefly investigated cycloadditions of indole with several pyridazines, but with only limited success.[24,38] Of the diazines and conditions tested, only tetramethyl pyridazine-3,4,5,6-tetracarboxylate **53** reacted with indole, but led only to rearranged phenanthridone **54** (41%) with no trace of carbazole detected (Figure 29).[38] The rearrangement presumably follows a pathway analogous to that observed in the pyrrole ring opening of the cycloadduct intermediates from the reactions of indole with tetrazines (see Figures 4, 5, 8, and 11) and triazines (see Figures 16, 17, and 18).

Haider, building on earlier work with pyridazines in inverse electron demand cycloadditions,[54] has shown that appropriately activated pyridazines will undergo cycloadditions with indole and its derivatives to give carbazoles, and has applied this chemistry to the synthesis of the *N*-methyl bis-trifluoromethyl analogue of ellipticene **58**.[31,55] *N*-Methylindole (**1b**) underwent a cycloaddition with pyrido[3,4-*d*]pyridazine **55a** to produce a mixture of products including the non-aromatized cycloadducts **56** and **57**, the latter two in a combined 37% yield (1:1,

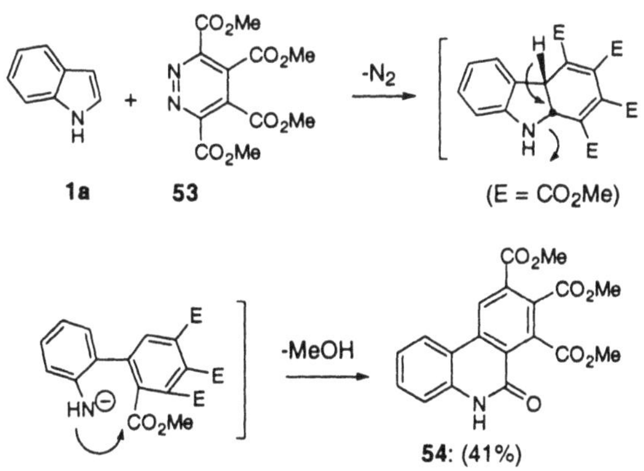

Figure 29. Inverse electron demand Diels–Alder reaction of indole (**1a**) with tetramethyl pyridazine-3,4,5,6-dicarboxylate **53** to produce phenanthridone **54**.[24,38]

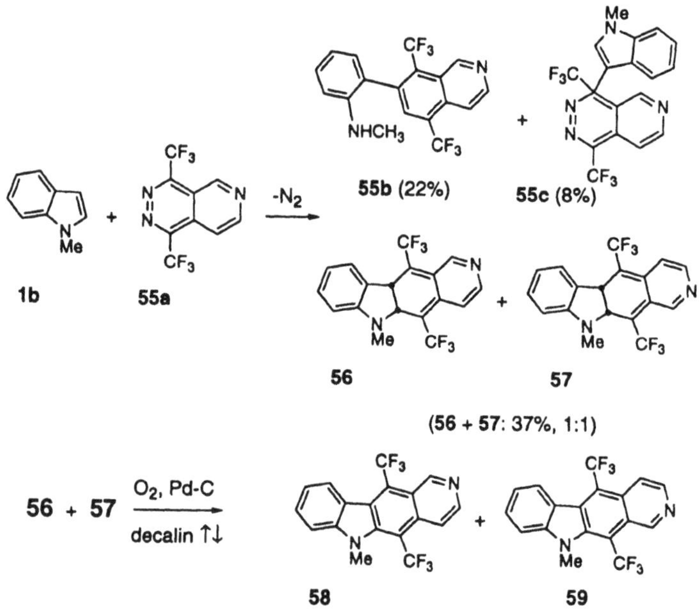

Figure 30. Cycloaddition of *N*-methylindole (**1b**) with 1,4-bis(trifluoromethyl) pyrido[3,4-*d*]pyridazine (**55a**) in Haider's synthesis of the bis(trifluoromethyl) analogue of ellipticene **58**.[31,55]

Figure 30). Also included in the product mixture was ring-opened adduct **55b** (22%) and simple addition product **55c** (8%). Aromatization of the inseparable mixture of **56** and **57** led to the desired **58** along with the isoellipticene analogue **59**. The reaction between **55a** and indole was similar, but relatively messy.

Carbazoles have also been produced from the cycloaddition of 1,4-dicyanophthalazine (**60a**) with *N*-methylindole (21%),[56] and from the reaction of 4,5-dicyanopyridazine (**61**) with both indole and *N*-methylindole[57] (Figure 31, Eqs. 1 and 2). No reaction ensued, however, between 1-cyanophthalazine (**60b**) and *N*-methylindole, emphasizing the importance of multiple electron-withdrawing substituents on the pyridazine ring in order for the cycloaddition to occur.

Pyridazines have also been linked to the terminal amino group of tryptamine for examination in intramolecular inverse reactions analogous to previously discussed investigations into similar chemistry with tryptamine bonded tetrazines (see Figure 14) and triazines (see Figures 27 and 28).[24,34b,37] No cycloaddition occurred, however, with the chloropyridazine **62** under any conditions (Figure 32, Eq. 1). Tethered tetrazine **64** was also prepared by S_NAr chemistry through the addition of the 3,6-bis(thiomethyl)-1,2,4,5-tetrazine (**2f**) to the lithiated *N*(1)-benzyl-N^{ω}-acylated tryptamine **63**.[34b] The cycloaddition of the terminal alkyne and tethered tetrazine occurred rapidly in refluxing CH_2Cl_2, and the cycloadduct **65a** was

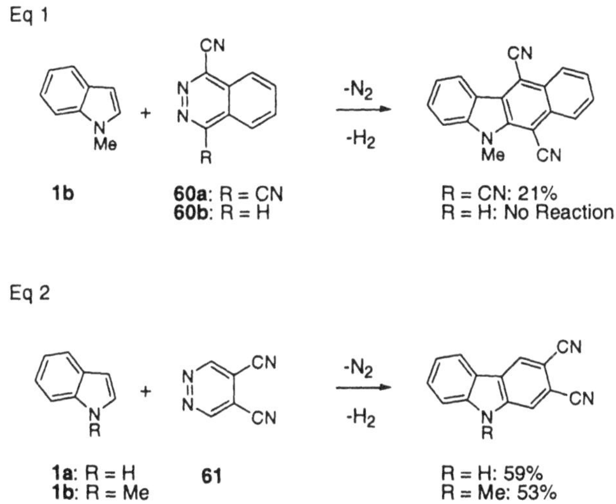

Figure 31. Intermolecular reactions of indole (**1a**) and *N*-methylindole (**1b**) with cyanopyridazines. **(Eq. 1)** Reaction of **1b** with dicyanophthalazine **60a** reported by Oishi;[56] no reaction ensued with monocyano derivative **60b**. **(Eq. 2)** Reactions of **1a** and **1b** with 4,5-dicyanopyridazine (**61**) reported by Nesi.[57]

isolated in quantitative yield.[34b] Thus not surprisingly, the better dienophile in this case was the terminal alkyne, not the 3-substituted indole. All efforts to effect the intramolecular cycloaddition of the pyridazine thioether with the indole dienophile by either thermal promotion or Lewis acid catalysis were unsuccessful. Since only highly activated pyridazines had been found to undergo inverse electron demand Diels–Alder reactions with indole, the thioether was oxidized with MCPBA to sulfoxide **65b** (92%). However, **65b** also failed to participate in intramolecular cycloadditions, and all attempts to further oxidize this sulfoxide to the sulfone produced only intractable mixtures.

From these studies with heteroaromatic azadienes it can be concluded that indole is a good dienophile for inverse electron demand Diels–Alder reactions, though not as reactive as traditional enamines. As expected, reducing the number of ring nitrogens (1,2,4,5-tetrazines > 1,2,4-triazines > 1,2-diazines), or reducing the number of electron-withdrawing substituents on the heteroaromatic azadiene resulted in increasingly harsher conditions necessary to achieve the desired chemistry. Other electron-deficient dienes have also been used in cycloaddition chemistry with indoles, as discussed in Section II.C.

Lewis Acid-Catalyzed Cycloadditions with Heteroaromatic Azadienes

The well-known Lewis acid catalysis of normal electron demand Diels–Alder reaction is often the critical difference between a successful synthesis and a search

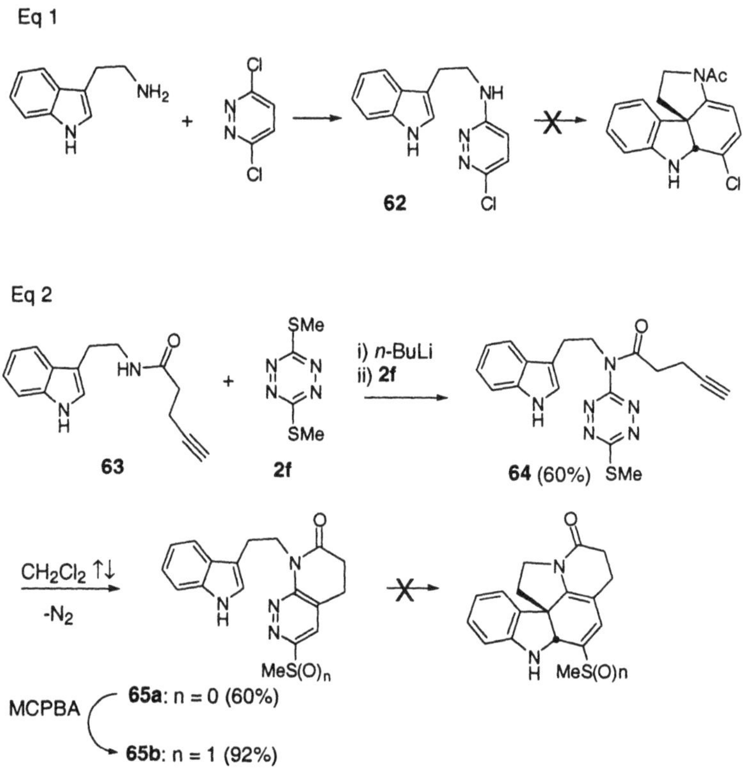

Figure 32. Attempted intramolecular cycloadditions between indoles and tethered pyridazines. **(Eq. 1)** Failed cycloaddition of tethered chloropyridazine **62**.[24,34b,37] **(Eq. 2)** Failed cycloaddition of tethered pyridazinosulfide **65a** and pyridazinosulfoxide **65b**.[34b]

for an alternative synthetic approach. Catalysis of inverse electron demand Diels–Alder reactions, by comparison, is extremely rare. Theoretically, Lewis acid catalysis should also be feasible if the Lewis acid would coordinate with a heteroatom on the diene, as in an azadiene, which would thereby lower the $LUMO_{diene}$ as desired. The difficulty in practice presumably lies with the competing, and often more favorable, interaction of the Lewis acid with the electron-rich dienophile. Indole itself, for example, is well-known to oligomerize in the presence of both Brönsted and Lewis acids.[58]

Mariano has reported, however, that the Diels–Alder reactions of 2-azadienes with vinyl ethers and enamines can be catalyzed by BF_3, though the yields of cycloadducts were rather poor.[59] Boger and Panek had also sought for Lewis acids that would catalyze the inverse electron demand cycloadditions between 1,2,4-triazines and enamines, but without success.[60]

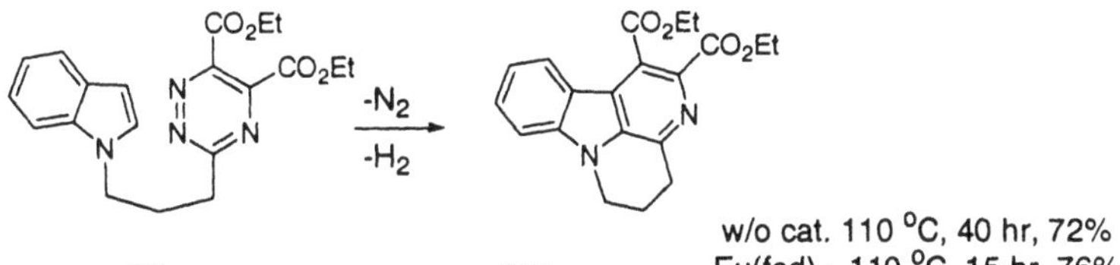

33a **35a**

Figure 33. Cycloaddition of trimethylene tethered triazine **33a** to give the canthine **35b** catalyzed by Eu(fod)3.[45]

Since tethered heteroaromatic azadienes offer several nitrogen lone pairs on the azadiene subunit which could function as Lewis bases, various intramolecular cycloadditions of indole were examined. We were also unable, however, to catalyze the inverse electron demand Diels–Alder reactions with Lewis acid catalysts with the few exceptions wherein the heteroaromatic azadiene bore substituents that enabled a chelation site to be formed. Numerous attempts were made to catalyze the cycloadditions of the trimethylene tethered triazines **33** (see Figure 21, $n = 1$) with various Lewis acids.[45b] In most cases this led only to decomposition with the exception of the diester triazine **33a**. With this tethered triazine, the use the NMR shift reagent Eu(fod)$_3$ (0.5 equiv) enabled shorter reaction times for the cycloaddition to canthine **35a** (Figure 33).[45a] Other Lewis acids ($ZnCl_2$, $LiClO_4$, $BF_3{\cdot}OEt_2$, $TiCl_4$) either had no effect or led only to decomposition of **33a**. We surmised that the Eu(fod)$_3$ chelates either the two ester carbonyls, or an ester carbonyl and a ring nitrogen, thereby lowering the azadiene LUMO and accelerating the reaction. Danishefsky had previously reported the use of Eu(fod)$_3$ as a catalyst for hetero Diels–Alder reactions.[61]

Table 1. Lewis Acid Catalyzed Cycloadditions of **20c** to **21c** (Figure 14, Eq. 1)

Item	*Catalyst (equiv)*	*Conditions*	*Yield (%)*[a]	*Recovered (%)* **20a**
1	none	TIPB or DMF, 60 °C, 24 h	0	100
2	$Ni(CN)_2$ (1.5)	DMF, 60 °C, 3 h	62	27
3	$Ni(acac)_2$ (1)	DMF, 60 °C, 3 h	40	44
4	$AlCl_3$ (1)	CH_2Cl_2, rt, 14 h	12	71
5	$MeAlCl_2$ (1)	TIPB, rt, 12 h	< 2	98
6	$MeAlCl_2$ (1)	TIPB, 80 °C, 4 h	0	0
7	$Eu(hfc)_3$ (0.05)	CH_2Cl_2, reflux, 8 h	10	82
8	$Eu(hfc)_3$ (0.05)	TIPB, 70 °C, 8 h	24	68
9	CuI (1)	DMF, 60 °C, 3 h	10	23
10	$BF_3{\cdot}OEt_2$	CH_2Cl_2, rt, 10 h	22	53

Note: [a] Isolated yields of **21c**.

Several Lewis acids were also surveyed for catalysis in the cycloaddition of **20c** to **21c** (Table 1, see Figure 14, Eq. 1) which also contains a sulfur as a potential site for Lewis acid coordination.[34b,37] While several Lewis acids did catalyze the cycloaddition in low yields at 60 °C or lower temperatures, the best yield obtained [with $Ni(CN)_2$, 62%, Item 2] was considerably lower than that achieved by heating **20c** to 160 °C in TIPB (95%). Nevertheless, this catalysis did enable the cycloaddition to proceed at a considerably lower temperature. Attempts to apply $Ni(CN)_2$ catalysis to the cycloaddition of other triazines as well as to the cycloadditions of sulfur-containing pyridazines **65a** and **65b** (see Figure 32) were unsuccessful. Thus, the successful catalysis of inverse electron demand Diels–Alder reactions remains relegated to very specific and isolated instances.

B. Cycloadditions with Other 1,3-Dienes

Reactions with Electron-Deficient 1,3-Diene Systems

In addition to the cycloadditions with heteroaromatic azadienes, other reports of cycloadducts from the reactions of indole and its derivatives with electron-deficient conjugated diene systems have appeared. In one of the earliest reports on the dienophilicity of indole, Raasch studied the exceptional ability of tetrachlorothiophene sulfone (**66**) to participate in cycloadditions with a vast array of dienophiles including indole (**1a**), which produced carbazole **67** in 77% yield (Figure 34).[62]

Omote and coworkers used the reaction between *o*-quinones generated in situ and 2,3-disubstituted indoles,[63] typically cycloalk[*b*]indoles **68**, initially as a means to prepare novel propellanes,[64] then later as an effective trap of unstable *o*-quinones proposed as intermediates in melanogensis[65] (Figure 35, Eq. 1). The formation of cycloadduct **69** was considered to proceed by a stepwise pathway beginning with formation of a charge-transfer complex. Interestingly, the cyclohex[*b*]indole cycloadduct **69a** existed in a equilibrium with the ring-opened taumeric form **69b** (Figure 35, Eq. 2), though for the corresponding cyclopent[*b*]indole adducts, which were typically produced in considerably higher yields, only the closed, cyclic tautomers were found.

Reactivity similar to that described by Omote has also been reported by Heine and coworkers in their studies of the cyclocondensation of *o*-quinone monoimides

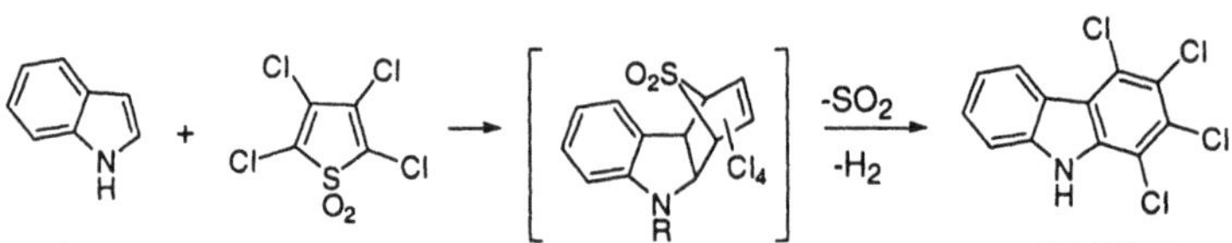

Figure 34. Cycloaddition of tetrachlorothiophene sulfone (**66**) with indole (**1a**) to produce 1,2,3,4-tetrachlorocarbazole (**67**) reported by Raasch.[62]

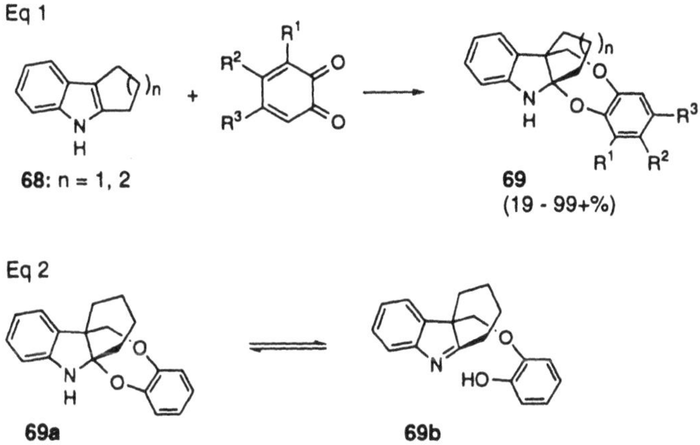

Figure 35. **(Eq. 1)** Reaction of indoles **68** with *o*-quinones to form cycloadducts **69** reported by Omote.[63–65] **(Eq. 2)** Tautomeric equilibrium between cyclic and ring-opened forms **69a** and **69b** of cyclohex[*b*]indole adduct.

70.[66] Either cyclic adducts **71** or acyclic products **72**, the latter thought to result from a cycloaddition followed by elimination ring opening,[67] were isolated depending upon the indole substitution (Figure 36). A very interesting reversal in regioselectivity was observed depending upon the electronic nature of the indole nitrogen substituent. Electron-withdrawing protecting groups (Ac, Bz, and SO_2Ph) favored addition of the *o*-quinone monoimide nitrogen to the indole C2 position leading to cyclic products **71**, while with indole and *N*-methylindole the imide nitrogen added to the indole C3 position when this site was also unsubstituted. The dominance of the ring-opened tautomer **72** in the latter adducts is reminiscent of Omote's observations with the cyclohex[*b*]indole/*o*-quinone cycloadduct **69a/69b** (see Figure 35).

Stepwise cyclocondensations between indole and 1,3-diene systems have also been reported. The acid-promoted reaction between indole and 1,4-diamino-2-azadienes **73** produced β-carbolines **75** in a stepwise reaction sequence (Figure 37, Eq. 1).[68] Best yields were obtained when the procedure was carried out in stepwise fashion, isolating intermediate **74** and resubjecting it to the acidic reaction conditions, though one-pot procedures were optimized to 79% yield in specific cases. The reaction proved to be completely regioselective with no γ-carbolines reported. The authors rationalized this exclusive regioselectivity from the addition of indole to the diprotonated diamino-2-azadiene. Substitution at the indole 3-position as with skatole (**1g**), however, diverted the chemistry to the pyrrolopyrimidine **76** (Figure 37, Eq. 2).[69]

Padwa has also reported the reaction between 1,3-bis(phenylsulfonyl)-1,3-butadiene (**77**) and indole (**1a**) to give tetrahydrocarbazole **78** (Figure 38).[70] In contrast,

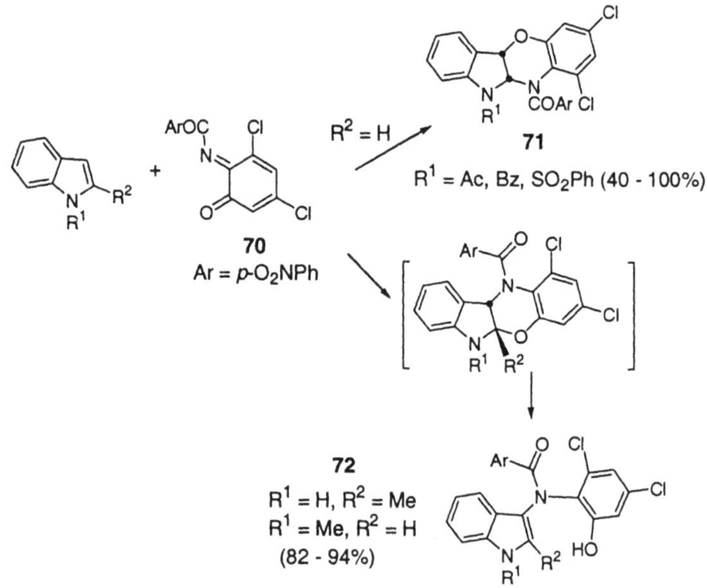

Figure 36. Reaction of indoles with *o*-quinone monoimide **70** to yield either cycloadducts **71**, or the regioisomeric noncyclized adducts **72** reported by Heine.[66]

the reaction between 2,3-bis(phenylsulfonyl)-1,3-butadiene **79** and indole was too slow, allowing **79** to rearrange to **77** which subsequently underwent the cycloaddition. This latter result contrasted with those using more reactive enamines which were able to undergo cycloadditions with **79** prior to its rearrangement.

Another cycloaddition strategy with a 1,3-diene was reported by Bäckvall, who reacted the indole Grignard salt **1i** with 2-phenylsulfonyl-1,3-dienes **80**[71] (Figure 39).[72] The reaction with (*E,E*)- and (*E,Z*)-1,4-disubstituted 2-phenylsulfonyl-1,3-dienes **80** produced mixtures of *cis*- and *trans*-cycloadducts **81**, establishing the stepwise nature of the chemistry. The cycloaddition with (*E,E*)-3-phenylsulfonyl-2,4-hexadiene was subsequently applied to the synthesis of the antitumor agents pyrido[4,3-*b*]carbazoles ellipticene and olivacine.[73]

Despite the numerous exploratory studies on the dienophilicity of indole in inverse electron demand and related cycloadditions, there have been relatively few applications of this methodology to the synthesis of predetermined targets. As mentioned in the previous paragraph, Bäckvall has completed the synthesis of ellipticene and olivacine using such a cycloaddition strategy. Ottenheijm has also developed an extensive synthetic program centered on the cycloaddition of vinylnitroso compound **82**, generated in situ from the base treatment of ethyl 3-bromo-2-hydroxyiminopropanoate, with indole and derivatives, a reaction first reported by Gilchrist,[74] ultimately producing the *N*-hydroxytryptophans **84** as a key syn-

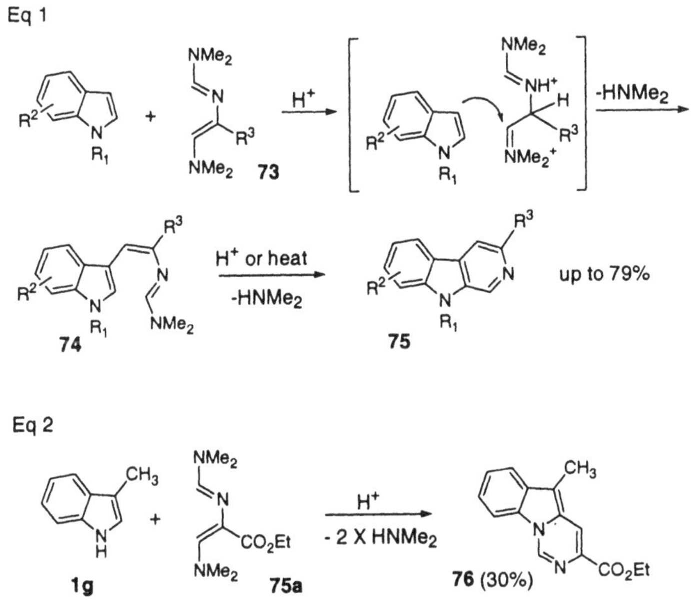

Figure 37. **(Eq. 1)** Stepwise cyclocondensation of indoles with 1,4-diamino-2-azadienes **73** to give β-carbolines **75** reported by Biere.[68] **(Eq. 2)** A similar reaction with 3-substituted indoles such as skatole (**1g**) leads to pyrrolopyrimidines **76**.[69]

thon[75] (Figure 40). A variety of targets have been addressed including the neoechinulins and sporidesmins,[76] C2-functionalized tryptophans including tryptothionine,[77] several of the eudistomin β-carbolines,[78] fumitremorgin C (**85**),[79] and verruculogen TR-2 (**86**),[80] though closure of β-carboline ring still required Pictet–Spengler chemistry on the *N*-hydroxytryptophans.[81] The initially formed cycload-

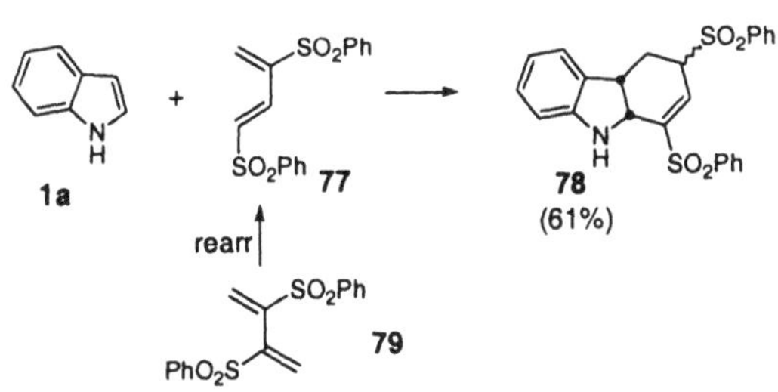

Figure 38. Reaction of indole (**1a**) with 1,3-bis(phenylsulfonyl)-1,3-butadiene (**77**) to give tetrahydrocarbazole **78** reported by Padwa.[70] A reaction between **1a** and 2,3-bis(phenylsulfonyl)-1,3-butadiene (**79**) was too slow, allowing rearrangement of **79** to **77** to occur first.

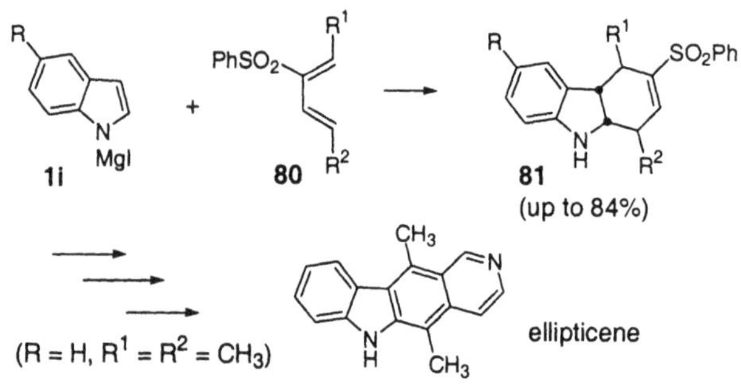

Figure 39. Stepwise cyclocondensation of the indole Grignard salt **1i** with 2-phenylsulfonyl-1,3-dienes **80** to produce tetrahydrocarbazoles **81** reported by Bäckvall.[72] This chemistry was then used by Bäckvall to complete the synthesis of ellipticene.[73]

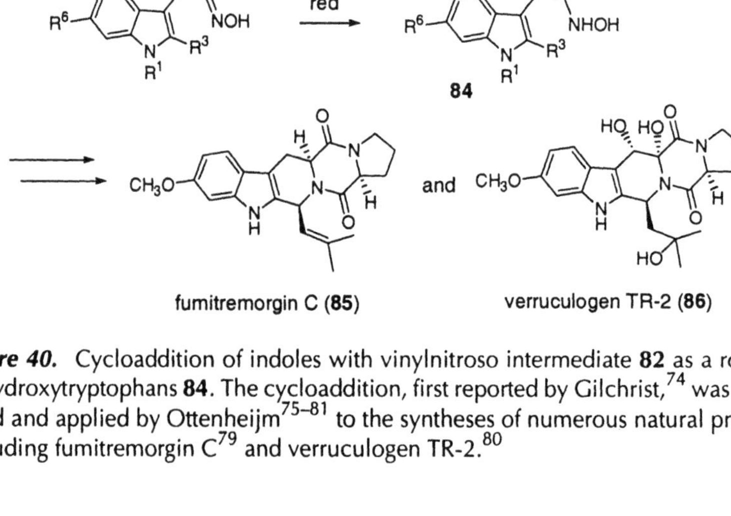

Figure 40. Cycloaddition of indoles with vinylnitroso intermediate **82** as a route to *N*-hydroxytryptophans **84**. The cycloaddition, first reported by Gilchrist,[74] was developed and applied by Ottenheijm[75–81] to the syntheses of numerous natural products including fumitremorgin C[79] and verruculogen TR-2.[80]

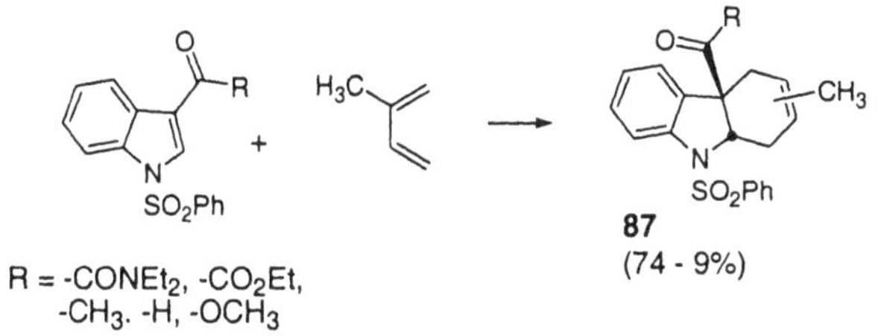

87
(74 - 9%)

Figure 41. Cycloadditions of *N*-benzenesulfonylindoles bearing C3 carbonyl substituents with isoprene to produce tetrahydrocarbazoles **87** as mixtures of regioisomers reported by Wenkert.[82]

duct **83** could not be isolated unless the indole derivative possessed a C3 substituent that resisted migration to the C2 position, such as an acetoxy group.[77c] Both C3 unsubstituted (**1:** R^3 = H), and C3 thioethers (**1:** R^3 = SR) indoles yielded only opened adducts upon product isolation.[77a]

Normal Electron-Demand Cycloadditions

Kraus and Wenkert have both demonstrated that electron-withdrawing C3 substituents along with electron-withdrawing *N*-protecting groups sufficiently lower the $LUMO_{dienophile}$ of indole to enable cycloadditions with electron-rich butadienes to proceed. Wenkert used a series of indole C3 carbonyl substituents along with the *N*-benzenesulfonyl protecting group to produce excellent yields of cycloadducts **87** with isoprene as mixtures of regioisomers (Figure 41).[82] Aluminum trichloride catalyzed the reactions, enabling the cycloadditions to proceed at lower temperatures, but the yields of cycloadducts were also greatly reduced. Lower yields were obtained with 1-benzenesulfonyl-3-nitroindole, and with butadiene as the diene, and no cycloadducts were produced when the electron-withdrawing substituent was located at the indole C2 position with C3 unsubstituted.

Kraus,[48,83] and later Padwa[84] have both utilized *N*-acylated indole derivatives in intramolecular, normal electron demand Diels–Alder reactions with tethered, electron-rich dienes. Kraus, using both amide[83] and urea[48] linkages to tether a variety of butadienes to the indole nitrogen, accomplished the cycloadditions with both 3-substituted and 3-unsubstituted indoles **88** (Figure 42, Eq. 1). The cycloadditions failed, however, when a carbamate (X = O) or trimethylene linkage to the butadiene was used, the latter failure emphasizing the importance of an electron-withdrawing group on the indole nitrogen. Padwa adapted his chemistry exploring the cycloadditions of 2-amidofuran dienes[85] to N^1-acetyltryptamine **90** producing cycloadduct **91** (Figure 42, Eq. 2). This latter strategy represents a novel approach to the Strychnos alkaloids.

Finally, Steckhan has investigated the cycloadditions of the indole radical cation, produced by photochemically induced electron transfer, with 1,3-cyclohexadienes

Eq 1

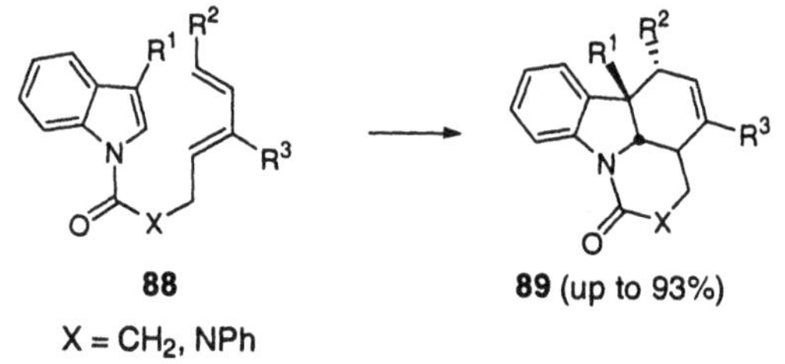

Eq 2

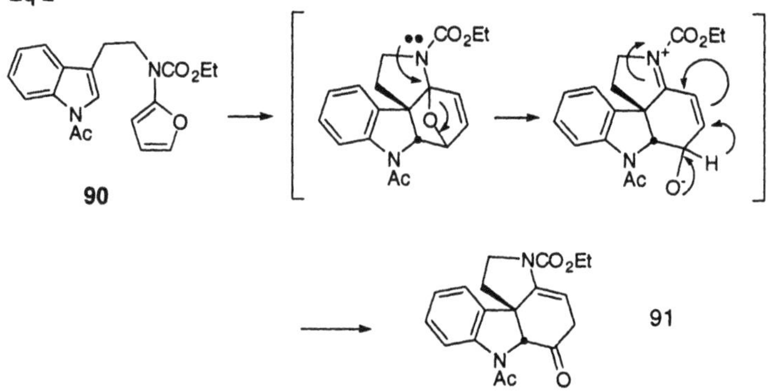

Figure 42. **(Eq. 1)** Kraus' use of amide (X = CH_2)[83] and urea (X = NPh)[48] linkages to tether butadienes with indole (**88**) for intramolecular cycloadditions. **(Eq. 2)** Intramolecular cycloaddition of indolyl amidofuran **90** reported by Padwa.[84]

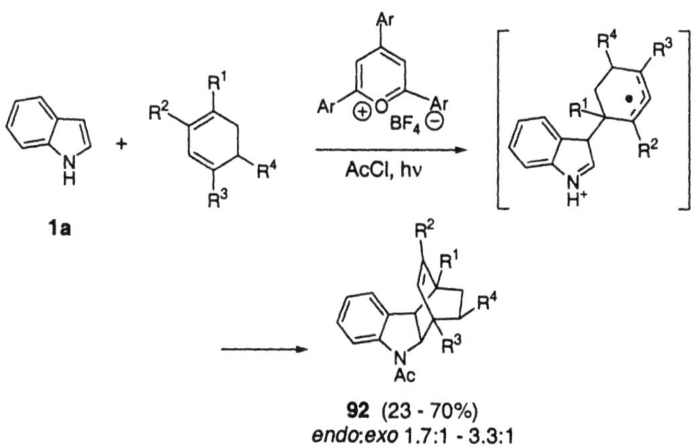

Figure 43. Stepwise cycloadditions of the indole radical cation, generated by photochemically induced electron transfer, with 1,3-cycohexadienes reported by Steckhan.[86–88]

(Figure 43), chemistry which does not proceed without the electron transfer.[86] With substituted 1,3-cyclohexadienes, excellent regioselectivity was observed. Due to the lower oxidation potential of the cycloadducts, which are produced as a mixture of *endo*- and *exo*-stereoisomers **92**, trapping by in situ acylation was necessary. Furthermore, the chemistry was limited to 2,3-unsubstituted indoles. Subsequent theoretical[87] and experimental[88] studies established the stepwise nature of this reaction that proceeds through initial radical addition of the indole radical cation to the 1,3-diene.

III. INDOLE AS A DIPOLAROPHILE IN DIPOLAR CYCLOADDITIONS

A. 1,3-Dipolar Cycloadditions

As an electron-rich subunit, the indole 2,3-double bond also functions as an excellent dipolarophile in 1,3-dipolar additions.[89] Many of the cyclic adducts are not stable, however, and noncyclized as well as other rearranged adducts are often among the product mixtures that result, making assessments of the overall yields of the cycloadditions difficult. For example, the reactions of sulfonyl azides with indoles[90] likely proceed via a 1,3-dipolar additions[91] to give 2-(*N*-sulfonylated)-aminoindoles **93a**, which are in equilibrium with their imino tautomers **93b** (Figure 44, Eq. 1), though no cyclic adducts were reported. The reactions with *N*-methylindole (**1b**) are considerably less regioselective, giving **93** as the indole tautomer and the 3-(*N*-sulfonylated) aminoindoles **94a** as well (Figure 44, Eq. 2). Ozonolysis, a well-understood reaction which proceeds through an initial dipolar addition,[92] of the indole 2,3-double bond has also been known in the literature for a long time.[93]

In general, the 1,3-dipolar cycloadditions of indole with simple 1,3-dipoles reported to date are characterized by modest to low yields of any single cyclic product. Table 2 lists the reported 1,3-dipolar cycloadditions of indole and its simple derivatives. Intermolecular reactions with nitrilimines (Items 1–4),[94] nitrile *N*-oxides (Items 5–10),[95] and nitrones (Item 11)[96] have all appeared in the literature, with an intramolecular dipolar addition with a nitrile oxide (Item 12)[97] also known.

The regioselectivity of the dipolar cycloadditions can usually be predicted from FMO theory with the electron-rich indole C3 position adding to the electropositive end of the 1,3-dipole as expected. This regioselectivity has been reversed with nitrilimine dipoles when indole-2-carboxylate esters are the dipolarophiles (Table 2, Item 4).[94d,94e] The production of arylpyrazolo[4,3-*c*]quinolines from these reactions is analogous to the ring openings and rearrangements observed in the inverse electron demand cycloadditions of indoles with 1,2,4,5-tetrazines **2a** and **2b** (see Figures 4, 5, 8, and 11), 1,2,4-triazines **22a** and **22b** (see Figures 16, 17, and 18), and pyridazines **53** and **55** (see Figures 29 and 30).

Dipolar cycloadditions with nitrile *N*-oxides also produce cycladducts (Table 2, Items 5–10), but the higher yielding reactions with the mesityl nitrile *N*-oxides are

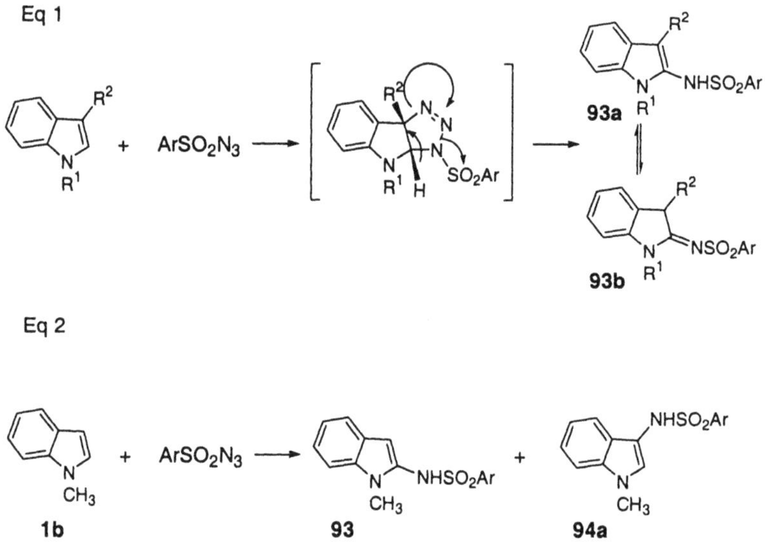

Figure 44. Reactions of indoles with sulfonyl azides via 1,3-dipolar additions.[90,91] **(Eq. 1)** Reactions giving equilibrium mixture of 2-(*N*-sulfonylated) aminoindoles **93a** and their imino tautomers **93b**. **(Eq. 2)** By comparison, the reaction of *N*-methylindole (**1b**) gives a mixture of the regioisomeric 2- and 3-(*N*-sulfonylated) aminoindoles, **93** and **94a**.

very slow, requiring several weeks (Items 5 and 6).[95a,95c] Protection of the indole nitrogen with an electron-withdrawing group also retards the reactions, and in some cases leads to the production of minor amounts of the alternative regioisomer (Item 7).[95a] Substitution at the indole C3 position has been reported to prevent the cycloaddition with nitrilimines,[95b] but the reaction of cycloalk[*b*]indoles with 2,6-dichlorobenzonitrile *N*-oxide gave modest to good yields of cycloadducts (Item 10).[95d] These latter cycloadditions, however, did not succeed when the indole nitrogen was unprotected.[98] Cycloadditions with nitrones also succeeded with 2- and 3-substituted, and 2,3-disubstituted indoles, but the yields were considerably lower than those with these positions open.[96]

In an attempt to prepare *N*-protected 2-nitroindoles by an ipso nitration/destannylation sequence, Gribble and Pelkey treated 2-stannylated indoles **95a,b** with tetranitromethane, producing isoxazolo[5,4-*b*]indole **97** instead (Figure 45).[99] A mechanism proposed for the formation of **97** proceeds through formation of a nitroindole (possible structures include **95c–e**) and the trinitromethane anion. The latter decomposes to nitroformonitrile oxide **96** which undergoes a 1,3-dipolar addition with the nitroindole, then leads to the observed product **97** following HNO_2 (or R_3SnNO_2) elimination.

Table 2. 1,3-Dipolar Cycloadditions of Indoles

Item	*Indole*	*Dipole*	*Products*		*Ref.*
1	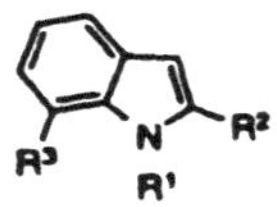	 R = Me, OEt	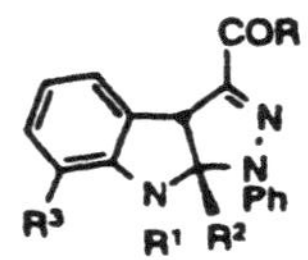 0 - 60%	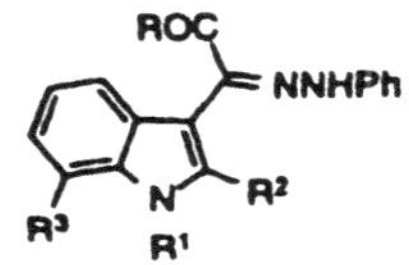 0 - 64%	94a 94b
2	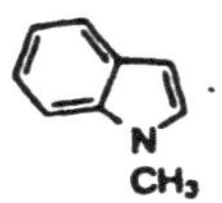	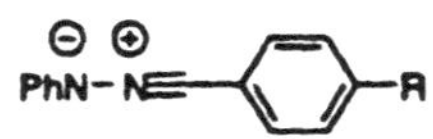 R = H, OCH_3, NO_2	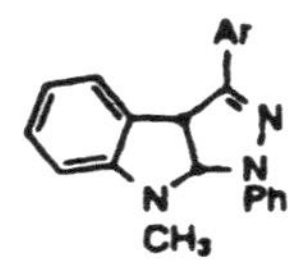 25 - 32%	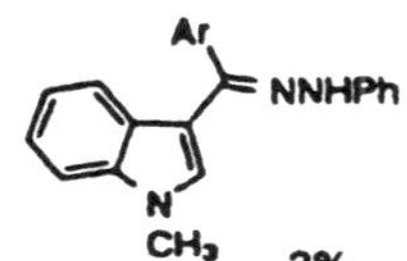 3%	94c
3	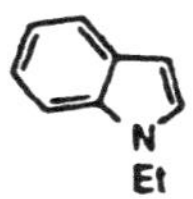		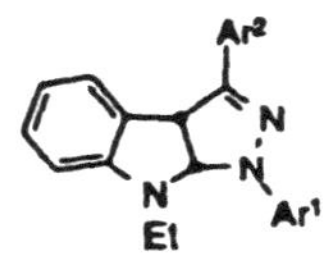 27 - 40%		94d 94e

Table 2. Continued

Item	*Indole*	*Dipole*	*Products*	*Ref.*
4	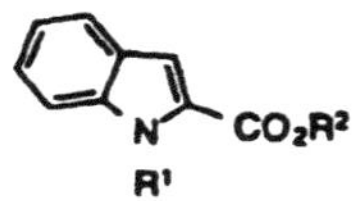 R^1 = Me, Et, Bn		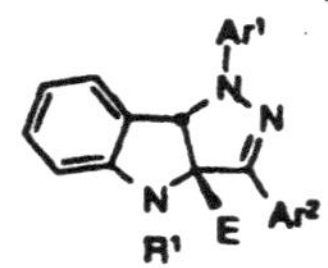 25 - 29% 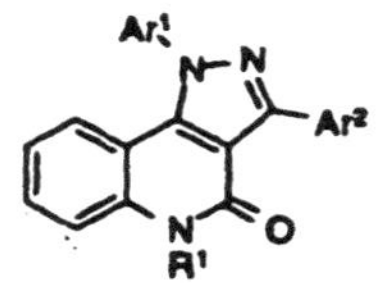 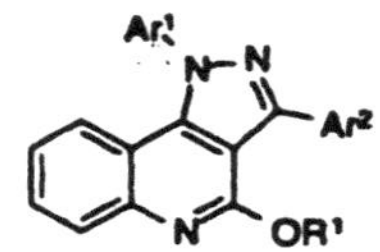	94d 94e
5	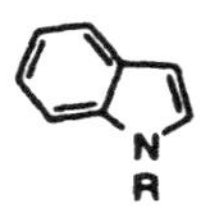 R = H, Me	 Ar = Ph, 2,4,6-$Me_3C_6H_2$	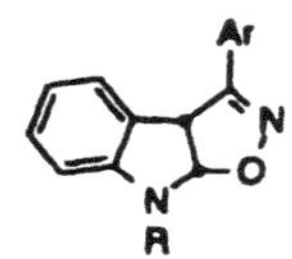 15 - 62% 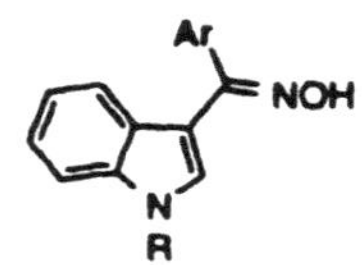	95a 95b
6	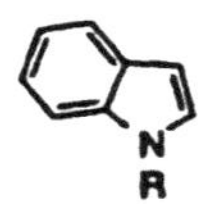 R = Ac, Bz, *p*-Ts, CO_2Me		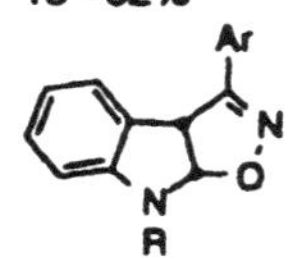 Ar = Ph, 1 - 5% Ar = 2,4,6-$Me_3C_6H_2$: 60 - 80%	95a 95b

7

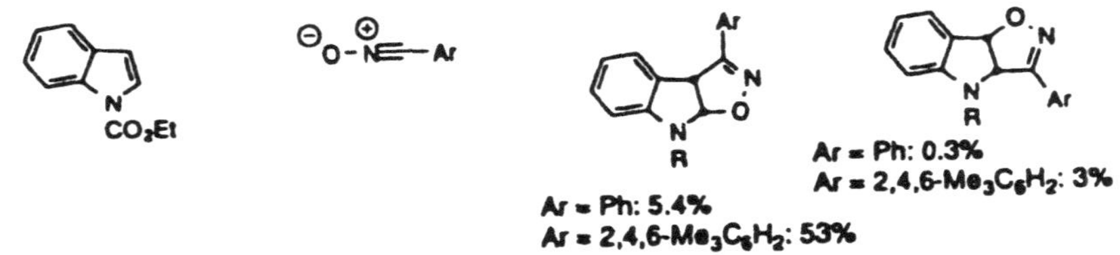

95a

8

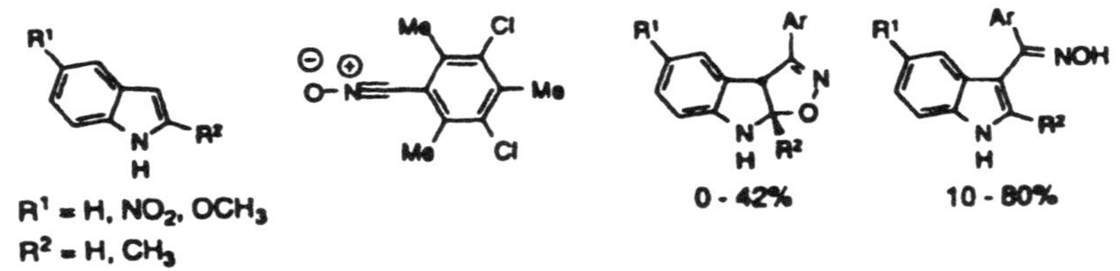

95b

9

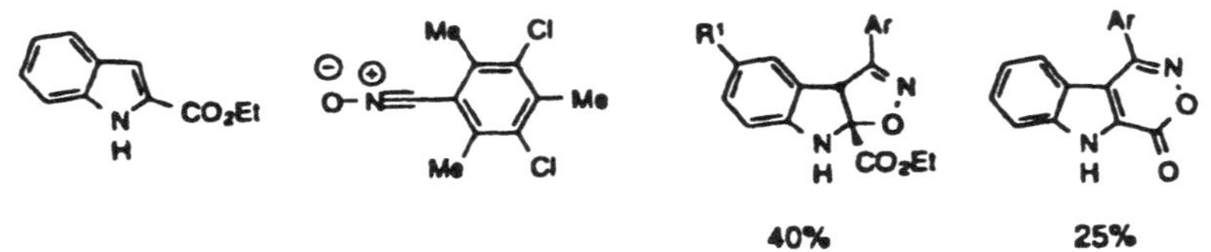

95b

10

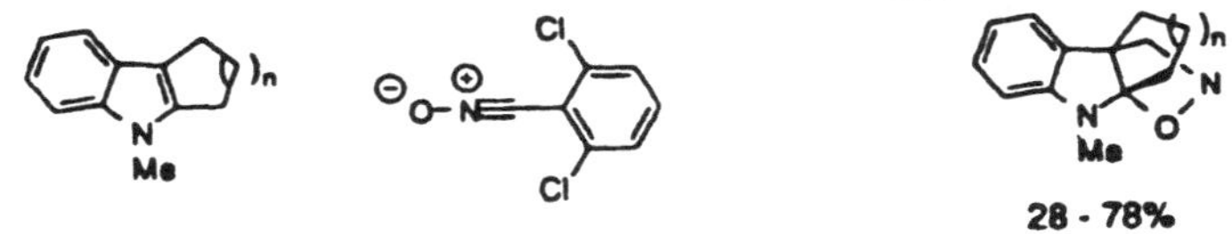

95d

Table 2. Continued

Item	Indole	Dipole	Products	Ref.
11	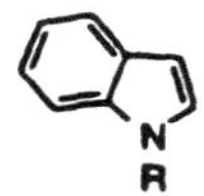 R = *n*-Bu, H, Me	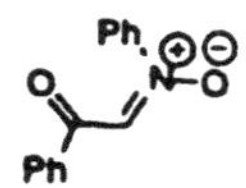	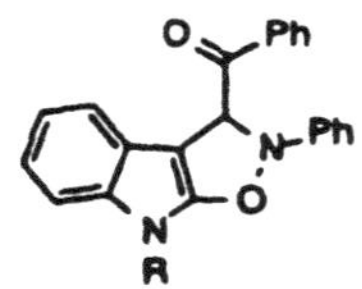 R = *n*-Bu: 9% R = Me:16% R = H: 34% 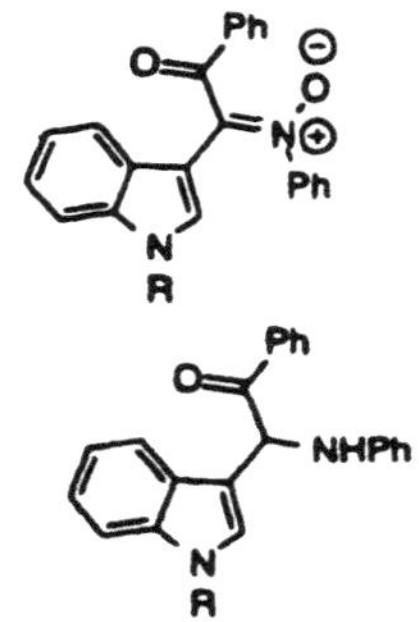	96
12	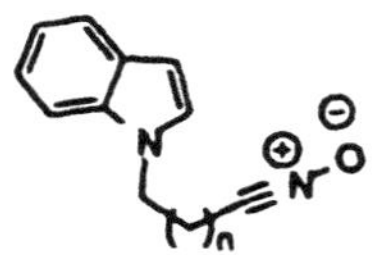 n = 1,2,3		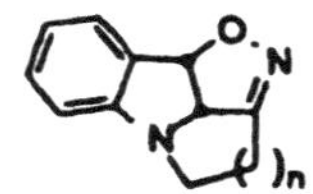 n = 1: no cycloaddition n = 2: 77% n = 3: 20%	97

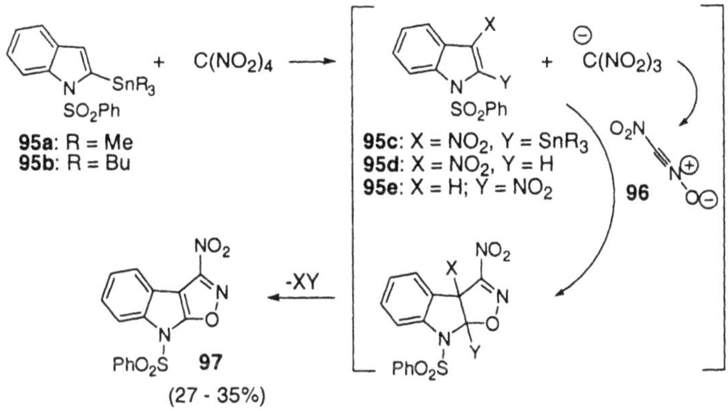

Figure 45. Reaction of 2-stannylated indoles **95a** and **95b** with tetranitromethane to produce isoxazolo[5,4-b]indole **97** through a 1,3-dipolar addition with nitroformonitrile oxide **96** followed by elimination discovered by Pelkey and Gribble.[99]

B. Other Dipolar Cycloadditions

Recent interest in the reactivity of indole with münchnones and related dipoles has paralleled increased activity investigating these reactive intermediates in general. Gribble has reported the cycloadditions of protected 2- and 3-nitroindoles with münchnones **98a,b**, which proceeded through initial cycloadducts **99** followed by loss of CO_2 and HNO_2, as a very facile route to pyrrolo[3,4-*b*]indoles **100** (Figure 46).[100] In general, yields of the pyrroloindoles were greater with R = Ph in comparison to R = Me.

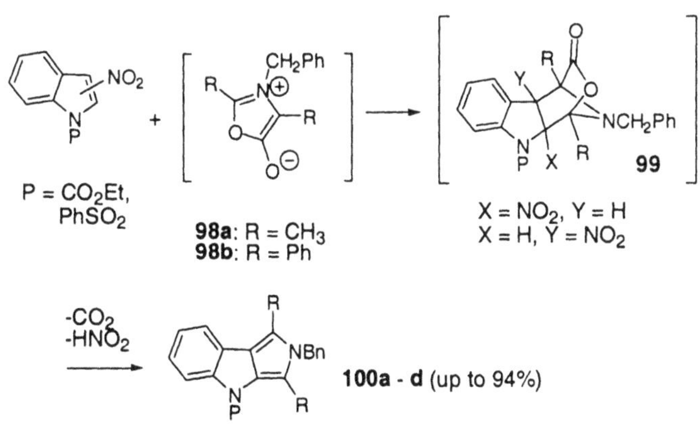

Figure 46. Cycloadditions of 2- and 3-nitroindoles with münchnones **98** to produce pyrrolo[3,4-*b*]indoles **100** reported by Gribble.[100]

With indole-2-carboxylate **101a**, the cycloadditions with nonsymmetrically substituted münchnones **98c** and **98d** gave predominantly the "anti-FMO" products **100e** and **100f**, respectively (Figure 47, Eq. 1).[99] This "reversed" regioselectivity is reminiscent of the dipolar adducts produced in the cycloadditions between indole-2-carboxylates and nitrilimines (Table 2, Item 4), and so may actually represent normal behavior of indole-2-carboxylates. In contrast, the reactions of 3-nitroindole **101b** with **98c** and **98d** were nonregiospecific (Figure 47, Eq. 2).[99]

Building on considerable work investigating the generation of mesoionic species from the intramolecular capture of carbenoid intermediates by carbonyls,[101] in a series of papers over the past few years, Padwa has outlined a strategically exemplary synthetic approach to a variety of indole alkaloids[102] using the intramolecular cycloaddition of indoles to these dipoles generated in situ.[103] Cycloaddition of the indole subunit across the carbonyl ylide dipole in **103**, generated by closure of the γ-carbonyl to a rhodium carbenoid intermediate, produced adduct **104** with the intact Aspidosperma alkaloid skeleton in a remarkable 95% yield (Figure 48, Eq. 1). Subsequent transformations led to desacetoxy-4-oxo-6,7-dihydrovindorosine (**105**).[104] An analogous intramolecular cycloaddition (95%!) across

Eq 1

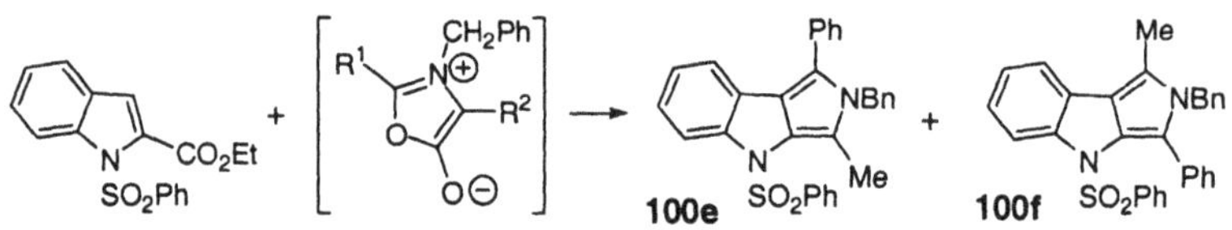

101a

98c: R^1 = Me, R^2 = Ph → **100e**:**100f** 9:1 (88%)
98d: R^1 = Ph. R^2 = Me → **100e**:**100f** 1:9 (85%)

Eq 2

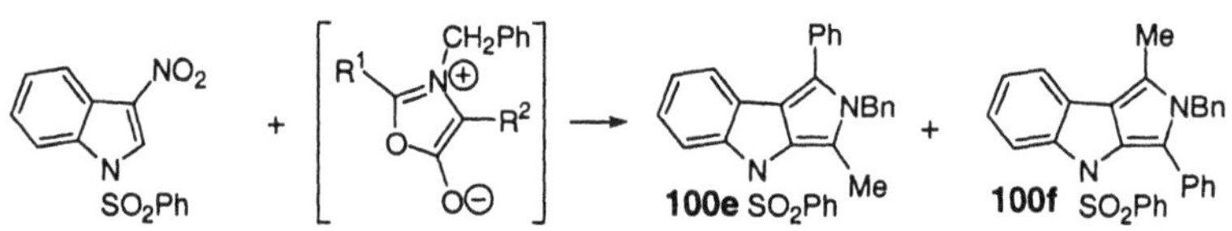

101b

98c: R^1 = Me, R^2 = Ph → **100e**:**100f** 7:3 (76%)
98d: R^1 = Ph. R^2 = Me → **100e**:**100f** 100:0 (74%)

Figure 47. **(Eq. 1)** Cycloadditions of ethyl-*N*-(benzenesulfonyl)indole-2-carboxylate (**101a**) with nonsymmetric münchnones **98c** and **98d** to give pyrrolo[3,4-*b*]indoles **101e** and **101f**, respectively, with "anti-FMO" regioselectivity. **(Eq. 2)** Nonregiospecific cycloadditions of *N*-(benzenesulfonyl)-3-nitroindole with **98c** and **98d**. Both examples were discovered by Pelkey and Gribble.[99]

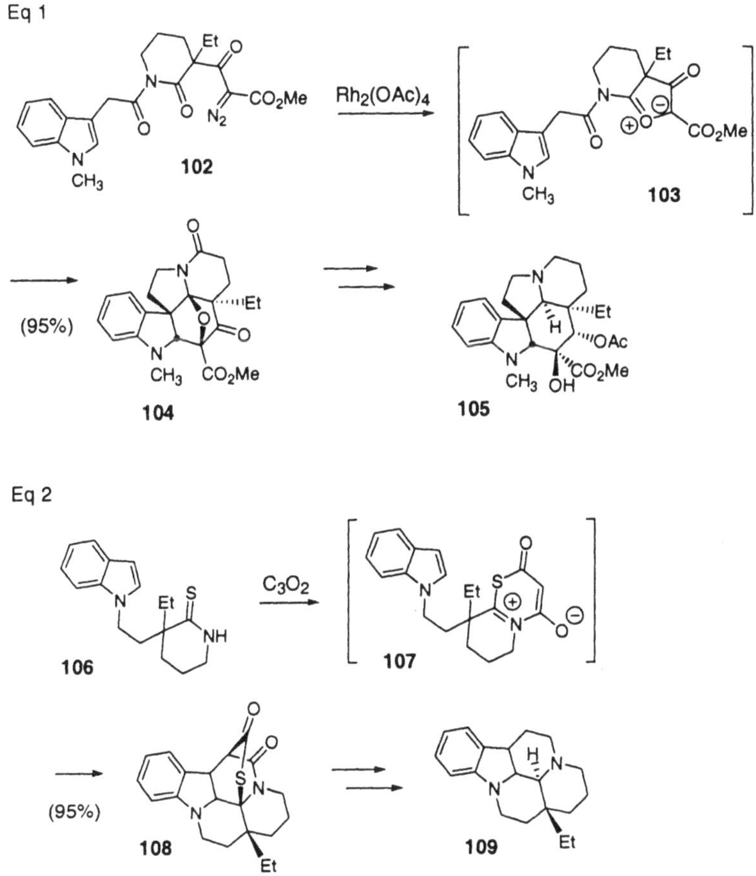

Figure 48. Intramolecular dipolar cycloadditions of indoles discovered and developed by Padwa.[101–105] **(Eq. 1)** Cycloaddition across the carbonyl ylide dipole in **103** gave adduct **104**, with the Aspidosperma alkaloidal skeleton.[104] **(Eq. 2)** Cycloaddition across the betaine in **107** gave adduct **108**, leading to the total synthesis of *epi*-eburnamenine (**109**).[105]

the betaine in **107** led to synthesis of epi-eburnamenine (**109**, Figure 48, Eq. 2).[105] In both of these cases, the cycloadditions were completely stereoselective, though the stereochemistry in adduct **108** was not reported since this is lost after conversion to epi-eburnamenine.

IV. CONCLUSION

Indole and its derivatives with their electron-rich 2,3-double bond function as very good dienophiles in inverse electron demand Diels–Alder reactions with electron-deficient dienes, particularly heteroaromatic azadienes such as 1,2,4,5-tetrazines and 1,2,4-triazines. Typically the reactions proceed with rapid loss of nitrogen gas and subsequent aromatization to annnulate a third aromatic ring onto the indole 2,3-double bond. Ring opening of the pyrrole ring, however, often intercedes as a secondary pathway leading to ring-opened or rearranged products. In addition, indoles also undergo facile 1,3-dipolar additions, though these latter reactions often lead to ring-opened products.

Despite the numerous examples of the cycloadditions involving the indole 2,3-double bond as a 2π component, there have been relatively few synthetic programs based on this chemistry. One notable effort is development of the cycloadditions of indoles with vinylnitroso compounds as a route to *N*-hydroxytryptophans with subsequent transformation to a variety of indole alkaloids by Ottenheijm. In addition, the emerging work of Padwa on the use of intramolecular cycloadditions of indole across in situ generated dipoles is really a beautifully conceived synthetic strategy. No doubt additional results will continue from these and other programs.

ACKNOWLEDGMENTS

To date, the work on the dienophilicity on indole in our group has been the subject of three Ph.D. dissertations by three excellent graduate students, Scott C. Benson, Jia-He Li, and Lily Lee, and by a superb Master's student, Wen-Hong Fan. In addition, two more are completing their doctoral studies on this theme: Zhao-Kui Wan and Rana Nomak. Numerous undergraduates have also participated in this project, many appearing as coauthors on published papers: Chris Palabrica, Urvashi Rangan, Jeanne DeCara, Jonathan Gross, Mamta Parikh, Kevin Daly, Wendy Hui-Yin Fu, Karina Macek, Marc Girardot, Lydie Yang, and Alan DiPesa. We are very grateful for the efforts of all these students.

Funding for this research has been provided in part by the Research Corporation, the Petroleum Research Fund, administered by the American Chemical Society, and the National Science Foundation.

REFERENCES AND NOTES

1. Remers, W. A. In *Indoles, Part 1*; Houlihan, W. J., Ed.; Chemistry of Heterocyclic Compounds Monograph Series; Wiley: New York, 1972, Chapter 1.
2. (a) Marian, L. In *The Alkaloids, Chemistry and Pharmacology*; Manske, R. H. F.; Holmes, H. L., Eds.; Academic: New York, 1952, Vol. 2, pp. 371–498. (b) Saxton, J. E. In *The Alkaloids, Chemistry and Pharmacology*; Manske, R. H. F., ed.; Academic: New York, 1960, Vol. 7, pp. 4–199. (c) Taylor, W. I. *Indole Alkaloids: An Introduction to the Enamine Chemistry of Natural Products*; Pergamon: London, 1966. (d) Neuss, N. In *Chemistry of the Alkaloids*; Pelletier, S. W., Ed.; van Nostrand Reinhold: New York, 1970, Chapter 9. (e) Husson, H.- P. In *The Alkaloids, Chemistry and Pharmacology*; Brossi, A., Ed.; Academic: New York, 1985, Vol. 26, Chapter 1. (f) *The Alkaloids, Part 4, Monoterpenoid Indole Alkaloids and Supplement to Part 4*, Saxton, J. E., Ed.; Chemistry of Heterocyclic Compounds Monograph Series; Wiley: New York, 1983, and 1994.

3. Reviews of indole chemistry: (a) Sundberg, R. J. *The Chemistry of Indoles*; Organic Chemistry Monograph Series, Blomquist, A. T. Ed.; Academic: New York, 1970. (b) Brown, R. T.; Joule, J. A.; Sammes, P. G. In *Comprehensive Organic Chemistry, The Synthesis and Reactions of Organic Compounds*; Sammes, P. G., Ed.; Pergamon: London, 1979, Vol. 4, pp. 411. (c) *Indoles, Parts 1, 2, and 3*; Houlihan, W. J., Ed.; Chemistry of Heterocyclic Compounds Monograph Series; Wiley: New York, 1972 and 1979.
4. Reviews: (a) Whaley, W.; Govindochari, T. *Org. React.* **1951**, *6*, 151. (b) Abramovich, R,; Spenser, I. In *Advances Heterocyclic Chemistry*; Katritzky, A. R.; Boulton, A. J.; Lagowski, J. M., Eds.; Academic: New York, 1964, Vol. 3, 79. (c) Stuart, K.; Woo-Ming, R. *Heterocycles* **1975**, *3*, 223. (d) Ungemach, F.; Cook, J. M. *Heterocycles* **1978**, *9*, 1089. (e) Kametani, T.; Fukumoto, K. In *Chemistry of Heterocyclic Compounds, Isoquinolines, Part I*; Weissberger, A.; Taylor, E. C., Eds.; Wiley: New York, 1981, 170. (f) Czerwinski, Y. M.; Cook, J. M. In *Advances in Heterocyclic Natural Products Synthesis*; Pearson, W., Ed.; JAI: Greenwich, CT, 1996, Vol. 3, p. 217.
5. Reviews: (a) Whaley, W.; Govindochari, T. *Org. React.* **1951**, *6*, 74. (b) Fodor, G.; Naguband, S. *Tetrahedron* **1980**, *36*, 1279. (c) Kametani, T.; Fukumoto, K. In *Chemistry of Heterocyclic Compounds, Isoquinolines, Part I*; Weissberger, A.; Taylor, E. C., Eds.; Wiley: New York, 1981, 142.
6. (a) Robinson, R.; Saxton, J. E. *J. Chem. Soc.* **1953**, 296. (b) Cockerill, D.A.; Robinson, R.; Saxton, J.E. *J. Chem. Soc.* **1955**, 4369. (c) Robinson, B.; Smith, G. F. *J. Chem. Soc.* **1960**, 4574. (d) Garnick, R. L.; Levery, S. B.; LeQuesne, P. W. *J. Org. Chem.* **1978**, *43*, 1226.
7. Harrington, P.; Kerr, M. A. *Can. J. Chem.-Rev. Can. Chim.* **1998**, *76*, 1256.
8. Knoelker, H.-J. In *Advances in Nitrogen Heterocycles*; Moody, C. J., Ed.; JAI: Greenwich, CT, 1995, Vol. 1, p. 173.
9. Grotjahn, D. B.; Vollhardt, K. P. C. *J. Am. Chem. Soc.* **1986**, *108*, 2091.
10. (a) Pindur, U. *Heterocycles* **1988**, *27*, 1253. (b) Pindur, U. In *Advances in Nitrogen Heterocycles*; Moody, C. J., Ed.; JAI: Greenwich, CT, 1995,Vol. 1, p. 121.
11. For an overview: (a) Overman, L. E.; Sworin, M. In *Alkaloids: Chemical and Biological Perspectives*; Pelletier, S. W., Ed.; Wiley: New York, 1985, Vol. 3, p. 275. For recent leading references from Kuehne's group on this approach: (b) Kuehne, M. E.; Bandarage, U. K.; Hammach, A.; Li, Y.-L.; Wang, T. *J. Org. Chem.* **1998**, *63*, 2172. (c) Kuehne, M. E.; Wang, T.; Seraphin, D. *J. Org. Chem.* **1996**, *61*, 7873. (d) Kuehne, M. E.; Wang, T.; Seaton, P. J. *J. Org. Chem.* **1996**, *61*, 6001. (e) Kuehne, M. E.; Bandarage, U. K. *J. Org. Chem.* **1996**, *61*, 1175.
12. Pindur, U.; Erfanian-Abdoust, H. *Chem. Rev.* **1989**, *89*, 1681.
13. (a) Magnus, P. In *Strategies and Tactics in Organic Synthesis*; Lindberg, T., Ed.; Academic: New York, 1984, p. 83. (b) Magnus, P.; Gallagher, T.; Brown, P.; Pappalardo, P. *Acc. Chem. Res.* **1984**, *17*, 35.
14. Gribble, G. W.; Keavy, D. J.; Davis, D. A.; Saulnier, M. G.; Pelcman, B.; Borden, T. C.; Sibi, M. P.; Olson, E. R.; Bel Bruno, J. J. *J. Org. Chem.* **1992**, *57*, 5878, and references therein.
15. Moody, C. J.; Rahimtoola, K. F.; Porter, B.; Ross, B. C. *J. Org. Chem.* **1992**, *57*, 2105.
16. Reviews: (a) Boger, D. L.; *Tetrahedron* **1983**, *39*, 2869. (b) Boger, D. L. *Chem. Rev.* **1986**, *86*, 781. (c) Boger, D. L.; Weinreb, S. M. *Hetero Diels-Alder Methodology in Organic Synthesis*; Organic Chemistry Monograph Series, Vol 47.; Academic: New York, 1987. (d) Kametani, T.; Hibino, S. In *Advances in Heterocyclic Chemistry*; Katritzky, A. R., Ed.; Academic: New York, 1987, Vol. 42, p. 245.
17. For a recent overview: Boger, D. L. *J. Heterocycl. Chem.* **1996**, *33*, 1519.
18. Taylor, E. C. *Bull. Soc. Chim. Belg.* **1988**, *97*, 599.
19. Sietz, G.; Kampchen, T. *Arch. Pharm. (Weinheim)* **1976**, *309*, 679.
20. Takahashi, M.; Ishida, H.; Kohmoto, M. *Bull. Chem. Soc. Jpn.* **1976**, *49*, 1725.
21. Acheson, R. M.; Bridson, J. N.; Cecil, T. R.; Hands, A. R. *J. Chem. Soc., Perk. Trans 1.* **1972**, 1569.
22. Seitz, G.; Mohr, R. *Chem. Zeit.* **1987**, *111*, 81.

23. Haider, N.; Wanko, R. *Heterocycles* **1994**, *38*, 1805.
24. Benson, S. C. Ph.D. Dissertation, Boston University, Boston, MA, 1992.
25. Benson, S. C.; Palabrica, C. A.; Snyder, J. K. *J. Org. Chem.* **1987**, *52*, 4610.
26. Daly, K.; Nomak, R.; Snyder, J. K. *Tetrahedron Lett.* **1997**, *38*, 8611.
27. (a) Boger, D. L.; Coleman, R. S.; Panek, J. S.; Huber, F. X.; Sauer, J. *J. Org. Chem.* **1985**, *50*, 5377. (b) Boger, D. L.; Panek, J. S.; Patel, M. *Org. Synth.* **1991**, *70*, 79.
28. Girardot, M.; Nomak, R.; Snyder, J. K. *J. Org. Chem.* **1998**, *63*, 10063.
29. (a) Basha, A.; Lipton, M. F.; Weinreb, S. M. *Tetrahedron Lett.* **1977**, 4171. (b) Lipton, M. F.; Weinreb, S. M. In *Org. Synth., Collect.*, Noland, W. E., Ed.; Wiley: New York, 1988, Vol. 6, pp. 492–495. (c) Sidler, D. R.; Lovelace, T. C.; McNamara, J. M.; Reider, P. J. *J. Org. Chem.* **1994**, *59*, 1231. (d) Chung, E.-A.; Cho, C.-W.; Ahn, K. H. *J. Org. Chem.* **1998**, *63*, 7590.
30. Nomak, R.; Snyder, J. K. Unpublished results.
31. Haider, N. *Acta Chim. Slov.* **1994**, *41*, 205.
32. Hansch, C.; Leo, A.; Unger, S. H.; Kim, K. H.; Nikaitani, D.; Lien, E. J. *J. Med. Chem.* **1973**, *16*, 1207.
33. (a) Pindur, U.; Kim, M.-H. *Tetrahedron Lett.* **1988**, *29*, 3927. (b) Pindur, U.; Pfeuffer, L.; Kim, M.-H. *Helv. Chim. Acta* **1989**, *72*, 65.
34. (a) Benson, S. C.; Lee, L.; Snyder, J. K. *Tetrahedron Lett.* **1996**, *37*, 5061. (b) Lee, L. Ph.D. Dissertation, Boston University, Boston, MA, 1999.
35. (a) Takano, S.; Yuta, K.; Hateyama, S.; Ogasawara, K. *Tetrahedron Lett.* **1979**, 369. (b) Takano, S.; Suzuki, Y.; Ogasawara, K. *Heterocycles* **1981, 16**, 1479. (c) Shin, K.; Ogasawara, K. *Chem. Lett.* **1995**, 289.
36. Reviews: (a) Shepherd, R. G.; Fedrick, J. L. In *Advances in Heterocyclic Chemistry*; Katritzky, A. R.; Boulton, A. J.; Lagowski, J. M., Eds.; Academic: New York, 1965, Vol. 4, p. 146. (b) Crampton, M. R. In *Organic Reactions Mechanisms*; Knipe, E.; Watts, W. E., Eds.; Wiley: New York, 1971, Chap. 7. (c) Crampton, M. R. In *Organic Reactions Mechanisms*; Knipe, E.; Watts, W. E., Eds.; Wiley: New York, 1985, Chap. 7. (d) Crampton, M. R. In *Organic Reactions Mechanisms*; Knipe, E.; Watts, W. E., Eds.; Wiley: New York, 1986, Chap. 7. (e) Crampton, M. R. In *Organic Reactions Mechanisms*; Knipe, E.; Watts, W. E., Eds.; Wiley: New York, 1993, Chap. 7.
37. Benson, S. C.; Lee, L.; Yang, L.; Snyder, J. K. *Tetrahedron*, **1999**. Submitted.
38. Benson, S. C.; Gross, J. L.; Snyder, J. K. *J. Org. Chem.* **1990**, *55*, 3257.
39. (a) Neunhoeffer, H.; Fruhauf, H. W. *Tetrahedron Lett.* **1970**, 3355. (b) Steigel, A.; Sauer, J. *Tetrahedron Lett.* **1970**, 3357. (c) Neunhoeffer, H.; Fruhauf, H. W. *Liebigs Ann. Chem.* **1972**, *758*, 125. (d) Burg, B.; Dittmar, W.; Reim, H.; Steigel, A.; Sauer, J. *Tetrahedron Lett.* **1975**, 2897. (e) Neunhoeffer, H.; Lehmann, B. *Liebigs Ann. Chem.* **1977**, 1413. (f) Neunhoeffer, H. In *Chemistry of 1,2,3-Triazines, 1,2,4-Triazines, Tetrazines, and Pentazines*; Neunhoeffer, H.; Wiley, P., Eds; Chemistry of Heterocyclic Compounds Monograph Series; Wiley: New York, 1978, p. 227.
40. (a) Boger, D. L.; Panek, J. S. *J. Org. Chem.* **1981**, *46*, 2179. (b) Boger, D. L.; Panek, J. S.; Meier, M. M. *J. Org. Chem.* **1982**, *47*, 895. (c) Boger, D. L.; Panek, J. S. *J. Org. Chem.* **1982**, *47*, 3763. (d) Boger, D. L.; Duff, S. R.; Panek, J. S.; Yasuda, M. *J. Org. Chem.* **1985**, *50*, 5782. (e) Neunhöffer, H.; Bachmann, M. *Liebigs Ann. Chem.* **1985**, 1263. (f) Taylor, E. C.; Macor, J. E. *Tetrahedron Lett.* **1985**, *26*, 2415. (g) Taylor, E. C.; McDaniel, K. F.; Warner, J. C. *Tetrahedron Lett.* **1987**, *28*, 1977. (h) Taylor, E. C.; Pont, J. L.; Warner, J. C. *J. Org. Chem.* **1988**, *53*, 3568. (i) Chenard, B. L.; Ronau, R. T.; Schulte, G. K. *J. Org. Chem.* **1988**, *53*, 5175. (j) Taylor, E. C.; Macor, J. E. *J. Org. Chem.* **1989**, *54*, 1249. (k) Neunhoeffer, H.; Phillipp, B.; Schildhauer, B. *Heterocycles* **1993**, *35*, 1089. (l) Richter, M.; Seitz, G. *Arch. Pharm.* **1995**, *328*, 175. (m) Rykowski, A.; Lipinska, T. *Pol. J. Chem.* **1997**, *71*, 83.
41. Macor, J. E.; Kuipers, W.; Lachicotte, R. J. *J. Chem. Soc., Chem. Comm.* **1998**, 983.
42. (a) Rykowski, A.; van der Plas, H. C.; van Veldhuizen, A. *Recl. Trav. Chim.* **1978**, *97*, 273. (b) Konno, S.; Sagi, M.; Yoshida, N.; Yamanaka, H. *Heterocycles* **1987**, *26*, 3111. (c) Rykowski, A.; Makosza, M. *Liebigs Ann. Chem.* **1988**, 627.

43. For a review of intramolecular Diels-Alder reactions: Ciganek, E. *Org. React.* **1984**, *32*, 1.
44. Ohmoto, T.; Koike, K. *The Alkaloids, Chemistry and Pharmacology*; Brossi, A., Ed.; Academic: New York, 1989, Vol. 36, Chap. 3.
45. (a) Benson, S. C.; Li, J.-H.; Snyder, J. K. *J. Org. Chem.* **1992**, *57*, 5285. (b) Li, J.-H. Ph.D. Dissertation, Boston University, Boston, MA, 1999. (c) Wan, Z.-K.; Snyder, J. K. Unpublished results.
46. Seitz, G.; Dietrich, S.; Gorge, L.; Richter, *J. Tetrahedron Lett.* **1986**, *27*, 2747.
47. (a) Li, J.-H.; Snyder, J. K. *Tetrahedron Lett.* **1994**, *35*, 1485. (b) Jager, H.; Luetolf, J.; Meyer, M. W. *Angew. Chem., Int. Ed. Eng.* **1979**, *18*, 786. (c) Schmidt, H.-J.; Schaefer, H. J. *Angew. Chem., Int. Ed. Eng.* **1979**, *18*, 787. (d) Bretherick, L. *J. Chem. Ed.* **1987**, *64 (2)*, A43. (e) Markgraf, H.; Sangani, P. K.; Finkelstein, M. *Syn. Comm.* **1997**, *27*, 1285.
48. Kraus, G. A.; Bougie, D.; Jacobson, R. A.; Su, Y. *J. Org. Chem.* **1989**, *54*, 2425.
49. Fan, W.-H.; Parikh, M.; Snyder, J. K. *Tetrahedron Lett.* **1995**, *37*, 6591.
50. Murata, O.; Shigemori, H.; Ishibashi, M.; Sugama, K.; Hayashi, K.; Kobayashi, J. *Tetrahedron Lett.* **1991**, *32*, 3539.
51. Wan, Z.-K.; Snyder, J. K. *Tetrahedron Lett.* **1998**, *39*, 2487.
52. Other S_NAr displacements of methylthiolate from 1,2,4-triazines, see ref. 46, also (a) Taylor, E. C.; Warner, J. C.; Pont, J. L. *J. Org. Chem.* **1988**, *53*, 800. S_NAr displacements of methylthiolate from 1,2,4,5-tetrazines: (b) Seitz, G.; Görge, L.; Dietrich, S. *Tetrahedron Lett.* **1985**, *26*, 4355. Taylor has also accomplished these S_NAr displacements from 1,2,4-triazines after oxidizing to the corresponding sulfone: (c) Taylor, E. C.; Macor, J. E. *Tetrahedron Lett.* **1986**, *27*, 431. (d) Taylor, E. C.; Macor, J. E. *Tetrahedron Lett.* **1986**, *27*, 2107. (e) Taylor, E. C.; Pont, J. L. *Tetrahedron Lett.* **1987**, *28*, 379. (f) Taylor, E. C.; Macor, J. E.; Pont, J. L. *Tetrahedron* **1987**, *43*, 5145. (g) Taylor, E. C.; Pont, J. L.; Warner, J. C. *Tetrahedron* **1987**, *43*, 5159. (h) Taylor, E. C.; Macor, J. E.; French, L. G. *J. Org. Chem.* **1991**, *56*, 1807. Also: (i) Konno, S.; Yokoyama, M.; Kaite, A.; Yamatsuta, I.; Ogawa, S.; Mizugaki, M.; Yamanaka, H. *Chem. Pharm. Bull.* **1982**, *30*, 152.
53. (a) Neunhöffer, H.; Werner, G. *Tetrahedron Lett.* **1972**, 1517. (b) Jojima, T.; Takeshiba, H.; Konotsune, T. *Chem. Pharm. Bull.* **1972**, *20*, 2191. (c) Neunhöffer, H.; Werner, G. *Liebigs Ann. Chem.* **1973**, 437. (d) Neunhöffer, H.; Werner, G. *Liebigs Ann. Chem.* **1973**, 1955. (e) Jojima, T.; Takeshiba, H.; Kinoto, T. *Chem. Pharm. Bull.* **1976**, *24*, 1581. (f) Jojima, T.; Takeshiba, H.; Kinoto, T. *Chem. Pharm. Bull.* **1976**, *24*, 1588. (g) Jojima, T.; Takeshiba, H.; Kinoto, T. *Chem. Pharm. Bull.* **1980**, *28*, 198. (h) Boger, D. L.; Coleman, R. S. *J. Org. Chem.* **1986**, *51*, 3250. (i) Boger, D. L.; Coleman, R. S.; Invergo, B. J. *J. Org. Chem.* **1987**, *52*, 1521. (j) Boger, D. L.; Patel, M. *J. Org. Chem.* **1988**, *53*, 1405. (i) Boger, D. L.; Sakya, S. M. *J. Org. Chem.* **1988**, *53*, 1415. (k) Boger, D. L.; Zhang, M. *J. Am. Chem. Soc.* **1991**, *113*, 4230.
54. (a) Haider, N. *Tetrahedron* **1991**, *47*, 3959. (b) Haider, N. *Tetrahedron* **1992**, *48*, 7173. (c) Haider, N.; Loll, C. *J. Heterocycl. Chem.* **1994**, *31*, 357. (d) Haider, N.; Mereiter, K.; Wanko, R. *Heterocycles* **1994**, *38*, 1845.
55. Haider, N.; Mereiter, K.; Wanko, R. *Heterocycles* **1995**, *41*, 1445.
56. Oishi, E.; Taido, N.; Iwamoto, K.; Miyashita, A.; Higashino, T. *Chem. Pharm. Bull.* **1990**, *38*, 3268.
57. Nesi, R.; Giomi, D.; Turchi, S.; Falai, A. *J. Chem. Soc., Chem. Comm.* **1995**, 2201.
58. See ref. 1, pp. 66–67. Also: Smith, G. F. In *Advances in Heterocyclic Chemistry*; Katritzky, A. T., Ed.; Wiley: New York, 1953, Vol. 2, p. 287.
59. Cheng, Y.-S.; Ho, E.; Mariano, P. S.; Ammon, H. L. *J. Org. Chem.* **1985**, *50*, 5678.
60. Boger, D. L.; Panek, J. S. *J. Am. Chem. Soc.* **1985**, *107*, 5745.
61. Bednarski, M.; Danishefsky, S. *J. Am. Chem. Soc.* **1983**, *105*, 3716.
62. Raasch, M. S. *J. Org. Chem.* **1980**, *45*, 856.
63. Komatsu, T.; Nishio, T.; Omote, Y. *Chem. Ind. (London)* **1978**, 95.
64. Omote, Y.; Harada, K.; Tomotake, A.;Kashima, C. *J. Heterocycl. Chem.* **1984**, *21*, 1841.

65. (a) Omote, Y.; Tomotake, A.; Kashima; C. *Tetrahedron Lett.* **1984**, *25*, 2993. (b) Omote, Y.; Tomotake, A.; Kashima, C. *J. Chem. Res. (S)* **1987**, 10.
66. (a) Heine, H. W.; Olsson, C.; Bergin, J. D.; Foresman, J. B.; Williams, E. A. *J. Org. Chem.* **1987**, *52*, 97. (b) Black, D. St. C.; Craig, D. C.; Heine, H. W.; Kumar, N.; Williams, E. A. *Tetrahedron Lett.* **1987**, *28*, 6691.
67. Heine, H. W.; LaPorte, M. G.; Overbaugh, R. H.; Williams, E. A. *Heterocycles* **1995**, *40*, 743.
68. Biere, H.; Russe, R.; Seelen, W. *Liebigs Ann. Chem.* **1986**, 1749.
69. Biere, H.; Russe, R. *Liebigs Ann. Chem.* **1987**, 491.
70. Padwa, A.; Gareau, B.; Harrison, B.; Rodriguez, A. *J. Org. Chem.* **1992**, *57*, 3540.
71. Bäckvall, J. E.; Chinchila, R.; Najera, C.; Yus, M. *Chem. Rev.* **1998**, *98*, 2291.
72. Bäckvall, J. E.; Plobeck, N. A.; Juntunen, S. K. *Tetrahedron Lett.* **1989**, *30*, 2589.
73. Bäckvall, J. E.; Plobeck, N. A. *J. Org. Chem.* **1990**, *55*, 4528.
74. Gilchrist, T. I.; Lingham, D. A.; Roberts, T. C. *J. Chem. Soc., Chem. Comm.* **1979**, 1089.
75. Ottenheijm, H. C. J.; Herscheid, J. D. M. *Chem. Rev.* **1986**, *86*, 697.
76. (a) Ottenheijm, H. J. C.; Plate, R.; Noordik, J. H.; Herscheid, J. D. M. *J. Org. Chem.* **1982**, *47*, 2147. (b) Plate, R.; Theunisse, A. W. G.; Nivard, R. J. F.; Ottenheijm, H. C. J. *Tetrahedron* **1986**, *42*, 6511. (c) Plate, R.; Akkerman, M. A. J.; Ottenheijm, H. C. J.; Smits; J. M. M. *J. Chem. Soc., Perk. Trans. 1* **1987**, 2481.
77. (a) Plate, R.; Ottenheijm, H. J. C.; Nivard, R. J. F. *J. Org. Chem.* **1984**, *49*, 540. (b) Plate, R.; Ottenheijm, H. J. C. *Tetrahedron Lett.* **1986**, *27*, 3755. (c) Plate, R.; Theunisse, A. W. G.; Ottenheijm, H. J. C. *J. Org. Chem.* **1987**, *52*, 370. (d) Plate, R.; Nivard, R. J. F.; Ottenheijm, H. C. J. *Tetrahedron* **1986**, *42*, 4503.
78. (a) Plate, R.; Merkens, P. H. H.; Smits, J. M. M.; Ottenheijm, H. J. C. *J. Org. Chem.* **1986**, *51*, 309. (b) Plate, R.; van Hout, R. H. M.; Behm, H.; Ottenheijm, H. J. C. *J. Org. Chem.* **1987**, *52*, 555.
79. (a) Plate, R.; Hermkens, P. H. H.; Behm, H.; Ottenheijm, H. J. C. *J. Org. Chem.* **1987**, *52*, 560. (b) Plate, R.; Hermkens, P. H. H.; Smits, J. M. M.; Nivard, R. J. F.; Ottenheijm, H. C. J. *J. Org. Chem.* **1987**, *52*, 1047. (c) Hermkens, P. H. H.; Plate, R.; Ottenheijm, H. J. C. *Tetrahedron* **1988**, *46*, 1991.
80. (a) Hermkens, R. H. H.; Plate, R.; Ottenheijm, H. C. J. *Tetrahedron Lett.* **1988**, *29*, 1323. (b) Hermkens, P. H. H.; Plate, R.; Kruse, C. G.; Scheeren, H. W.; Ottenheijm, H. J. C. *J. Org. Chem.* **1992**, *57*, 3881.
81. (a) Plate, R.; Nivard, R. J. F.; Ottenheijm, H. C. J.; Fardes, J.; Simonyi, M. *Heterocycles* **1986**, *24*, 3105. (b) Hermkens, P. H. H.; van Maarseveen, H.; Cobbon, P. L.; Ottenheijm, H. C. J.; Kruse, C. G.; Scheeren, H. W. *Tetrahedron* **1990**, *46*, 833. For a review: (c) Hino, T.; Nakagawa, M. *Heterocycles* **1998**, *49*, 499.
82. Wenkert, E.; Moeller, P. D. R.; Piettre, E. R. *J. Am. Chem. Soc.* **1988**, *110*, 7188.
83. Kraus, G. A.; Raggan, J.; Thomas, P. J.; Bougie, D. *Tetrahedron Lett.* **1988**, *29*, 5605.
84. Padwa, A.; Brodney, M. A.; Dimitroff, M. *J. Org. Chem.* **1998**, *63*, 5304.
85. (a) Padwa, A.; Dimitroff, M.; Waterson,, A. G.; Wu, T. *J. Org. Chem.* **1997**, *62*, 4088. (b) Padwa, A.; Dimitroff, M.; Waterson,, A. G.; Wu, T. *J. Org. Chem.* **1998**, *63*, 3986.
86. Gieseler, A.; Steckhan, E.; Wiest, O.; Knoch, F. *J. Org. Chem.* **1991**, *56*, 1405.
87. Wiest, O.; Steckhan, E.; Grein, F. *J. Org. Chem.* **1992**, *57*, 4034.
88. Wiest, O.; Steckhan, E. *Tetrahedron Lett.* **1993**, *34*, 6391.
89. *1,3-Dipolar Cycloaddition Chemistry*; Padwa, A., Ed.; Wiley: New York, 1984, Vols. 1 and 2.
90. (a) Bailey, A. S.; Buckley, A. J.; Warr, W. A.; Wedgwood, J. J. *J. Chem. Soc., Perk. Trans. 1* **1972**, 2411, and references therein. (b) Hannon, R. E.; Wellman, G.; Gupta, S. K. *J. Org. Chem.* **1973**, *38*, 11.
91. Lwowski, W. In *1,3-Dipolar Cycloaddition Chemistry*; Padwa, A., Ed.; Wiley: New York, 1984, Vol. 1, Chap. 5, pp. 616–617.
92. Bailey, P. S. *Ozonation in Organic Chemistry;* Academic: New York, 1978 and 1982, Vols. 1 and 2.

93. (a) Witkop, B. *Liebigs Ann. Chem.* **1944**, 103, and references therein. (b) Mentzer, C.; Molho, D.; Berguer, Y. *Bull. Soc. Chim. Fr.* **1950**, 555.
94. (a) Ruccia, M.; Vivona, N.; Piozzi, F.; Averssa, M. C. *Gazz. Chim. Ital.* **1969**, *99*, 588. (b) Ruccia, M.; Vivona, N.; Cusmano, G.; Marino, M. L.; Piozzi, F. *Tetrahedron* **1973**, *29*, 3159. (c) Laude, B.; Soufiaoui, M.; Arriau, J. *J. Heterocycl. Chem.* **1977**, *14*, 1183. (d) Daou, B.; Soufiaoui, M.; Carrie, R. *J. Heterocycl. Chem.* **1989**, *26*, 1485. (e) Daou B.; Soufiaoui, M. *Tetrahedron* **1989**, *45*, 3351.
95. (a) Caramella, P.; Corsico, A. C.; Corsaro, A.; Del Monte, D.; Albini, F. M. *Tetrahedron* **1982**, *38*, 173. (b) Brucke, L.; Zecchi, G. *J. Org. Chem.* **1983**, *48*, 2772. (c) Albini, F. M.; Albini, E.; Bandiera, T.; Caramella, P. *J. Chem. Res. (S)* **1984**, 36. (d) Malamidou-Xenikaki, E.; Coutouli-Argyropoulou, E. *Liebigs Ann. Chem.* **1992**, 75.
96. Fisera, L.; Mesko, P.; Lesko, J.; Dandarova, M.; Kovac, J.; Goljer, I. *Coll. Czech. Chem. Commun.* **1983**, *48*, 1854.
97. Dahaen, W.; Hassner, A. *J. Org. Chem.* **1991**, *56*, 896.
98. Coutouli-Argyropoulou, E.; Malamidou-Xenikaki, E.; Mentzafos, D.; Terzis, A. *J. Heterocycl. Chem.* **1990**, *27*, 1185.
99. Pelkey, E. T. Ph.D. Dissertation, Dartmouth College, Hanover, NH, 1999.
100. Gribble, G. W.; Pelkey, E. T.; Switzer, F. L. *Synlett.* **1998**, 1061.
101. Padwa, A.; Hornbuckle, S. F. *Chem. Rev.* **1991**, *91*, 4731–309.
102. Padwa, A. *J. Chem. Soc., Chem. Comm.* **1998**, 1417.
103. (a) Herzog, D. L.; Austin, D. J.; Nadler, W. R.; Padwa, A. *Tetrahedron Lett.* **1992**, *33*, 4731. (b) Osterhout, M. H.; Nadler, W. R.; Padwa, A. *Synthesis* **1994**, 123.
104. (a) Padwa, A.; Price, A. T. *J. Org. Chem.* **1995**, *60*, 6258. (b) Padwa, A.; Price, A. T. *J. Org. Chem.* **1998**, *63*, 556.
105. (a) Padwa, A.; Semones, M. A. *Tetrahedron Lett.* **1996**, *37*, 335. (b) Padwa, A. Harring, S. R.; Semones, M. A. *J. Org. Chem.* **1998**, *63*, 44.

ASPECTS OF THE INTRAMOLECULAR DIELS–ALDER REACTION OF A FURAN DIENE (IMDAF) LEADING TO THE FORMATION OF 1,4-EPOXYDECALIN SYSTEMS

Brian A. Keay and Ian R. Hunt

Advances in Cycloaddition
Volume 6, pages 173–210.

ISBN: 0-7623-0531-2

ABSTRACT

Twenty-three examples of intramolecular Diels–Alder reactions between a furan ring and an olefin (IMDAF) connected by a four-carbon-atom chain are described in detail. Attempts to promote the IMDAF reaction in aqueous solution with salt or β-cyclodextrin, high-pressure conditions (12.5 kbar), and Lewis acids (LA) are discussed. The best results were obtained with catalytic quantities of $MeAlCl_2$. In the majority of cases studied, sub-stoichiometric quantities of $MeAlCl_2$ provided better starting material : product ratios than when 1 equiv of $MeAlCl_2$ was employed. This observation is rationalized by comparing the relative basicity of the functional groups in the starting materials and products and invoking that the LA prefers to bind to the more basic unsaturated ketone in the starting material. This postulate is supported by detailed investigations involving: (a) ^{1}H NMR competitive binding studies, (b) using ^{1}H NMR to follow an IMDAF reaction catalyzed by 10 mol% $MeAlCl_2$, (c) AM1 semi-empirical calculations to allow the calculation of ΔH_f and ΔH_R for IMDAF reactions, (d) examination of four possible reaction coordinate diagrams, (e) a mathematical simulation of the equilibria involved in the IMDAF reaction, and (f) 3-D plots that illustrate how K_{obs} increases as less LA is used. Comparisons are made of the results from these studies with other IMDAF and LA promoted reactions reported in the literature. Finally, the synthesis of 1,4-epoxycadinane and the C-15 to C-23 segment of venturicidins A, B, and X are presented, which utilize the highly stereoselective IMDAF reaction as the key step.

I. INTRODUCTION

The intramolecular version of the Diels–Alder[1] reaction (IMDA) is one of the most powerful reactions in an organic chemist's repertoire for the creation of two or more rings with good to excellent stereo- and regiocontrol. The IMDA reaction has been extensively studied, used, and reviewed within the past century.[2] Among the wide variety of IMDA reactions that have been reported, those with furan rings (IMDAF) as the diene component are among the most intriguing. For the IMDAF reaction to proceed favorably, the aromatic character of the furan ring and the strain associated with the oxa-bicyclic adduct must be overcome. Experimentally, one cannot simply swamp the reaction mixture with dienophile in the *intramolecular* version to get the reaction to proceed as one can in the *intermolecular version*. Thus, interesting methods for promoting IMDAF reactions have been reported over the years.[3,4] Our endeavors[5,6] with the IMDAF reaction, in which the tether between furan diene and olefin of the dienophile has four carbon atoms, will be presented in this chapter.

Although the first report of an IMDA reaction with a furan diene was in 1945 by Herz[7] (**1**→**4**, Scheme 1) only four other groups had reported successful IMDAF reactions by 1970. Reports by Cram[8] (1961, **7**→**8**, Scheme 1), Wasserman[9] (1962, **9**→**10**, Scheme 1), Bilovic[10] (1964, **2**→**5** and **3**→**6**, Scheme 1), and Katz[11] (1968, **9**→**10**, Scheme 1) set the groundwork for a resurgence of activity in the 1970s, which continued on into the 1980s and 1990s. In 1979, De Clercq[12] reported the

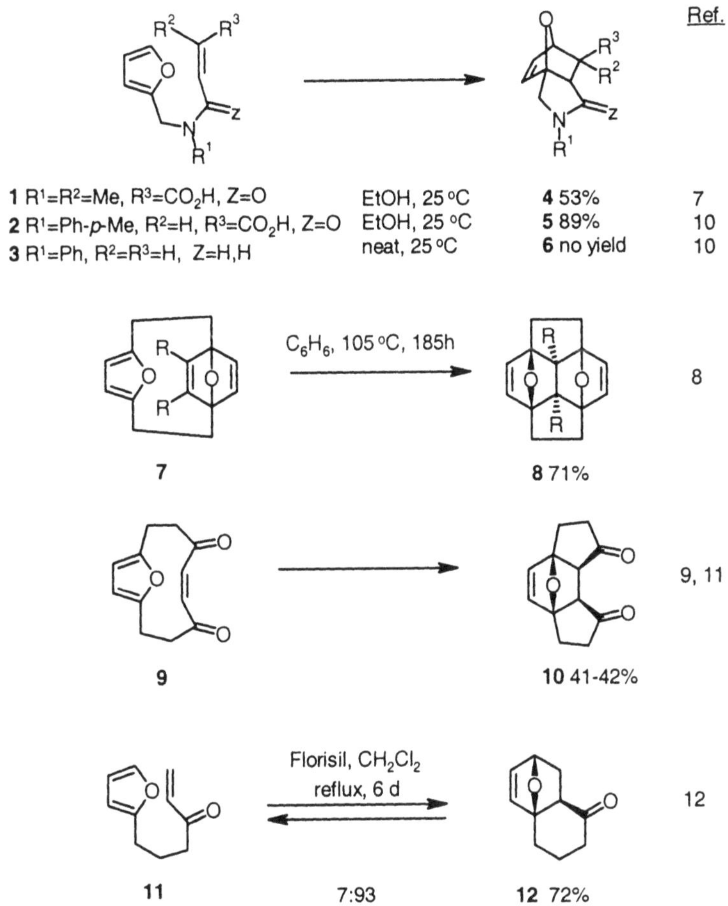

Scheme 1.

first example of an IMDAF reaction with a tether containing four carbon atoms (**11**→**12**, Scheme 1). Refluxing **11** in CH_2Cl_2 in the presence of Florisil® for 6 days provided a 7:93 ratio of **11**:**12** in 72% yield. Interestingly, only the adduct in which the side arm was *exo* with respect to the bridge was detected and isolated. In 1985, we became intrigued with this reaction for several reasons. First, the reaction took 6 days to reach equilibrium while examples containing only three carbon atoms in tether (none of which were a carbonyl group) reached equilibrium within 1–2 days.[13] Second, the best results were reported in the presence of Florisil®, which is a somewhat weak Lewis acid. Finally, only the *exo*-adduct was detected in the reaction mixture. Thus, we decided to investigate this reaction further to: (1) determine if alternative conditions could be found to reduce the time taken for the

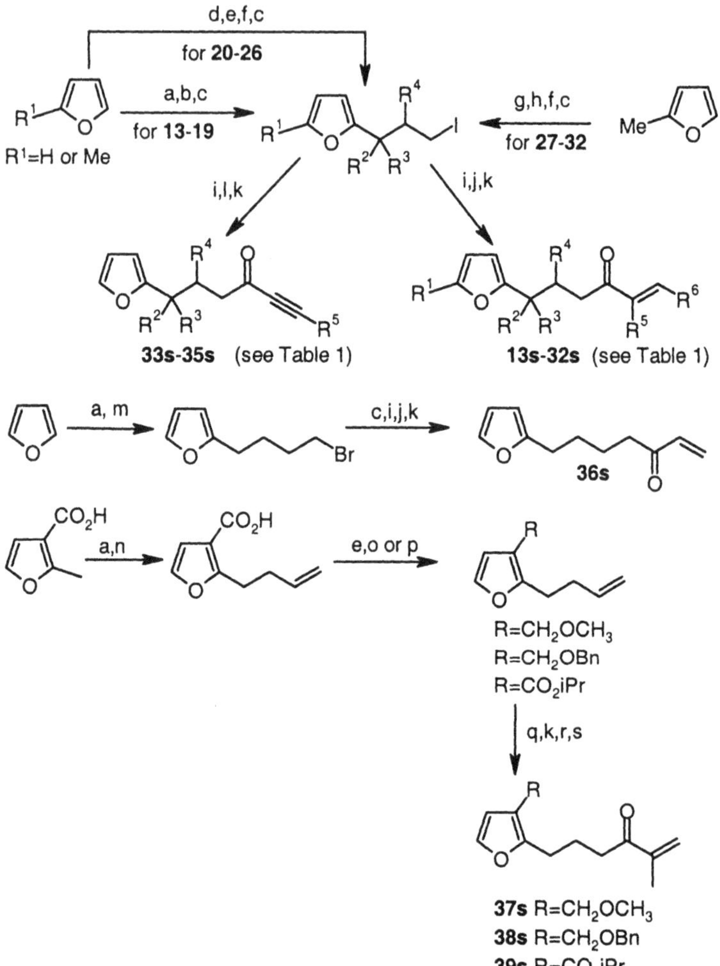

Conditions: a) n-BuLi, THF, 0 °C, 1.5 h; b) 1-bromo-2-chloropropane; c) NaI, acetone, reflux; d) 4-methyl-2-oxo-3-pentenenitrile, $AlCl_3$, then MeOH; e) $LiAlH_4$, Et_2O; f) TsCl, DMAP, CH_2Cl_2; g) methacrolein or crotonaldehyde, H_2SO_4; h) $NaBH_4$, EtOH; i) 2.2 equiv. t-BuLi, Et_2O, -78 °C; j) acrolein, methacrolein, crotonaldehyde or tiglic aldehyde; k) Swern [O]; l) trimethylsilylpropynal or 2-butynal; m) 1,4-dibromobutane; n) allyl bromide; o) NaH, MeI or benzyl bromide; p) K_2CO_3, 2-iodopropane, DMF, rt; q) $BH_3.Me_2S$, Et_2O, 0 °C; then H_2O_2, NaOH; r) 2-lithiopropene; s) Ag_2CO_3 on Celite, benzene.

Scheme 2.

reaction to reach equilibrium; (2) how substituents on the furan ring, tether, and dienophile would affect the equilibrium and stereochemistry of the product(s); (3) to use the somewhat rigid 1,4-functionalized epoxydecalin skeleton to introduce further groups with high stereo- and regiocontrol[14]; and (4) to attempt to develop an asymmetric version of the IMDAF reaction for further application to the synthesis of natural products.

This chapter will focus on our studies of this reaction over a 10-year period. The syntheses of the various precursors will be briefly described, followed by the various methods that we have used in attempts to drive the reaction to completion

Table 1. Structures of Compounds **13–35s** and p

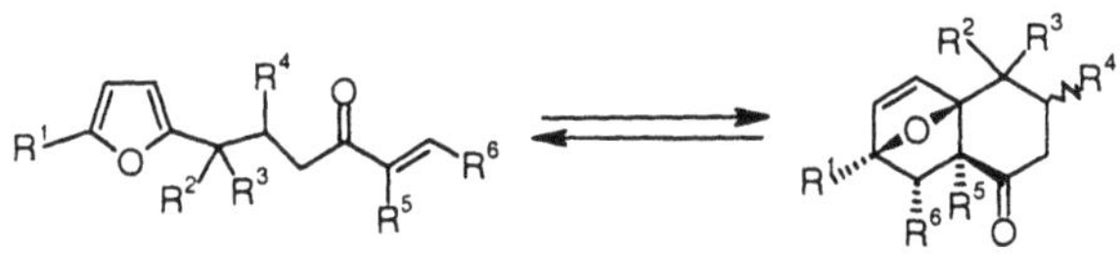

s **p**

Compound	*R¹*	*R²*	*R³*	*R⁴*	*R⁵*	*R⁶*
13	H	H	H	H	H	H
14	H	H	H	H	Me	H
15	H	H	H	H	H	Me
16	H	H	H	H	Me	Me
17	Me	H	H	H	H	H
18	Me	H	H	H	Me	H
19	Me	H	H	H	H	Me
20	H	Me	Me	H	H	H
21	H	Me	Me	H	Me	H
22	H	Me	Me	H	H	Me
23	H	Me	Me	H	Me	Me
24	Me	Me	Me	H	H	H
25	Me	Me	Me	H	Me	H
26	Me	Me	Me	H	H	Me
27	Me	Me	H	H	H	H
28	Me	Me	H	H	Me	H
29	Me	Me	H	H	H	Me
30	Me	H	H	Me	H	H
31	Me	H	H	Me	Me	H
32	Me	H	H	Me	H	Me

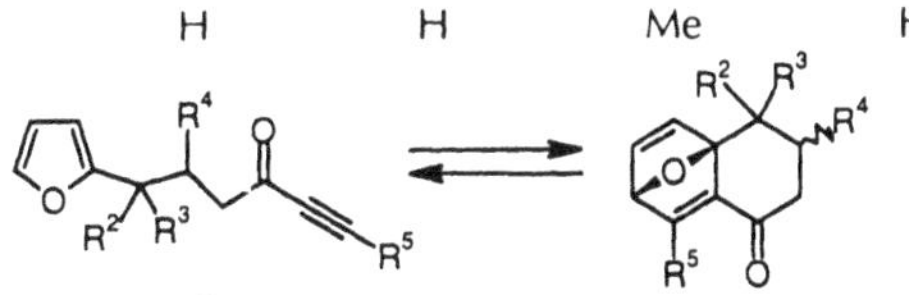

s **p**

Compound	*R¹*	*R²*	*R³*	*R⁴*	*R⁵*	*R⁶*
33		Me	Me	H	Me	
34		Me	Me	H	TMS	
35		H	H	Me	TMS	

in a shorter period of time. During these studies, Lewis acids (LA) emerged as the most efficient method for promoting these reactions, thus our experience with LA will be presented in detail. From our work we determined that ***less is really better!***

Finally, two synthetic applications that utilize the IMDAF reaction as the key step in generating a number of stereogenic centers of known relative stereochemistry will be described.

II. SYNTHESIS OF THE PRECURSORS

Precursors **13s–35s**,[15–18] **36s**[17] and **37s–39s**[19] were prepared as outlined in Scheme 2 and the structures for **13s–35s** are shown in Table 1. Full details of the various syntheses[15–19] have already been published and will not be repeated in this article.

III. METHODS TO PROMOTE THE IMDAF REACTION

A. Use of Salts, Water, and β-Cyclodextrin

As mentioned above,[12] compounds **13s** and **17s** have been reported to provide a 1:9 and 1:7 ratio of **S:P** when stirred at room temperature for 6 and 14 days, respectively, in the presence of Florisil®/CH_2Cl_2 (Table 2). In an attempt to shorten the time to equilibrium and to determine if the equilibrium could be shifted more toward products, we undertook a study to determine the effect water and/or salts would play on the IMDAF reaction.[15a] At the time, there had been some reports that aqueous solutions containing salts could be used to promote DA reactions due to

Table 2. Some IMDAF Reactions Run in Florisil®, Water and 2.0 M $CaCl_2$[a]

	Conditions (% Yield)[b,c]		
S:P	*Florisil*	*Water*[f]	*2.0 M $CaCl_2$*[f]
13s:13p	1:9[d] (71)	1:1 (69)	1:1 (66)
14s:14p	NR[e]	1:2 (70)	1:4 (73)
15s:15p	NR[e]	1:2 (70)	1:2 (69)
17s:17p	1:7[e] (65)	3:1 (72)	2:1 (68)
18s:18p	NR[e]	1:1 (71)	1:2 (78)
19s:19p	NR[e]	10:1 (—)	2:1 (61)

Notes: [a]All reactions performed at atmospheric pressure.
[b] Yield based on recovered starting material.
[c]Ratio obtained from ^1H NMR integration.
[d]Stirred 6 days at rt.
[e]Stirred 14 days at rt.
[f]Stirred 4 days at rt.

the hydrophobic effect.[13b,20] The results are summarized in Tables 2 and 3. Precursor **18s** was used as a standard in the optimization as it did not provide any product with Florisil® (Table 2). As Table 3 indicates, the best **S:P** ratio was obtained using a 2.0 M $CaCl_2$ solution. This provided a 1:2 ratio of **18s:18p** after 4 days at rt and at atmospheric pressure. Other concentrations of $CaCl_2$, using LiCl or water alone, did not improve the ratio. Interestingly, β-cyclodextrin failed to produced any product even though others had reported its usefulness in promoting Diels–Alder reactions[13d,20c] by forming hydrophobic inclusion complexes that bring the diene and dienophile into close proximity. Although the 2.0 M $CaCl_2$ gave a modest improvement in the IMDAF reaction and also provided products for systems that did not react in the presence of Florisil®, we were not satisfied with the overall yields and investigated other methods of improving the **S:P** ratio.

B. Use of High Pressure (12.5 kbar)

In 1985, Harwood and Issacs reported that high pressure could be used to promote an IMDAF reaction in which the side arm connecting the furan ring to the dienophile contained five carbon atoms.[21,22] In 1987, when we started investigating the IMDAF reaction,[23] high pressure had not yet been reported on systems containing a four-carbon-atom tether. Thus we prepared compounds **13s–15s** and **17s–19s** and subjected them to 12.5 kbar for 24 h at rt.[16] Removal of the CH_2Cl_2 provided only the tricycloadducts; no starting material was detectable by ^1H NMR (Table 4). The isolated yields were moderate at best and it was thought that performing the reaction at lower pressure may reduce the tendency for the dienophile to polymerize under the high-pressure conditions. Although the yield improved from 51 to 65% with the unsubstituted precursor **13s** at the lower pressure of 5.2 kbar, the addition of a methyl group α to the carbonyl resulted in no observed reaction at 5.2 kbar.

Table 3. Optimization of the IMDAF Reaction of Compound 18 with Water, Salts, and Additives[a]

Conditions	**18s:18p**[b]
Water	1:1
2 M LiCl	1:1
3 M LiCl	3:2
4.86 M LiCl	2:1
1.0 M $CaCl_2$	1:1
2.0 M $CaCl_2$	1:2
4.0 M $CaCl_2$	1:1
2.0 M $CaCl_2$ (90 °C)	NR
β-cyclodextrin	NR

Notes: [a]All reactions stirred at rt for 4 days at atmospheric pressure.
[b]Ratio obtained from ^1H NMR integration.

Table 4. S:P Ratios of IMDAF Reactions at High Pressure

S:P	*Pressure (% Yield)*[a,b]	
	5.2 kbar[c]	12.5 kbar[d]
13s:13p	0:100 (65)	0:100 (51)
14s:14p	100:0	0:100 (55)
15s:15p	—	0:100 (56)
17s:17p	—	0:100 (56)
18s:18p	—	0:100 (40)
19s:19p	—	0:100 (43)

Notes: [a]Yield based on recovered starting material.
[b]Ratio obtained from ^{1}H NMR integration.
[c]0.16 M solutions in CH_2Cl_2, 24 h, rt.
[d]0.16 M solutions in CH_2Cl_2, 12 h, rt.

Although the application of high pressure provided excellent **S:P** ratios, the low yields and limited access to a high-pressure apparatus[24] resulted in our turning our attention to the potential application of Lewis acids to promote the IMDAF reaction.

C. Use of Lewis Acids

Introduction

Although Lewis acids (LA) have been known to promote Diels–Alder reactions since 1960,[25] it was not until 1984 that they were reported to promote an intermolecular DA reaction with a furan diene.[26] In 1991, we discovered that IMDAF reactions with an internally activated ketone dienophile were promoted by using 1.1 equiv of methylaluminum chloride (MAC) at –78 °C (Table 5, Scheme 3).[17–19,27] Amazingly, the reaction time was shortened to a maximum of 8 h[28] with no noticeable decomposition. Subsequently, and totally unexpectedly, we found that *catalytic conditions* (0.1 equiv of Lewis acid) gave better conversion to products than with 1.1 equiv in good to excellent yields after only 2 h at –65 °C.

Interestingly, and in contrast, we found that the very closely related IMDAF reactions containing acetylenic dienophiles **33s–35s** were more efficient with 1.1 equiv of dimethylaluminum chloride (DMAC) (Table 5). Intrigued by these seemingly discrepant findings, we decided to seek an explanation for these observations.

To Use Catalytic or Stoichiometric Amounts of Lewis Acids?

In the LA literature, as recently as 1990, it has been stated that "the choice of the appropriate Lewis acid and the amount in which it should be used continue to tax the ingenuity and intuition of the chemist. As a rule, trial and error remains the best way to answer these questions"[29] This encouraged us to seek a rationalization for our observations and try to overcome the need for "trial and error."

Table 5. Enthalpies of Formation, ΔH_f, and Reaction, ΔH_R, From AM1 Level Calculations and Experimentally Observed S:P Ratios[a]

S	*S* ΔH_f	*P* ΔH_f	ΔH_g	*0.1 Equiv LA*[bc]	*1.1 Equiv LA*[bc]
13	−115.5	−159.4	−43.9	0:100	0:100
14	−145.2	−164.8	−19.6	11:89	35:65
15	−158.6	−177.8	−19.2	31:69	78:22
16	−183.3	−176.1	7.2	95:5	100:0
17	−144.8	−176.1	−31.3	0:100	0:100
18	−174.5	−181.6	−7.1	0:100	19:81
19	−187.3	−193.4	−6.1	24:76	82:18
20	−132.2	−178.7	−46.5	(Florisil 0:100)[d]	
21	−161.1	−181.2	−20.1	40:60	68:32
22	−176.1	−196.6	−20.5	27:73	73:27
23	−200.8	−191.6	9.2	100:0	100:0
24	−161.9	−195.4	−33.5	(Florisil 0:100)[d]	
25	−192.0	−197.5	−5.5	23:77	78:22
26	−208.4	−212.1	−3.7	69:31	78:22
33	−20.5	−40.9	−20.4	—	0:100[e]
34	−206.9	−181.4	25.5	—	12:88[e]
35	−202.8	−185.2	17.6	99:1	13:87[e]
37	−128.4	−130.5	−2.1	80:20	—
38	−81.2	−86.8	−5.5	7:93	—
39	−53.1	−58.0	−4.9	5:95	—
40	−474.9	−654.8	−179.9	—	—
41	−699.1	−868.6	−169.5	—	—
42	−20.2	−205.5	−185.3	—	—
				(S : Peq. : Pax)	
27	−157.3	eq −191.6	−34.3	—	0:90:10
		ax −188.3	−31.0		
28	−185.4	eq −197.1	−11.7	—	9:88:3
		ax −192.0	−6.6		
29	−203.3	eq −208.8	−5.5	17:75:4	79:16:5
		ax −205.4	−2.1		
30	−159.4	eq −197.5	−38.1	(Florisil 0:100:0)[d]	
		ax −191.2	−31.8		
31	−187.4	eq −202.9	−15.5	—	11:89:0
		ax −196.2	−8.8		
32	−202.5	eq −215.1	−12.6	11:87:2	61:31:8
		ax −208.4	−5.9		

Notes: [a] ΔH_f and ΔH_R in kJ mol^{-1}.

[b] Unless indicated otherwise, reaction conditions are 1.1 equiv MAC, CH_2Cl_2, 8 h, −78 °C and 0.1 MAC, CH_2Cl_2, 2 h, −65 °C.

[c] Isolated yields are generally very close to these figures indicating that there is no significant decomposition of either S or P, hence these ratios can be regarded as yields.

[d] We have found that reactions generally give better yields of product with 0.1 equiv of MAC than with Florisil, therefore, it is reasonable to assume that these reactions will go to completion with 0.1 equiv MAC.

[e] 1.1 or 0.1 equiv of DMAC, −50 °C, 2.5 h. Note that DMAC was used to reduce aromatization of adducts that was seen with MAC.

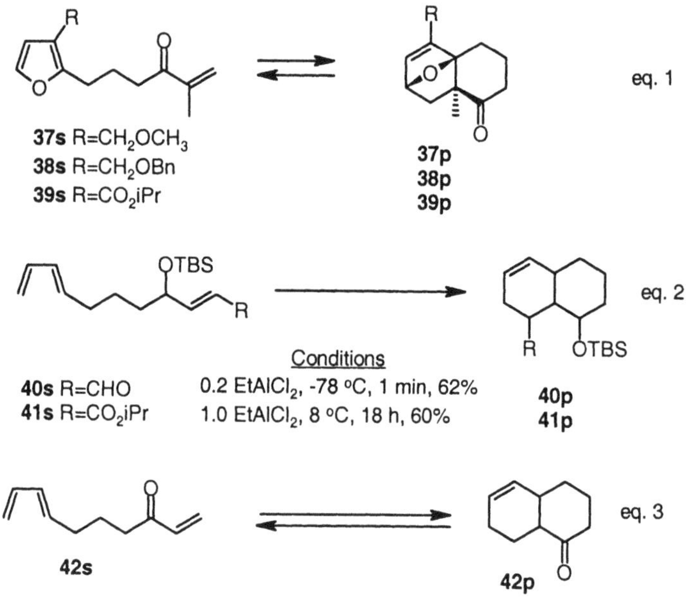

Scheme 3.

Similar experimental observations had been reported by Snider[30] for the LA-catalyzed ene reaction and by others for intermolecular DA reactions.[31] In the IMDA literature, we found that LA-promoted IMDA reactions, with ester- or aldehyde-activated dienophiles generally used 0.95 equiv of the LA,[32] but there were a few examples of truly catalytic IMDA reactions when the dienophile was an enal (e.g. compare **40s** and **41s**, Scheme 3).[33]

This indicated to us that the nature of the components of the dienophile (activating functional group with alkene or alkyne) was critical in determining the amount of LA required to promote the IMDA reaction most efficiently. Empirically, we noted that catalytic quantities of LA (defined by us as type "A" reactions) can be used with olefinic ketones or aldehydes dienophiles (e.g. **13s–32s**). In contrast, stoichiometric LA conditions (defined as type "B" reactions) are used with olefinic ester dienophiles[32] (e.g. **41s**), or acetylenic ketone dienophiles (e.g. **33s–35s**). The question about the origin of this difference with respect to the quantity of LA remained unanswered at this stage. Our experimental observations indicated that the relative amount of the LA employed influenced the position of the IMDAF equilibrium. Therefore, as a referee reminded us, in a strict chemical sense the LA should not be described as a "catalyst." However, since the LA increases the rate of the reaction, is not consumed during the actual IMDAF reaction (though we destroy it during work-up), and it can be employed in low concentration relative to the reactant, there is no better word in the English language to describe the effect of the LA in these reactions and thus the terms catalyst and catalytic will be used throughout this chapter.

Lewis acids increase reactivity by influencing the frontier molecular orbitals of the dienophile. These IMDAF reactions are examples of normal DA reactions in that they are diene–HOMO/dienophile–LUMO controlled.[34] It is generally accepted that the LA activates the carbonyl group by coordination at the carbonyl oxygen, thus lowering the LUMO of the dienophile and making it more electrophilic.[35] The overall reduced HOMO–LUMO gap increases the rate of the reaction in question. Since all the systems under discussion here are activated by carbonyl oxygens, and so have a similar mode of activation, this then cannot explain the observed differences in reactivity with respect to the quantity of LA required to facilitate the type "A" or "B" IMDA reactions. To solve the enigma, one must take a careful look at the relative basicity of the all the functional groups present in both the starting materials and the Diels–Alder products.

Defining the Equilibrium

Initially we postulated that the success of the simple catalytic IMDAF reactions was because the reaction was actually a series of three equilibria as shown in Scheme 4, requiring that the nonconjugated ketone of the adduct was less basic than the conjugated ketone in the dienophile of the starting material (i.e. $K_1 > K_3$). Note that K_3 is defined in the direction of formation of the product–Lewis acid complex. For the IMDAF reactions, it also requires that the etheral oxygen atoms of the furan and the product oxabicyclic bridge are not coordinating with the LA. Although it is known that MAC exists as a dimer, we have ignored this in our scheme. With catalytic quantities of LA, the overall reaction equilibria most closely resembles the relative free energies of the *uncomplexed* starting material (**S**) and the DA product (**P**). The LA preferentially coordinates with the more basic enone and facilitates the forward IMDAF reaction. Once the P–LA complex is formed, dissociation occurs and the catalytic cycle can repeat. This effectively shifts the reaction towards completion.

This conclusion in its own right is revealing as it indicates that the IMDAF reactions studied *must actually be energetically favorable!* This in itself contradicts many literature statements,[12] including our own,[17] that IMDAF reactions are intrinsically unfavorable due to a combination of the aromaticity of furan and the inherent product ring strain. These factors contribute to making the activation barrier high, and therefore, often make the equilibrium thermally unobtainable *rather than* unfavorable. Modern computing power has provided us with at least one other method of verifying this realization (see discussion below).

Returning to the LA issue, and the situation with 1.1 equiv of LA, the observed reaction equilibrium then tends to reflect the relative free energies of the two complexes, **S**–LA and **P**–LA, and the equilibrium between them. The experimental

$$S + LA \underset{k_{-1}}{\overset{k_1}{\rightleftharpoons}} S......LA \underset{k_{-2}}{\overset{k_2}{\rightleftharpoons}} P.......LA \underset{k_{-3}}{\overset{k_3}{\rightleftharpoons}} P + LA$$

$$K_1 = k_1 / k_{-1} \qquad K_2 = k_2 / k_{-2} \qquad K_3 = k_{-3} / k_3$$

$$K_{eq} = [P] / [S] = K_1 K_2 / K_3$$

$$K_{obs} = \frac{[P] + [P...LA]}{[S] + [S...LA]}$$

Scheme 4.

results thus indicate that for these IMDAF reactions the equilibrium lies towards the S side. However, we have to realize that the observed **S:P** ratio reported in Table 5 corresponds to the isolated yields after workup, and the experimentally observed equilibrium, K_{obs}, is therefore defined in Scheme 4.

If correct, the three equilibria models shown in Scheme 4 should also be able to explain the reactivity observed with other LA reaction systems. The crux for type "A" behavior is that the activating carbonyl group in the starting material needs to be more basic that the resulting carbonyl group in the product (i.e. $K_1 > K_3$). For type "B" behavior, the product carbonyl group is more basic than the activating carbonyl group in the starting material (i.e. $K_3 > K_1$), and P will remain complexed to the catalyst and inhibit the forward DA reaction. So an equivalent of LA is required to ensure that there is sufficient LA present to complex the starting material.

The relative basicity of the reaction-controlling functional groups represent only two of the three steps in the model. The final contribution, which could also be a controlling factor, is the thermodynamics of the DA reaction itself (i.e. K_2)

Interpreting the model therefore requires information on the relative basicity of the activating groups.

Relative Basicities of the Activating Groups in IMDAF Reactions: Competitive ^1H NMR Binding Studies

Experimental data on the relative basicity of the functional groups was obtained by carrying out a series of low-temperature NMR, competitive complexation studies using MAC and $BF_3{\cdot}Et_2O$ with a representative selection of Lewis bases to "model" fragments of our IMDAF reactions. Our model compounds were 2-cyclohexenone (**43**), cyclohexanone (**44**), 2-methylfuran (**45**), THF (**46**), methyl propionate (**47**), and methyl acrylate (**48**) (Scheme 5). In model studies, a comparison of the ^1H or ^{13}C NMR spectra of the free base and the complex formed when 1 equiv of the appropriate LA was present, produced shifts that were consistent with those from $EtAlCl_2$ with similar bases reported by Childs et al.[36] These results are shown

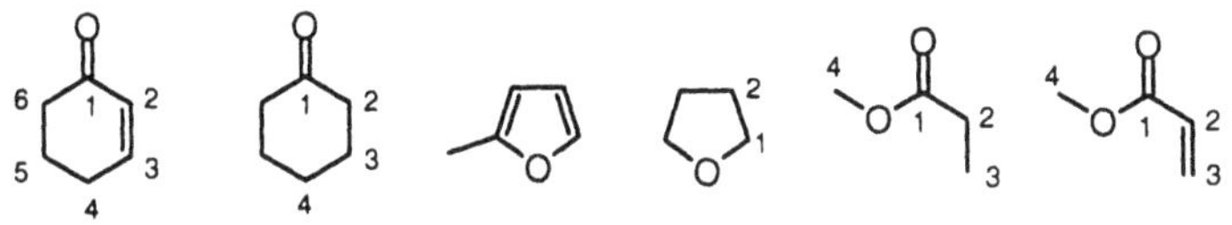

43 44 45 46 47 48

Scheme 5.

in Tables 6 and 7 as the relative shifts, Δδ = δ (complex) – δ (free base), and were used to identify the complexed species in the subsequent competitive complexation studies.

At –60 °C, fast exchange conditions were observed for the MAC complexes. This temperature was selected to represent the typical MAC-promoted IMDAF reaction conditions. Slow exchange conditions for the MAC complexes were observed by –90 °C. As expected cyclohexenone (**43**), methyl propionate (**47**), and methyl acrylate (**48**) existed as both *syn*- and *anti*-complexes, consistent with complexation at the carbonyl oxygen.[37]

In the competitive complexation studies with 1:1:1 mixtures of base X:base Y:LA, we were able to determine which functional group was the more basic by interpretation of peak shifts and line shapes (see Table 8). Experiment 1 confirms that the enone is more basic than the ketone as required by the type "A" hypothesis (see Figure 1[38]), whereas experiment 2 indicates that MAC complexes to the saturated

Table 6. Change in the Chemical Shift, Δδ for Nucleus *i* of each of the Model Compounds Complexed in the Presence of an Equivalent of MAC, Relative to the Uncomplexed Base, in $CDCl_3$ at –60 °C[a]

		Δδ/ppm[b]					
Base	*i*	*1*	*2*	*3*	*4*	*5*	*6*
43	^1H	—	0.60	0.76	0.27	0.14	0.45
	^{13}C	13.5	–0.9	19.3	–0.9	–1.2	0.9
44	^1H	—	0.52	0.25	0.13	—	—
	^{13}C	2.7	1.3	0.9	–1.2	—	—
47	^1H	—	0.41	0.32	0.22	—	—
	^{13}C	8.3	0.7	5.4	–0.5	—	—
48	^1H	—	0.39	0.43	0.29	—	—
	^{13}C	7.4	–3.1	10.2	5.7	—	—
46	^1H	0.63	0.36	—	—	—	—
	^{13}C	5.5	–0.1	—	—	—	—

Notes: [a]2-Methylfuran (**45**) polymerized under the experimental conditions and has been omitted from the table.
[b]Column headings refer to the positions defined in the structural diagrams in Scheme 5.

Table 7. Change in the Proton Chemical Shift, Δδ, and Percent Complex for Each of the Model Compounds Complexed in the Presence of an Equivalent of $BF_3 \cdot Et_2O$ in $CDCl_3$ at –60 °C[a]

		$\Delta\delta_H$/ppm[b]					
Base	*% Complex*	*1*	*2*	*3*	*4*	*5*	*6*
43	72	—	brd[c]	0.94	0.34	0.14	0.58
44	26	—	0.55	0.25	nd[d]	—	—
47	9	—	0.64	nd[e]	0.51	—	—
48	0[f]	—	—	—	—	—	—
46	79	0.61	0.37	—	—	—	—

Notes: [a]2-Methylfuran (**45**) was not included in this study.
[b]Column headings refer to the positions defined in the structural diagrams.
[c]The H2 peak for the complex is broad due to exchange between *syn* and *anti* forms at -60 °C. At -90 °C, both the H2 and H3 peaks are resolved into two signals of approximately equal intensity with $\Delta\delta_{H2}$ = 0.84, 0.47 ppm and $\Delta\delta_{H3}$ = 0.97, 0.88 ppm for the *syn* and *anti* forms.
[d]The shift of this peak could not be accurately determined due to overlap with H2 of the free base.
[e]The shift of this peak could not be accurately determined due to overlap with peaks due to Et_2O.
[f]No evidence of complexation (even in th presence of 5 equiv of $BF_3 \cdot Et_2O$).

ester rather than the conjugated ester as for type "B." Experiment 3 is important since it shows that the enone is more basic than the furan ether oxygen. Note that the presence of the enone also prevents polymerization of the 2-methylfuran (**45**) but that no intermolecular DA reaction was observed. The greater basicity of THF (**46**) in experiments 4 and 5 indicates that the ether oxygen of the DA adduct should complex the MAC and inhibit the IMDAF reaction. This was not unexpected, but it does contradict the hypothesis and cannot be ignored. Evidence for the reduced basicity of the ethereal oxygen in the oxatricyclic DA adducts is presented below.

Similar studies with $BF_3 \cdot Et_2O$ are shown in Table 9. In contrast to MAC, the BF_3 complexes were under slow exchange at –60 °C with separate signals for free and complexed base that could be integrated to determine the percent complexation and hence relative basicity. The order of THF (**46**) > cyclohexenone (**43**) > cyclohexanone (**44**) > methyl propionate (**47**) > methyl acrylate (**48**) is in good agreement with the results with MAC. Simple $BF_3 \cdot Et_2O$ "titration" NMR experiments also allowed us the measure the K for cyclohexenone ($K = 3.5 \pm 1.5$) and cyclohexanone ($K = 0.09 \pm 0.02$), indicating that the enone is an order of magnitude more basic than the ketone toward $BF_3 \cdot Et_2O$.[39] Thus, in the competitive binding experiments, $BF_3 \cdot Et_2O$ was found to show the same behavior as MAC.

Literature studies have been reported on many substrates complexed by BF_3. Low-temperature ^{19}F NMR chemical shifts[40] (used to measure the strength of the interaction of the LA and substrate), thermodynamic measurements[41] of the enthalpy of complexation, and ab initio[42] calculations indicate the following general order of basicity (in CH_2Cl_2): ethers > alcohols > amides > esters > enones >

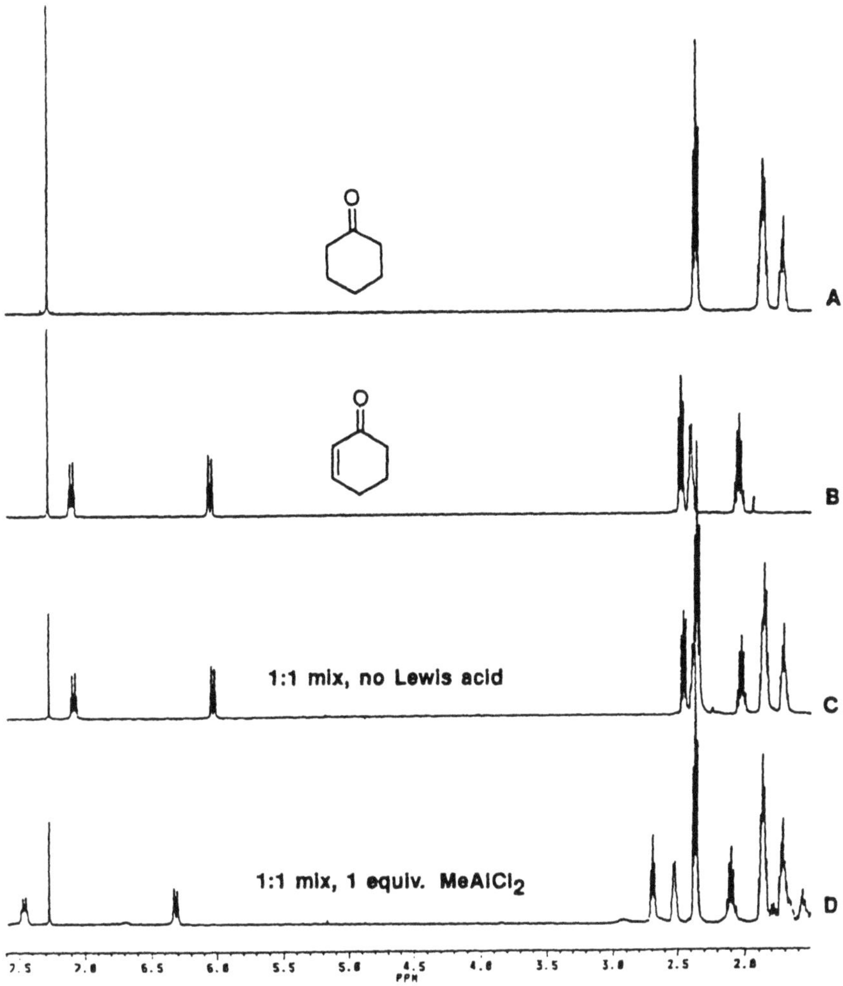

Figure 1. [38] ^{1}H NMR Spectra of competitive complexation experiment 1, Table 8, between cyclohexenone (**43**) and cyclohexanone (**44**) (–60 °C in $CDCl_3$). A = **43** alone; B = **44** alone; C = 1:1 mixture of **43**:**44**; D = 1:1:1 mixture of **43**:**44**:MAC.

ketones > aldehydes. However, this trend for BF_3, is sensitive to both steric and electronic effects, and the complexation of trialkylaluminiums have also been reported to be sensitive to steric effects.[43] Experimental evidence for the low basicity of furan is also provided by an experimental study of the complexation of obacunone (**49**, Scheme 6) with BF_3 in $CDCl_3$ which preferentially coordinates at the A and D ring lactones.[40c] Experimentally, it is also known that complexation to

Table 8. Shifts Observed for Each of the Bases in the Competitive Complexation Experiments with an Equivalent of MAC in $CDCl_3$ at −60 °C

Expt.	*Base X*	*H*	*Δδ/ppm*[a]	*Base Y*	*H*	*Δδ/ppm*[a]
1	**43**	3	0.36	**44**	2	0.00
2	**47**	2	0.38	**48**	3	0.04
3	**43**	3	0.66	**45**	—	—[b]
4	**43**	3	0.01	**46**	1	0.16
5	**44**	2	0.01	**46**	1	0.36

Notes: [a]Values are for the shift of the ^1H peak that exhibited the largest shift on complexation in the model studies.

[b]No figure is available for 2-methylfuran (**45**) because of the polymerization observed during the model studies.

an ethereal oxygen is gradually weakened due to steric effects: dimethyl ether > methyl *n*-butyl ether > di-*n*-butyl ether.[40a] In the IMDAF systems we feel that it is the steric requirements at the bridgehead etheral oxygen that prevent complexation of the LA at the P ether site which could otherwise inhibit the catalytic process.

The Effect of Other Lewis Acids on the IMDAF Reaction

While we found that the LA effect is not specific to MAC, neither is it general to all LA as can be seen in Table 10. Me_3Al and Ti(O-*i*Pr)$_4$ are both too weak as LA[29,36] to promote the IMDAF reaction; an equivalent of $BF_3{\cdot}Et_2O$ is required for reasonable adduct formation, while $AlCl_3$ and $TiCl_4$ both show the same type of behavior as MAC. We have attributed the differences between MAC and $BF_3{\cdot}Et_2O$ to the following factors: (1) stronger binding by BF_3 as indicated by the higher coalescence temperatures, (2) competitive inhibition by Et_2O, a strong Lewis base, and (3) the ability of BF_3 to complex at the **P** ether oxygen of the bridge, vide infra, that causes **P** to inhibit the catalytic process.

Table 9. Percentage of Each Base Complexed in the Competitive Complexation Experiments with an Equivalent of $BF_3{\cdot}Et_2O$ in $CDCl_3$ at −60 °C

Expt.	*Base X*	*% Complex*[a]	*Base Y*	*% Complex*
6	**43**	65	**44**	5
7	**47**	9	**48**	0
8	**43**	70	**45**	0
9	**43**	6	**46**	83
10	**44**	1	**46**	87

Note: [a]Values are based on peak integrals of free and complexed species.

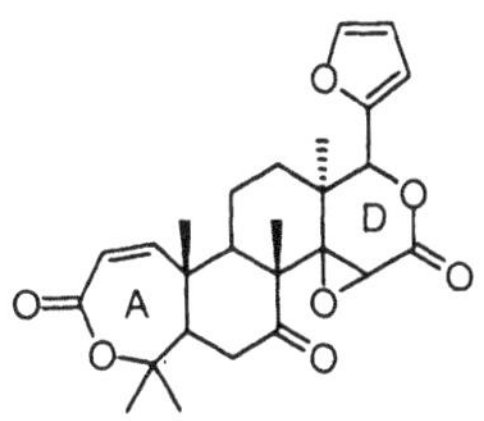

49

Scheme 6.

We know that the MAC IMDAF reactions are under thermodynamic control since the same **S:P** ratios are observed adding MAC to either pure **S** or **P**. The observation of the retro-IMDAF reaction is, in accord with the law of microscopic reversibility,[44] an indicator that MAC can complex to the **P** ketone. Of course, the forward or reverse reaction can be promoted by just a fraction of the total MAC complexed at the appropriate reactive center in a pre-equilibrium for the DA reaction.

A ^1H NMR Study of an IMDAF Reaction Promoted with MAC

We have also performed a direct ^{1}H NMR study of the IMDAF reaction itself despite the difficulties associated with the facile nature of the DA reaction that MAC promotes (Figure 2[38]). With 0.1 equiv of MAC we have observed a downfield shift of the dienophile olefinic protons in **29s** that increases as the reaction progresses due to the increase in the fraction of **S** that is complexed, but no evidence of a shift for the **P** protons. In the presence of 1.1 equiv of MAC there is sufficient MAC to complex the carbonyl groups of both **S** and **P**. The **S**–enone complex is readily observed but the **P** complex is generally more difficult due to overlap with the **S** protons, which is generally the major species. However for **13s**→**13p**, the product is stable to MAC (see Table 5), and in the presence of MAC a downfield shift of

Table 10. Isolated S:P Ratios for **16s**→**16p** with 0.1 and 1.0 Equivalents of Various Lewis Acids[a]

Lewis acid	*0.1 Equiv LA*	*1.1 Equiv LA*
$AlCl_3$	8:92	68:32
$MeAlCl_2$	24:76	82:28
Me_3Al	100:0	—
$BF_3{\cdot}Et_2O$	96:4	63:37
$TiCl_4$	27:73	89:11
Ti (OiPr)$_4$	100:0	100:0

Note: [a]All reactions performed for 2 h at –78 °C in CH_2Cl_2.

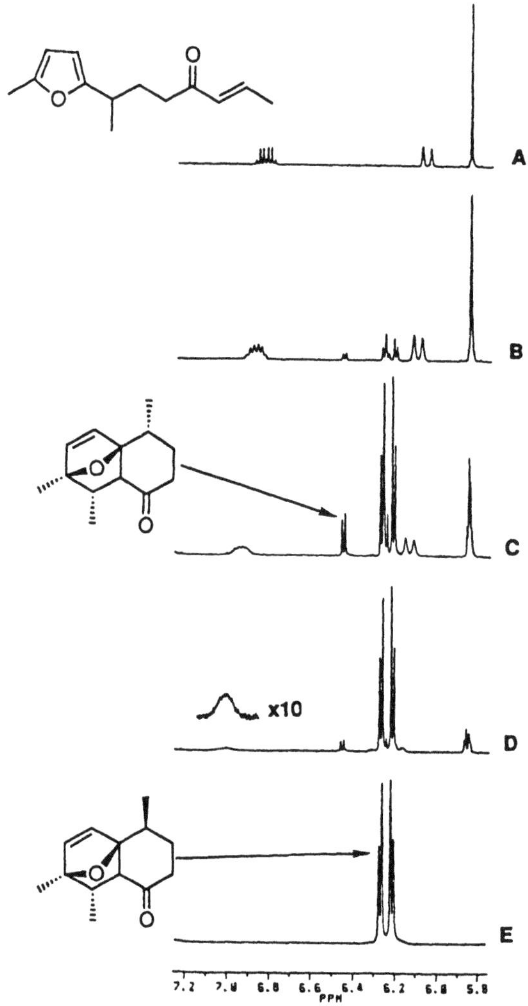

Figure 2. [38] ^{1}H NMR Spectra (–60 °C in $CDCl_3$) obtained for the IMDAF reaction of **29s**→**29p** in the presence of 0.1 equiv of MAC. A = **29s**; B = **29s** + 0.1 MAC, 5 min; C = **29s** + 0.1 equiv MAC, 15 min; D = **29s** + 0.1 equiv MAC, 3 h; E = **29p**.

the α-protons and upfield shifts for other protons in the cyclohexanone ring (presumably due to anisotropy effects caused by the ligands in MAC[37]) are observed. There was no evidence of complexation at the bridgehead ethereal oxygen.

A similar study with $BF_3{\cdot}Et_2O$ showed downfield shifts in the α-protons and small downfield shifts in the olefinic *and* bridgehead protons. During the course of the study, 1-tetralone was formed due to aromatization of the oxabicyclic DA product, presumably via complexation at the oxatricyclic ethereal oxygen. Ab initio calculations[42] have indicated that 7-oxanorbornene is more basic than either a ketone or an enone, and should complex preferentially with the BF_3 in an *anti* position with respect to the double bond.

AM1 Level Calculations[45,46]

A careful analysis of the experimental observations for the IMDAF reactions in Table 5 has provided more information on these reactions. The relative energies of **S** and **P** can be established by considering the position of the equilibrium of **S:P** under catalytic conditions (Table 5). With 0.1 equiv of MAC the ratio of **S:P** typically favors **P**, with a few exceptions when the dienophile is methylated (**16s**, **23s**, **35s**). This indicates that the IMDAF reaction is in fact a *favorable* equilibrium (i.e. $K_{eq} > 1$, Scheme 4). Since the reactions are notoriously difficult to perform under thermal conditions (vide supra) there must a high activation barrier. Similarly, there must be a high activation barrier to the reverse reaction as the IMDAF adducts can usually be distilled without cyclo-reversion. The relative energies of the **S**–LA and **P**–LA complexes can be approximated on the basis of the isolated **S:P** yields in the presence of 1.1 equiv of MAC (e.g. 80:20 for the reaction of **19s**) where the forward IMDAF has to compete with the retro DA reaction (k_2[**S**–LA] = k_2[**P**–LA], Scheme 4). The yields indicate that in many cases the **S**–LA complex is the more stable species. This is unusual for LA-promoted DA reactions, which are typically highly favorable (i.e. $K_{eq} >> 1$).[47] In this case, however, the aromaticity of the furan and high product ring strain make the equilibrium of IMDAF reactions much closer to unity.

Once we realized that the experimental observations with catalytic MAC were indicating to us that the IMDAF reactions leading to cyclohexanone derivatives were generally favorable equilibria despite their literature reputation, we decided to consider the energetics of the reaction scheme of these LA-promoted reactions. Similar work had been reported by Harwood and Dolata[48] using WIZARD and MM2 to calculate the transition states of IMDAF reactions performed under kinetic control at high pressure. Houk et al.[49] has also calculated transition states of IMDA reactions.

Since the IMDAF reactions with catalytic MAC are under thermodynamic control, these reactions proved to be amenable to study as only the relative ground states of **S** and **P** needed to be evaluated. We found that AM1 semiempirical calculations of ΔH_f for the addends and adducts allowed us to calculate ΔH_R that could be correlated with the experimentally observed yields with reasonable accuracy given that we ignored entropic and solvation effects (Table 5). Comparisons with closely related IMDA reactions showed that the IMDAF reactions were indeed significantly less favorable by approximately 140 $kJmol^{-1}$ (compare reaction **42s-p**, $\Delta H_R = -185.3$ $kJmol^{-1}$, with **13s-p**, $\Delta H_R = -43.9$ $kJmol^{-1}$) of which 67

$kJmol^{-1}$ can be attributed to the aromaticity of the furan.[50] Most of the reactions were found to be enthalpically favorable—the only exceptions being the reactions of the dimethylated enones (**16s** and **23s**) and the trimethylsilyl acetylenes (**34s** and **35s**), presumably due to steric effects. The AM1 calculations were also in accord with other experimental observations. These LA-promoted IMDAF reactions give only the *exo* orientation; for **13p** the *endo* orientation is calculated to be approximately 20 $kJmol^{-1}$ less favorable. For the diastereoselective reactions of **27s–32s**, AM1 predicts the equatorial methyl group to be favored by 3.3 to 6.7 $kJmol^{-1}$ over the axial adducts, with the greatest preference being seen with a methyl β to the furan (R^4 = Me, Tables 1 and 5). Experimentally we have observed that both products are formed initially but equilibration to the equatorial isomer then occurs. Further analysis is possible by comparing related pairs of reactions (e.g. for the effect of an α-methyl group on the furan compare **13s-p** with **17s-p** etc.). The major trend is that alkyl substitution of the dienophile or the furan diene makes the IMDAF reaction less favorable. On the furan, the 5-methyl costs +14 $kJmol^{-1}$, and the 3-ether or ester substituents (**37s–39s**) +15 $kJmol^{-1}$. On the dienophile, a single α- or β-Me substituent on the enone contributes +26 $kJmol^{-1}$. In contrast, alkyl substituents on the carbon tether have only a small, but generally slightly favorable effect on ΔH_R (9 of 13 reactions: +2.2 to −8.4 $kJmol^{-1}$), with a β-Me having a greater effect than an α-Me. In the context of the *gem*-dialkyl effect,[48b,51] it is interesting to note that the α,α-Me_2 is not calculated to be significantly different to the single α-Me, suggesting that the *gem*-dialkyl effect is not due to enthalpic effects in the ground states. It is also useful to note that these enthalpic trends seem to be additive.

The IMDAF reaction for acetylenic dienophiles containing a TMS group are calculated to be unfavorable (Table 5) and this is reflected when **35s** was treated with catalytic amounts of MAC. However, when 1.1 equiv of MAC is used, the adducts **33p–35p** are formed with excellent **S:P** ratios. The improved yields are due to the **P**–LA complex being more favorable than the **S**–LA complex when esters are employed on the dienophile. Ab initio level calculations[42] showed that saturated esters are more basic than unsaturated esters, thus the LA preferentially binds to the **P**, and thus >1.0 equiv of LA are necessary to effect a complete reaction. These results with the acetylenic esters **33s–35s** show how dramatically the **S:P** ratios can change depending on the amount of LA used in the reaction.

Of course it is really the change in free energy, ΔG_R, that governs the outcome of reactions. However, since we found that trying to evaluate the contribution of the entropy, ΔS_R, was complicated by the conformational flexibility of the carbon tether leading to wide variations in the entropic terms that provided no significant improvements, we decided that it was more advantageous to consider ΔH_R alone. The contribution of ΔS_R, can be assumed to be negative (i.e. $\Delta S_R < 0$)[52] and therefore an unfavorable contribution to ΔG_R. Although the activation entropies of similar IMDAF reactions have been reported,[51b] we have not found any measured ΔS_R. It is very tempting to assume that the change in ΔS_R across the series of IMDAF reactions is negligible (i.e. $\Delta\Delta S_R \cong 0$). However, not only is there no literature precedent for this, but an investigation of the *gem*-dialkyl effect in alkylcyclohexanes concluded that entropy and enthalpy effects are of comparable importance and that neither can be neglected.[53]

Despite this limitation, however, the AM1-calculated ΔH_R does correlate, at least in a *qualitative manner*, with the yields of the IMDAF reactions in the presence of

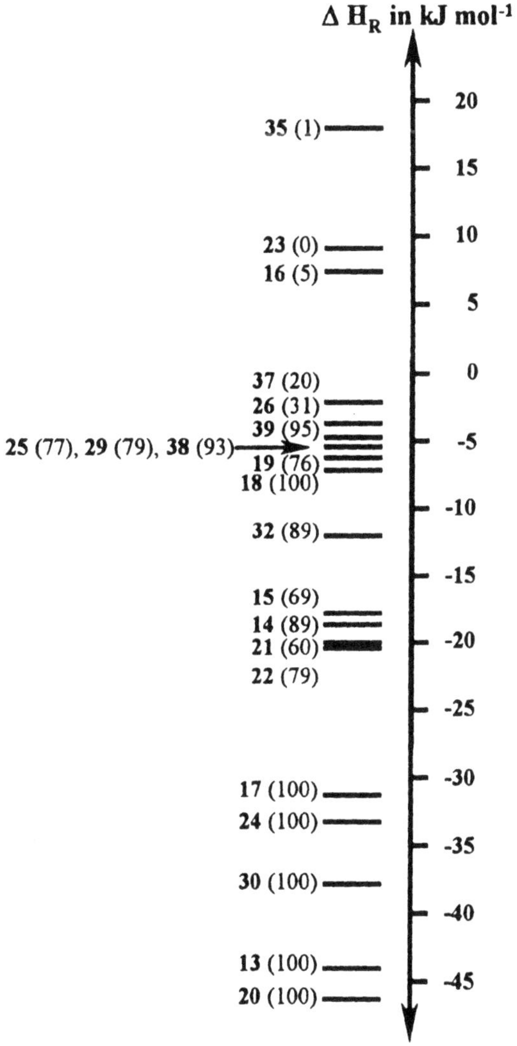

Figure 3. An "energy ladder"[48] displaying the calculated AM1 ΔH_R (kJ mol^{-1}) for each of the IMDAF reactions (Table 5) that were treated with 10 mol% MAC. The value in brackets is the experimentally observed yield of product (%).

catalytic quantities of LA (see Figure 3). Given that $\Delta S_R \neq 0$, and that other factors such as solvation effects have been neglected, an **S:P** ratio of 50:50 does not necessarily correspond to $\Delta H_R = 0$. However, the *general* trend is that the more successful reactions have increasingly negative ΔH_R and thus the AM1-calculated ΔH_R provides a potential tool for assessing the feasibility of the LA-mediated IMDAF reactions.

In comparison, the IMDA reactions of **40** and **41** (Table 5) are calculated to be so favorable ($\Delta H_R = -179.9$ and -169.5 kJmol^{-1}, respectively) that the energetics for the complexation of the LA is minor in comparison to that for the DA itself. The implication for these IMDA reactions is that the change in **S:P** with stoichiometric compared to catalytic quantities of LA would be experimentally unobservable in synthetically motivated research. This also indicates that there should not be a significant difference in the thermodynamic **S:P** ratio with different activating groups (e.g. **40** = aldehyde or **41** = ester). The application of >0.95 equiv of LA for the ester **41** is probably just a kinetic effect, reducing the time required for the reaction to reach equilibrium[32c] (vide infra).

Effect of Tether Length on the IMDAF Reaction with MAC

In the course of our synthetic studies with IMDAF reactions, we have also varied the number of carbon atoms in the furan–dienophile tether (from three to six carbon atoms, including the carbonyl) in an attempt to give access to other ring systems

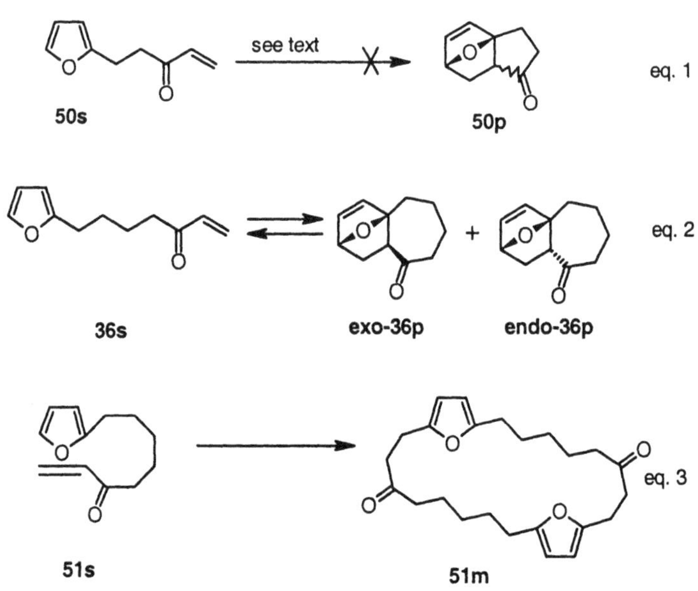

Scheme 7.

Table 11. Effect of Tether Length on the IMDAF Reaction

SM	*Equiv*	*LA*	*Time*	*Temp (°C)*	*Results*
50s	0.1	MAC	1 h	–65	polymer
50s	1.1	MAC	1 h	–65	polymer
					36s:endo:exo-**36p**
36s	0.1	MAC	1 h	–78	8:84:8
36s	0.1	MAC	1 h	–65	8:81:11
51s	1.1	MAC	0.5 h	–78→rt	polymer
51s	1.1	$BF_3 \cdot Et_2O$	10 min	–78	**51m** (67%)
51s	1.1	$BF_3 \cdot Et_2O$	10 min	–65→rt	polymer
51s		Florisil/CH_2Cl_2	7 d	rt NR	

with varying degrees of success (Scheme 7, Table 11). For precursor **50s** which leads to the formation a 6,5 ring system **50p**, neither catalytic nor stoichiometric LA gave anything but polymer.[54] However, the AM1 calculations indicate that the reaction should be favorable (Table 12, *exo*-adduct, $\Delta H_R = -26.6$ kJ mol^{-1}). It is interesting to note that Harwood[22] has also reported that **50s** fails to undergo the IMDAF reaction even at high pressure (19.5 kbar). Presumably, this reaction is prohibited by an extremely unfavorable transition state. Preliminary AM1 calculations indicate that the $\Delta H^{\ddagger}$ is approximately 85 kJ mol^{-1} less favorable for **50s**→**50p** than that for the homologous 6,6 system **13s**→**13p**.[55] It is interesting to note that the reaction related to that of **50s**→**50p** in which the carbonyl group is replaced by a methylene is known,[13] suggesting that the sp^2 carbon of the carbonyl is involved in preventing the IMDAF reaction of **50s** presumably due to geometric constraints.

In contrast, the cycloheptane homologue **36p** can be prepared in good yield with either catalytic or stoichiometric MAC,[17c] with the *endo-* isomer being produced as the major product. Harwood[21,22] has reported that at 12 kbar a 1:1 *endo*:*exo*

Table 12. AM1 Calculated ΔH_f, ΔH_R (kJ mol^{-1}) for IMDAF Reactions of **36**, **50**, and **51**

	ΔH_f		
Compound	**S**	**P**	ΔH_R
50	–87.8	*exo* –114.4	–26.6
36	–146.8	*exo* –174.7	–27.9
		endo –174.0	–24.2
51	–167.2	*exo* –187.2	–20.0
		endo –190.2	–21.7

mixture of products is formed and that the adducts cycloreversed at atmospheric pressure. In our hands, neither *exo*- nor *endo*-**36p** showed any signs of cycloreversion on isolation and the observed *endo*:*exo* ratio is close to that predicted by Dolata.[48a]

The IMDAF reaction of the next higher homologue cyclooctane system **51s**→**51p** is also calculated to be favorable (Table 12, $\Delta H_R = -21.7$ kJ mol^{-1}) but experiment gave only polymer with 1.1 equiv of MAC.[56] Interestingly, the 24-membered macrocycle **51m** was formed in good yield (67%) when **51s** was treated with 1.1 equiv of $BF_3{\cdot}Et_2O$ at −65 °C. Presumably, **51m** was formed via two successive Friedel–Crafts type alkylation reactions.

Reaction Coordinate Diagrams

Once we realized that the IMDAF reactions were in fact exothermic, although less exothermic than analogous IMDA or intermolecular DA reactions, we began an investigation of the reaction profiles for these IMDAF reactions.

Typically DA reactions have very favorable ΔG_R due to the enthalpy change associated with the formation of two σ-bonds at the expense of two π-bonds. If stoichiometric quantities of LA are used in a solvent of low basicity, then **S** and/or **P** may exist mainly in the complexed forms **S**–LA and **P**–LA respectively. Typically, the energy differences associated with the differences in basicity of **S** and **P** will be small compared to that of the DA reaction itself, so the energy change for **S**–LA to **P**–LA will also be very favorable. Thus the presence of either catalytic or stoichiometric LA serves only to enhance the observed rate at which the reaction comes to equilibrium but does not significantly alter the position of the equilibrium (as the traditional definition of a catalyst implies).

An appreciation for the general principles controlling the IMDAF reactions can be gained by contemplating the reaction coordinate diagrams shown in Figure 4.[57] Four scenarios have been selected for discussion; others are of course possible, but are not discussed here. For IMDAF reactions, the energy changes associated with the differences in the basicity of **S** and **P** are similar in magnitude to those of the IMDAF reaction, ΔG_R. The differences in the free energies of the LA complexes, ΔG_C (which is of course related to K_2) could be larger or smaller, and even have the opposite sign compared to ΔG_R. Therefore, the catalyst can significantly alter the experimental outcome of the reaction. In all four of the cases depicted $|\Delta G_C| > |\Delta G_R|$, but the relative basicities of **S** and **P** (i.e. K_1 and K_3) are varied to create the four situations. Note that experimental yields for catalytic LA reactions are reflected by the relative stabilities of **S** and **P** (as shown by the outer energy levels), while the yields from the stoichiometric LA reactions are reflected by the relative stabilities of **S**–LA and **P**–LA (as shown by the inner energy levels).

Figure 4a depicts a favorable overall reaction (i.e. $\Delta G_R < 0$) with **P** more basic than **S**, so stoichiometric LA will give a better yield than catalytic LA. The IMDAF reaction of **33s** is an example of this type of reaction. If **S** is more basic than **P**, as is the case in the enone/ketone systems, then case Figure 4b is observed, where catalytic LA provides **P** formation (i.e. $K_{obs} > 1$) while stoichiometric LA decreases the yield of **P** (i.e. $K_{obs} < 1$). This is the general situation for the IMDAF reactions that we have labeled as type "A" (e.g. the reaction of **15s**). Figure 4c depicts a slightly unfavorable overall reaction (i.e. $\Delta G_R > 0$) and where **P** is more basic than **S** (e.g. the ynone/enone systems). Here the yield of **P** is improved by using stoichiometric LA. Our type "B" reactions, such as that of **35s**, are described by this diagram. Finally, if **S** is the more basic species (or even if the relative basicities are very similar) the case shown in Figure 4d is encountered where neither

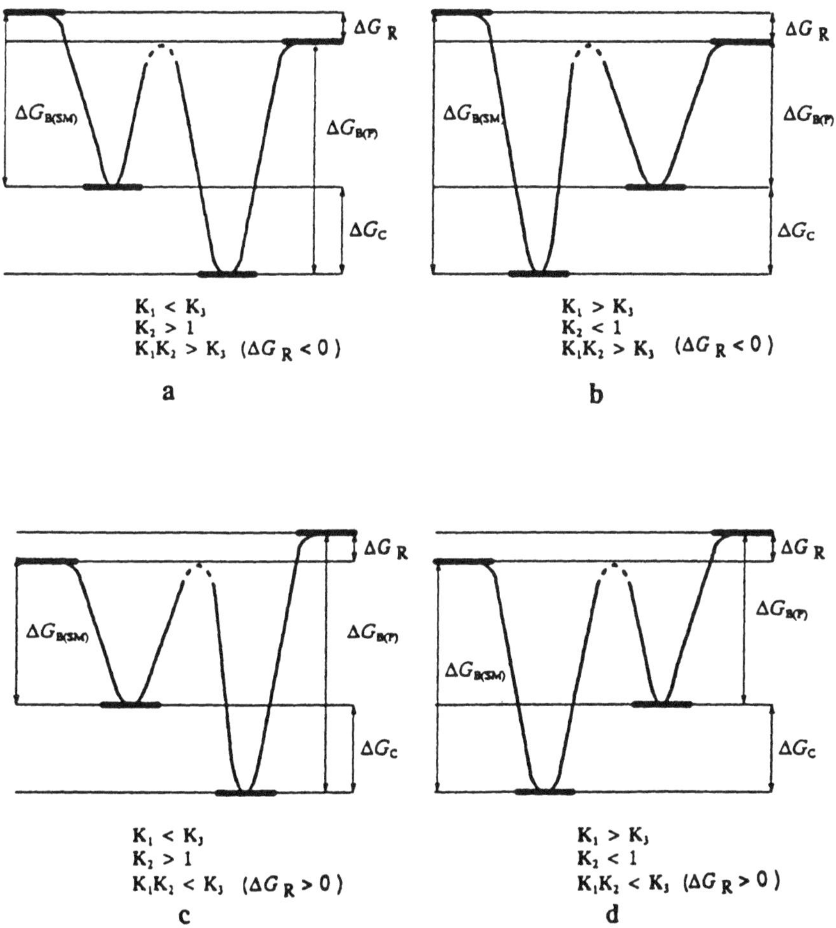

Figure 4. [57] Reaction coordinate diagrams for different scenarios in the LA-promoted IMDAF reaction.

stoichiometric nor catalytic LA provide an efficient means of generating **P**. The reaction of **16s** fits this scenario.

Mathematical Simulation of the IMDAF Equilibria

Both the AM1 calculations (Table 5) and the analysis of the reaction coordinate diagrams (Figure 4) provide a qualitative overall picture of the role of the LA and the equilibria of the catalyzed IMDAF reactions (Scheme 4). Since the principles behind the observed behavior should also extend to other LA-catalyzed reactions, we investigated the reaction equilibria in a more quantitative fashion by performing a mathematical simulation for Scheme 4. This has allowed us to calculate **S:P** ratios based on input "experimental conditions" (i.e. quantity of LA, relative basicity of **S** and **P**, DA reactivity), which can be very conveniently visualized and interpreted by using 3-D representations.

In order to do this, an expression for the observed **S:P** ratio in terms of experimental variables (which are the initial LA and **S** concentrations) is required. The equilibrium constants in Scheme 4 are defined in terms of the chemical species present as:

$$K_1 = [\mathbf{S}\text{–LA}] / [\mathbf{S}]\,[\text{LA}] \tag{1a}$$

$$K_2 = [\mathbf{P}\text{–LA}] / [\mathbf{S}\text{–LA}] \tag{1b}$$

$$K_3 = [\mathbf{P}\text{–LA}] / [\mathbf{P}]\,[\text{LA}] \tag{1c}$$

The equilibrium constants K_1 and K_3 measure the basicity, of **S** and **P**, respectively, toward the LA *relative to solvent*. The effective values of K_1 and K_3 could therefore be less than or greater than unity. The former situation applies when the solvent competes favorably for the LA (e.g. THF). In our studies with MAC in CH_2Cl_2, K_1 and K_3 are expected to be greater than unity. The equilibrium constant K_2 describes the relative stability of the LA complexes of **S** and **P**, and is greater than unity if the complex **P**–LA is more stable than the complex **S**–LA. The equilibrium constant for the uncatalyzed DA reaction in terms of **S** and **P** is K_{DA}:

$$K_{DA} = [\mathbf{P}] / [\mathbf{S}] = K_1\, K_2 / K_3 \tag{2}$$

Note that the DA reaction is intrinsically favorable in the absence of LA if $K_1\, K_2 > K_3$. The mathematical analysis is complicated to some degree because in our experiments the reaction mixture, including any added LA, has been quenched and **P** is therefore isolated from the combined amounts of **P** and **P**–LA. Likewise, recovered **S** originates both from **S** and **S**–LA. Therefore it is necessary to define K_{obs}, which is the more useful term:

$$K_{obs} = [\mathbf{P}]_{tot} / [\mathbf{S}]_{tot} \tag{3a}$$

$$= ([\mathbf{P}] + [\mathbf{P}\text{–LA}]) / ([\mathbf{S}] + [\mathbf{S}\text{–LA}]) \tag{3b}$$

By substituting for [**S**], [**S**–LA], [**P**], [**P**–LA], in terms of the three equilibrium constants, K_1, K_2, and K_3, and the free catalyst, [LA], then K_{obs} can be expressed as:

$$K_{obs} = K_1\, K_2\, (1 + K_3\, [\mathrm{LA}]) \,/\, K_3\, (1 + K_1\, [\mathrm{LA}]) \qquad (4)$$

Since the experimental variables are the initial amounts of addend, $[\mathbf{S}]_o$, and catalyst $[\mathrm{LA}]_o$, an expression for the free catalyst, [LA] is required:

$$[\mathrm{LA}]_{tot} = [\mathrm{LA}]_o = [\mathrm{LA}] + [\mathbf{S}\text{–LA}] + [\mathbf{P}\text{–LA}] \qquad (5)$$

By similar substitutions to those used above, the following expressions can be obtained:

$$[\mathrm{LA}] = 1/2\, ((\mathrm{B}^2 + 4\,\mathrm{D}\, [\mathrm{LA}]_o)^{0.5} - \mathrm{B}) \qquad (6)$$

where

$$\mathrm{B} = [\mathbf{S}]_o - [\mathrm{LA}]_o + \mathrm{D}$$

and

$$\mathrm{D} = (K_3 + K_1\, K_2) \,/\, K_1\, K_3\, (1 + K_2)$$

Equation 6 can be substituted into Eq. 4 to generate the complete equation for K_{obs} in terms of K_1, K_2, K_3, $[\mathbf{S}]_o$ and $[\mathrm{LA}]_o$. There is no merit in reproducing this awkward equation here. Note that under equilibrium conditions, if **S** and **P** are equally basic, i.e. $K_1 = K_3$, then from Eq. 4, $K_{obs} = K_2 = K_{DA}$, and that in the absence of added LA that Eq. 4 collapses to $K_{obs} = K_1\, K_2 \,/\, K_3 = K_{DA}$ as required by Eq. 2. The complex, nonlinear equation for K_{obs} may be readily solved and visualized using *Mathematica*[58] to generate graphical representations of the expression to illustrate the features under investigation.

Figures 5a–d[57] are 3-D surface representations of the dependence of the experimental yield (i.e. K_{obs}) on the relative basicity of **S** and **P** (i.e. K_1 and K_3 respectively) for unfavorable and favorable reactions (i.e. K_2 = 0.25 and 4, respectively) with catalytic and stoichiometric quantities of LA. The values of K_2 were chosen because we felt that they were reasonably representative of the general IMDAF reactions of most interest to us. These figures are a little intimidating, so to aid the interpretation of these figures, note that: (i) that if the reaction yields $[\mathbf{P}]_{tot} > [\mathbf{S}]_{tot}$, then $K_{obs} > 1$, and (ii) that the plane defined by the rear corner through the forward corner (see Figure 5a) corresponds to **S** and **P** being equally basic (i.e. $K_1 = K_3$). Therefore points to the right of that plane where $K_3 > K_1$ correspond to **P** being more basic than **S**. This would correspond to the reactions of the acetylenic IMDAF systems (Table 1). Similarly, points to the left of that plane have $K_1 > K_3$, so **S** is more basic than **P**, the situation corresponding to the reactions of the enone IMDAF systems. These 3-D surfaces are related to the coordinate diagrams in Figure 4 via the relationships between the terms in Eq. 2 with ΔG_R and ΔG_C.

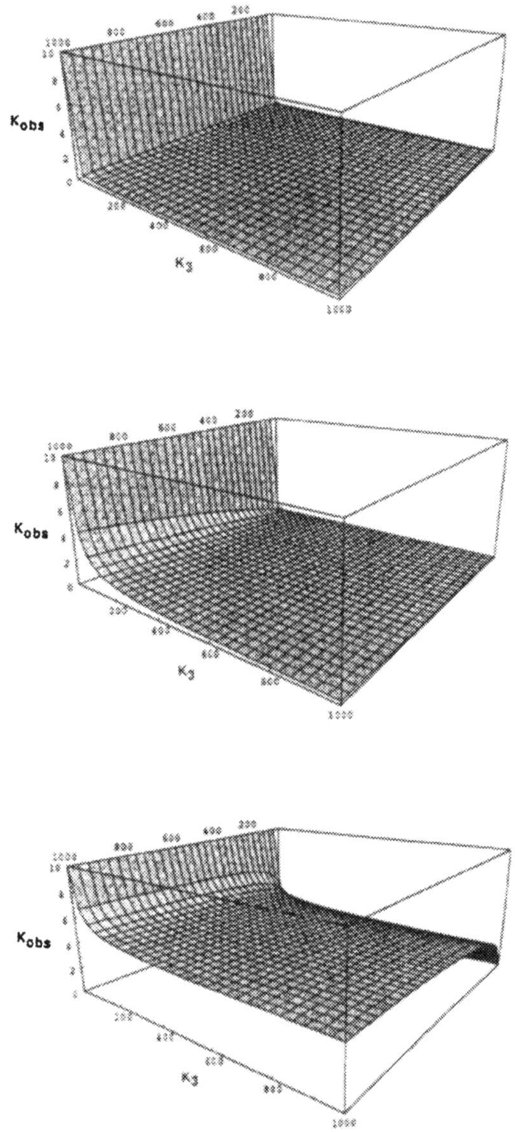

Figure 5.[57] 3-D plots of Eq. 4 obtained using Mathematica to show the general characteristics of the equilibria shown in Scheme 4. The following conditions were used: (**a**) $[SM]_o$, $[LA]_o = 0.1$ M; K_1, $K_3 = 0.1 - 1000$; $K_2 = 0.25$; (**b**) $[SM]_o = 0.1$ M; $[LA]_o = 0.01$ M; K_1, $K_3 = 0.1 - 1000$; $K_2 = 0.25$; (**c**) $[SM]_o$, $[LA]_o = 0.1$ M; K_1, $K_3 = 0.1 - 1000$; $K_2 = 4$; (**d**) $[SM]_o = 0.1$ M; $[LA]_0 = 0.001$ M; K_1, $K_3 = 0.1 - 1000$; $K_2 = 4$.

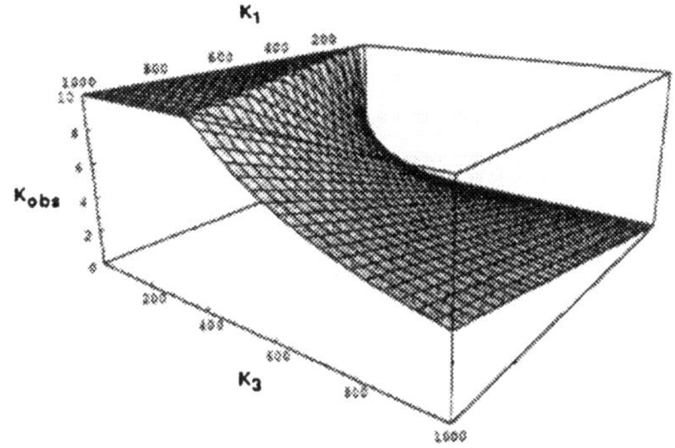

Figure 5. Continued

In Figures 5a and 5b, the reaction is set to be unfavorable with $K_2 = 0.25$, (i.e. [**S**–LA] > [**P**–LA]). Figure 5a shows that with an equivalent of LA, the reaction will only generate **P** efficiently when **S** is much more basic than **P** (i.e. $K_1 >> K_3$). In contrast, Figure 5b, which represents the reaction in the presence 0.1 equiv of LA, **P** is generated over a slightly wider range of relative basicities but with minimal **P** formed when **P** is the more basic species (i.e. $K_3 > K_1$).

Figures 5c and 5d, represent a favorable reaction with $K_2 = 4$. (i.e. [**P**–LA] > [**S**–LA]). For stoichiometric LA, Figure 5c clearly shows that efficient conversion to **P** is obtained over a large range of relative basicities but particularly when **S** is much more basic than **P** (i.e. $K_1 >> K_3$) but not if **S** is of low basicity (i.e. K_1 is small) when K_{obs} favors **S**. In contrast, Figure 5d clearly illustrates that with just 0.01 equiv of LA, the *same* equilibrium is dramatically shifted in favor of **P** particularly when **S** is more basic than **P** (i.e. $K_1 >> K_3$).

In general, IMDA reactions (and other LA-promoted organic reactions) will be characterized by significantly more favorable reaction energetics so K_2 will be much larger than those K_2 representative of the IMDAF reactions. The behavior of such reactions (e.g. **42**) has been investigated using a simple numerical evaluation of the expression for K_{obs}59 by using values of $[S]_o = 0.1$ M, and $[LA]_o = 0.1$ and 0.01 M. This has shown that, for more favorable reactions, catalytic quantities of LA will give higher conversion to **P** than stoichiometric LA provided **S** is more basic (e.g. with $K_1 = 10$, $K_2 = 10{,}000$, $K_3 = 1$, then $K_{obs} \approx 57{,}000$ for stoichiometric and $K_{obs} \approx 92{,}000$ with catalytic quantities of LA). If, however, **P** is more basic, then stoichiometric LA gives higher conversion to **P** (e.g. with $K_1 = 10$, $K_2 = 10{,}000$, $K_3 = 20$, then $K_{obs} \approx 6700$ for stoichiometric and $K_{obs} \approx 5200$ with catalytic quantities of LA). These observations indicate the importance of the relative basicity hypothesis for determining the amount of LA required for promoting organic reactions most effectively. In these cases, though, where the reactions are intrinsically very favorable, the yields are already so high that the calculated changes in K_{obs} will in all likelihood not be experimentally observable, especially in synthetic applications. However, from a practical perspective, the use of catalytic quantities of LA is likely preferred due to reduced cost, simplified experimental workup, and fewer problems with side reactions. These issues are particularly significant in the drive to develop chiral LA catalysts. The implication then is that the practical organic chemist should consider the relative Lewis basicity of the functional groups in their molecules as they decide how much LA to utilize, notably for IMDAF reactions, but for LA-catalyzed organic reactions in general.

These 3-D figures reproduce the experimental characteristics of both the type "A" reactions (i.e. increasing **P** with reduced catalyst) when **S** is more basic than **P** and type "B" reactions (i.e. increasing **P** with stoichiometric catalyst) when **P** is more basic than **S**. It is important to note that this occurs *regardless of whether the reaction (i.e.* K_2*) is favorable or not.* This accord with the experimental evidence implies that the LA-promoted IMDAF reaction is well represented by the general Scheme 4. The figures also serve to demonstrate how the outcome of a reaction can be dramatically effected by the relative concentration of **S** and LA. Since the surfaces are derived for a general scheme, the results should be applicable to other organic reactions that follow the same scenario.

Conclusions

IMDAF precursors containing a four-carbon-atom chain that is internally activated with a carbonyl group and contain no substituents on the dienophile or tether undergo a Diels–Alder reaction when stirred in a Florisil/CH_2Cl_2 mixture. Introducing methyl substituents onto the dienophile resulted in no reaction with Florisil; however, in a 2.0 M $CaCl_2$ solution for 6–14 days some product was isolated. If the same precursors are treated at high pressure (12.5 kbar) in CH_2Cl_2, only products are observed by ^{1}H NMR, albeit isolated in poor overall yield (~50%). In addition, specialized equipment is necessary to effect the reaction. Lewis acids, like $MeAlCl_2$ (MAC) and Me_2AlCl (DMAC), are by far the best reagents to use for effecting IMDAF reactions on precursors that have four or five carbon atoms in the tether (with an internally activated carbonyl group).

More interestingly was the observation that *catalytic quantities of LA provided better S:P ratios than when 1.1 equiv of LA were used.* In the majority of cases reported, ratios like 70:30 in favor of starting material could be reversed to ~30:70 using sub-stoichiometric amounts of LA. This observation has been rationalized by establishing experimental and theoretical evidence for the relative Lewis basicity of the functional group that activates the dienophile. Competitive complexation NMR studies established the relative basicities of saturated and unsaturated ketones and esters, furan rings, and THF. NMR experiments indicated that MAC in catalytic quantities promotes the IMDAF reactions as the most basic site in the system is the reactive site in **S** (unsaturated ketone) so that **P** (ketone or the oxygen bridged atom) does not inhibit the MAC. Increasing the number of equivalents of MAC decreases the conversion to **P**, since the complex **S**–LA is most favorable of the complexed forms. Reactions with acetylenic esters required a minimum of 1.1. equivalents of LA as the **P**–LA complex is more stable than the **S**–LA complex. Thus the LA cannot dissociate from **P** to recomplex with **S**, so >1.0 equiv of LA is necessary to drive the reaction to completion.

Correlation of AM1 calculated ΔH_R with experimental yields has shown that AM1 is a reasonable tool for assessing the feasibility of executing a desired IMDAF reaction. Reasonable qualitative agreement was observed between the experimentally determined yields and the AM1 calculated ΔH_R in systems where catalytic quantities of MAC were used. A reaction scheme involving three equilibria was postulated and ultimately verified via simulation using *Mathematica*. The 3-D plots of K_{obs} vs. K_1 and K_3 clearly indicated that K_{obs} will increase over a large range of K_1 and K_3 even if K_2 is unfavorable (i.e. $K_2 = 0.25$) when moving from 1.1 equiv of LA to catalytic quantities. This effect should also hold true for any reaction that is under thermodynamic control and in which the LA preferentially coordinates to the starting material. Work is continuing towards the development of chiral LA that will promote the IMDAF reaction of these systems.

IV. SYNTHETIC APPLICATIONS

A. Introduction

The IMDAF precursors that have a single substituent on the tether have been shown to undergo a highly diastereoselective IMDAF reaction providing only *exo*-adducts in which the substituent preferentially adopts an equatorial position on the newly formed cyclohexane ring. This type of IMDAF reaction can create adducts with up to 5 stereogenic centers and of the 16 diastereomers that are possible, only one is preferentially formed. We have used this diastereoselective IMDAF reaction to prepare 1,4-epoxycadinane,[18] and the C-15–C-23 portion of venturicidin A.[27,60] These synthetic endeavors are presented in this section.

B. Synthesis of 1,4-Epoxycadinane

(±)-1,4-Epoxycadinane (**52**) was isolated from the brown alga *Dilophus fasciola* in 1979 (Scheme 8).[61] It possesses a unique oxygen bridge between carbons 1 and 4 of the cadinane carbon skeleton. Since the C-6–C-7 bond and the methyl group at C-10 are *syn* with respect to the oxygen bridge at C-1 we felt that a diastereoselective IMDAF reaction would be an ideal method for generating 4 of the 5 stereogenic centers in one step. It was also thought that the rigidity of the oxatricyclo system could be used to an advantage to introduce the group at C-7. This group is situated in the equatorial position on the six-membered cyclohexane portion of **52**. This ring is conformationally locked due to the oxygen bridge, thus if the C-7 ketone that is generated from the IMDAF reaction (**27s**→**27p**) can be extended by one

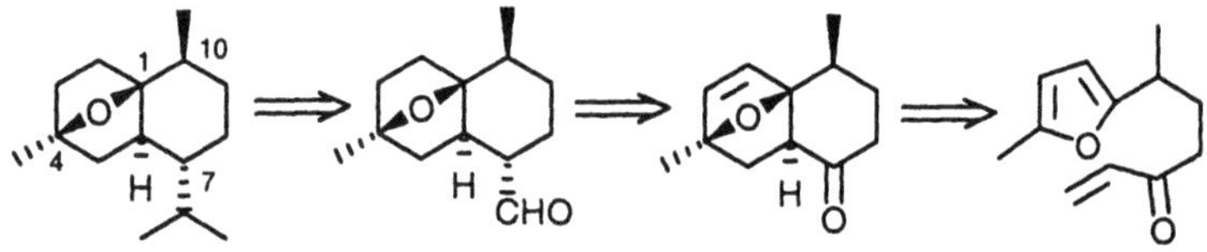

52
(+/-)-1,4-epoxycadinane

53

27p

27s

Scheme 8.

carbon to an aldehyde (**27p**→**53**), it was felt that the aldehyde would preferentially adopt the equatorial position through an equilibration via an enol or enolate.

Compound **27s** was prepared as previously described and when treated with 1.1 equiv of $MeAlCl_2$ (−78 °C, CH_2Cl_2) provided a 9:1 mixture of adducts **27p-eq** and **27p-ax** (98%, Scheme 9).[62] Adducts **27p-eq** and **27p-ax** were easily separated and the unwanted isomer **27p-ax** could be re-equilibrated into **27p-eq** by further treatment with $MeAlCl_2$. Compound **27p-eq** was converted into aldehyde **53** by reduction of the double bond by catalytic hydrogenation (H_2, Pd/C, 95%) followed by a Wittig reaction that was worked up with 10% HCl (63%).[63] Only isomer **53**

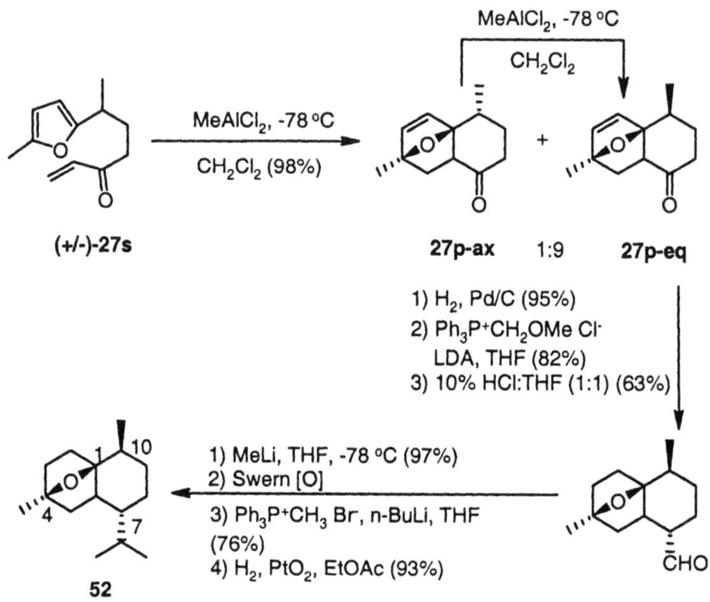

(+/-)-27s

27p-ax

27p-eq

52
(+/-)-1,4-epoxycadinane

53

Scheme 9.

with the aldehyde in the equatorial position was detected (by ^{1}H NMR) and isolated. Conversion of the aldehyde in **53** into an isopropyl group was done in four steps. Treatment of **53** with MeLi (THF, −78 °C, 97%) provided a mixture of alcohols that were immediately oxidized into a methyl ketone via a Swern oxidation[64] (90%). A Wittig reaction ($PPh_3P{=}CH_2$, THF, 76%) followed by another catalytic hydrogenation (H_2, PtO_2, EtOAc, 93%) furnished (±)-1,4-epoxycadinane (**52**).

C. Synthesis of the C-15 to C-23 Segment of Venturicidins A, B, and X

Venturicidins A[65] (**54**) and B[66] (**55**) were isolated from the soil actinomycetes and *Streptomyces aureofaciens*, respectively, while venturicidin X (**56**),[67] the aglycone of venturicidins A and B, was recently isolated from an unidentified species of *Streptomyces* (Scheme 10). In addition to being active against a variety

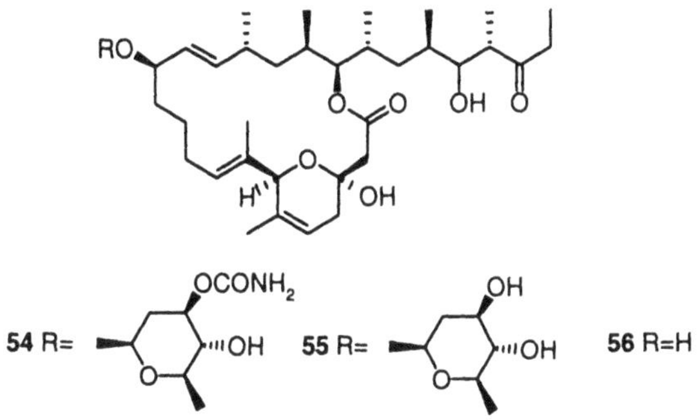

Scheme 10.

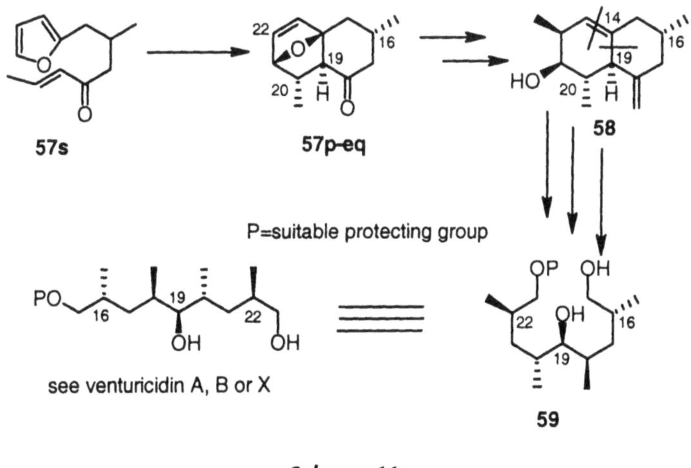

Scheme 11.

of fungi (cucumber, apple mildew, barley, apple scab, and grey mold),[65–67] they all inhibit the ATP synthetase of mitochondria.[68] In 1990, Akita et al. reported the first, and to date the only, total synthesis of venturicidin X.[69] Since then, two partial syntheses of the C-1–C-14[70] and C-15–C-27[71] portions of the venturicidins have been reported.

Our strategy towards the C-15–C-23 segment is outlined in Scheme 11. The highly diastereoselective IMDAF reaction of **57s** would produce **57p-eq** selectively in which five stereogenic centers are created. The rigidity of the oxatricyclo adduct would be used to introduce the C-22 Me group (**57p-eq→58**[27,60] and then both rings would be cleaved via an oxidative cleavage of the double bond at C-14–C-23 and

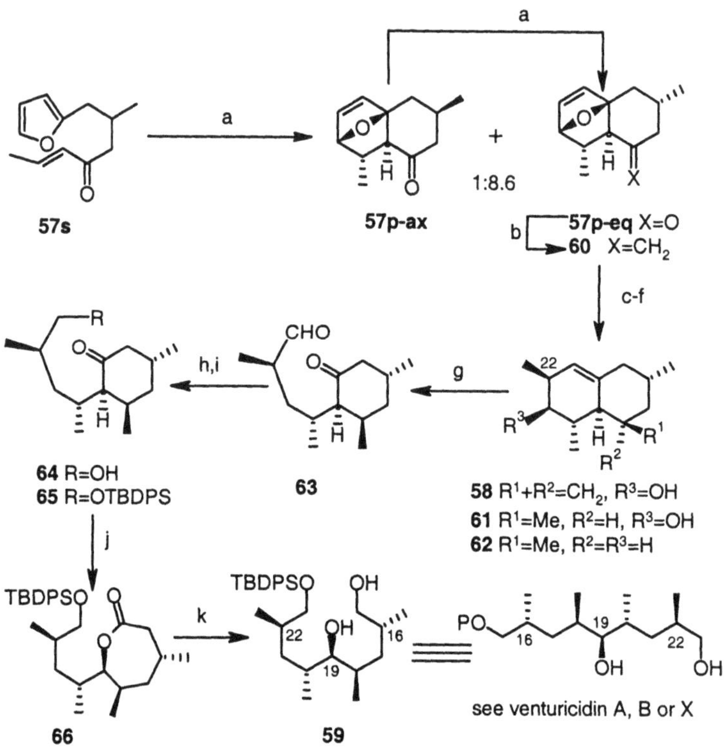

a) 0.1 equiv. $MeAlCl_2$, CH_2Cl_2; -40 °C (92%); b) $Ph_3P{=}CH_2$, 0 °C (95%); c) 30 equiv. MeLi, DME, 24 h, rt (68%); d) H_2, PtO_2, EtOH/benzene, 2h, rt (88%); e) KH, THF, CS_2, 2 h, then MeI, 12 h; f) TTMSS, AIBN, toluene, 90 °C, 2h (73%, 2 steps); g) $RuO_2{\cdot}H_2O$, $NaIO_4$ (77%); h) $NaBH_4$, EtOH:CH_2Cl_2, -78 °C, 2h; i) TBDPSCl, DMAP, CH_2Cl_2, 76 h, rt (96%, 2 steps); j) MCPBA, CH_2Cl_2, 76 h, rt (92%); k) LAH, Et_2O, 1h, rt (90%)

Scheme 12.

Baeyer–Villiger oxidation[72] (at C-14–C-19 bond) to yield a 10-carbon fragment **59**. Compound **59** has the correct relative stereochemistry of the C-15–C-23 portion of the venturicidins with both ends of the chain differentiated for further elaboration.

Treatment of **57s** with 0.1 equiv of $MeAlCl_2$ provided a 8.6:1 mixture of **57p-ax** and **57p-eq** in 92% yield (Scheme 12). These isomers were easily separated and **57p-ax** could be further equilibrated into **57p-eq** by treatment with catalytic amounts of $MeAlCl_2$. A Wittig reaction on **57p-eq** furnished diene **60** which when treated with excess MeLi in DME underwent an S_N2' ring opening of the oxygen bridge to provide **58** in 68% yield and surprisingly, 11% of a compound in which an *ethyl group*[73] was introduced at C-22 in **58**.[27,60] A highly chemo- and stereoselective catalytic hydrogenation of the exocyclic double bond in **58** afforded **61** which after a Chatgilialoglu[74]-modified Barton[75] deoxygenation gave **62**. Treatment of **62** with 1 equiv of RuO_4 (generated in situ with RuO_2 and 2 equiv of $NaIO_4$) in acetone gave the unstable aldehyde **63**, which was selectively reduced immediately using $NaBH_4$ at –78 °C giving alcohol **64**. The alcohol in **64** was protected as its TBDPS ether **65** and treated with MCPBA to give lactone **66**. Reduction of the lactone with LAH afforded diol **59**, which contained the correct relative stereochemistry for the C-15–C-23 portion of the venturicidins (Scheme 10). Although we reported that furan **57s** can be prepared in 98% ee,[76] we did not repeat the sequence shown in Scheme 12, which would have allowed us to form diol **59** in optical pure form.

ACKNOWLEDGMENTS

We thank our colleagues, Dr. P.W. Dibble, Dr. J.A. Nieman, Dr. A. Rauk, Dr. C. Rogers, Dr. S. Woo, Dr. S. Yu, and Giovanna Beese, who performed some of the chemistry mentioned in this chapter. We also thank the Natural Sciences and Engineering Research Council of Canada (NSERC), Merck Frosst (Dorval, PQ), and the University of Calgary for financial support over the years. We also thank and the Royal Society of (Great Britain)/NSERC Bilateral Exchange program (1992) and the Alberta Foundation for Medical Research (AHFMR) (1993) for postdoctoral fellowships (I.R.H.) and NSERC for postgraduate scholarships (C.R. and S.W.).

REFERENCES AND NOTES

1. Diels, O.; Alder K. *Liebigs. Ann. Chem.* **1928**, *460*, 98.
2. Reviews of the intramolecular Diels–Alder reaction: (a) Carlson, R. G. *Ann. Rep. Med. Chem.* **1974**, *9*, 270. (b) Oppolzer, W. *Angew. Chem., Int. Ed. Engl.* **1977**, *16*, 10. (c) Oppolzer, W. *Synthesis* **1978**, 793. (d) Brieger, G.; Bennett, J. N. *Chem. Rev.* **1980**, *80*, 63. (e) Ciganek, E. *Org. Reactions* **1984**, *32*, 1. (f) Funk, R. L.; Vollhardt, K. P. C. *Chem. Soc Rev.* **1980**, *9*, 41. (g) Fallis, A. G. *Can. J. Chem.* **1984**, *62*, 183. (h) Taber, D. F. *Intramolecular Diels–Alder and Alder Ene Reactions*; Springer-Verlag: Berlin, 1984. (i) Craig, D. *Chem. Soc. Rev.* **1987**, *16*, 187. (j) Jung, M. E. *Synlett* **1990**, 186. (k) Roush, W. R. *Adv. In Cyclo. Add.* **1990**, *2*, 91. (l) Oppolzer, W. In *Comprehensive Organic Synthesis*; Trost, B. M., Fleming, I., Eds.; Pergamon, Oxford, 1991. Vol.

5, p. 315. (m) Roush, W. R. In *Comprehensive Organic Synthesis*; Trost, B. M., Fleming, I., Eds.; Pergamon, Oxford, 1991. Vol. 5, p. 513.

3. For an excellent review on the inter- and intramolecular Diels–Alder reaction with furan dienes, see: Kappe, C. O.; Murphree, S. S.; Padwa, A. *Tetrahedron* **1997**, *53*, 14179.
4. For other reviews and papers mentioning IMDAF reactions, see ref. 2 and: Lipshutz, B. *Chem. Rev.* **1986**, *86*, 795.
5. Hunt, I. R.; Rogers, C.; Woo, S.; Rauk, A.; Keay, B. A. *J. Am. Chem. Soc.* **1995**, *117*, 1049.
6. Hunt, I. R.; Rauk, A.; Keay, B. A. *J. Org. Chem.* **1996**, *61*, 751.
7. Herz, W. *J. Am. Chem. Soc.* **1945**, *67*, 2272.
8. (a) Cram, D. J.; Montgomery, C. S.; Knox, G. R. *J. Am. Chem. Soc.* **1961**, *83*, 2204 and ibid. **1966**, *88*, 515.
9. (a) Wasserman, H. H.; Doumanux, A. R. *J. Am. Chem. Soc.* **1962**, *84*, 4611. (b) Wasserman, H. H.; Doumanux, A. R. *Tetrahedron Lett.* **1969**, 5315.
10. (a) Bilovic, D.; Strojanac, Z.; Hahn, V. *Tetrahedron Lett.* **1964**, 2071. (b) Bilovic, D. *Crot. Chem. Acta* **1966**, *38*, 293. (c) Bilovic, D.; Hahn, V. *Crot. Chem. Acta* **1967**, *39*, 189. (d) Bilovic, D. *Crot. Chem. Acta* **1968**, *40*, 15.
11. Katz, T. J.; Balogh, V.; Schulman, J. *J. Am. Chem. Soc.* **1968**, *90*, 734.
12. (a) De Clercq, P. J.; Van Royen, L. A. *Syn. Commun.* **1979**, *9*, 771. (b) Van Royen, L. A.; Mijngheer, R.; De Clercq, P. J. *Tetrahedron Lett.* **1982**, *23*, 3283. (c) Van Royen, L. A.; Mijngheer, R.; De Clercq, P. J. *Tetrahedron Lett.* **1983**, *24*, 3145. (d) Van Royen, L. A.; Mijngheer, R.; De Clercq, P. J. *Bull. Soc. Chim. Belg.* **1984**, *93*, 1019. (e) Van Royen, L. A.; Mijngheer, R.; De Clercq, P. J. *Tetrahedron* **1985**, *41*, 4667. (f) Missiaen, P.; De Clercq, P. J. *Bull. Soc. Chim Belg.* **1987**, *96*, 105. (g) Cauwberghs, S. G.; De Clercq, P. J. *Tetrahedron Lett.* **1988**, *29*, 6501. (h) Missiaen, P.; De Clercq, P. J. *Bull. Soc. Chim. Belg.* **1990**, *99*, 271. (i) Nuyttens, F.; Hoflack, J.; Appendino, G.; De Clercq, P. J. *Synlett* **1995**, 105.
13. (a) Sternbach, D. D.; Rossana, D. M. *Tetrahedron Lett.* **1982**, *23*, 303. (b) Sternbach, D. D.; Rossana, D. M. *J. Am. Chem. Soc.* **1982**, *104*, 5853. (c) Sternbach, D. D.; Rossana, D. M. Onan, K. D. *J. Org. Chem.* **1984**, *49*, 3427. (d) Sternbach, D. D.; Rossana, D. M. Onan, K. D. *Tetrahedron Lett.* **1985**, *26*, 591. (e) NcNelis, B. J.; Sternbach, D. D.; MacPhail, A.T. *Tetrahedron* **1994**, *50*, 6767.
14. For a review on the stereo- and regiocontrolled opening of oxygen-bridged systems, see: Woo, S.; Keay, B. A. *Synthesis* **1996**, 669.
15. (a) Keay, B. A. *J. Chem. Soc., Chem. Commun.* **1987**, 419. (b) Keay, B. A.; Rogers, C.; Bontront, J.-L. J. *J. Chem. Soc., Chem. Commun.* **1989**, 1782.
16. Keay, B. A.; Dibble, P. W. *Tetrahedron Lett.* **1989**, *30*, 1045.
17. (a) Rogers, C.; Keay, B. A. *Synlett* **1991**, 353. (b) Rogers, C.; Keay, B. A. *Tetrahedron Lett.* **1991**, *32*, 6477. (c) Rogers, C.; Keay, B. A. *Can. J. Chem.* **1992**, *70*, 2929.
18. (a) Rogers, C.; Keay, B. A. *Tetrahedron Lett.* **1989**, *30*, 1349. (b) Rogers, C.; Keay, B. A. *Can. J. Chem.* **1993**, *71*, 611.
19. (a) Yu, S.; Beese, G.; Keay, B. A. *J. Chem. Soc., Perkin Trans. I* **1992**, 2729. (b) Beese, G.; Keay, B. A. *Synlett* **1991**, 33. (c) Yu, S.; Keay, B. A. *J. Chem. Soc., Perkin Trans. I* **1991**, 2600. (d) Bures, E.; Nieman, J. A.; Yu, S.; Spinazze, P. G.; Bontront, J.-L.J.; Hunt, I. R.; Rauk, A.; Keay, B. A. *J. Org. Chem.* **1997**, *62*, 8750. (e) Bures, E.; Spinazze, P. G.; Beese, G.; Hunt, I. R.; Rogers, C.; Keay, B. A. *J. Org. Chem.* **1997**, *62*, 8741.
20. (a) von Hippel, P. H.; Schleich, T. *Acc. Chem. Res.*, **1969**, *9*, 257. (b) von Hippel, P. H.; Wong, K.-Y. *J. Biol. Chem.* **1965**, *240*, 3909. (c) Rideout, D. C.; Breslow, R. *J. Am. Chem. Soc.* **1980**, *102*, 7817.
21. (a) Burrell, S. J.; Derome, A. E.; Edenborough, M. S.; Harwood, L. M.; Leeming, S. A.; Isaacs, N. S. *Tetrahedron Lett.* **1985**, *26*, 2229. (b) Isaacs, N. S.; Van der Beeke, P. *Tetrahedron Lett.* **1982**, *23*, 2147. (c) Isaacs, N. S.; George, A. V. *Chem. Brit.* **1987**, *23*, 47.

22. (a) Harwood, L. M.; Leeming, S. A.; Isaacs, N. S.; Jones, G.; Pickard, J.; Thomas, R. M.; Watkin, D. *Tetrahedron Lett.* **1988**, *29*, 5017. (b) Harwood, L. M.; Jones, G.; Pickard, J.; Thomas, R. M.; Watkin, D. *J. Chem. Soc., Chem. Commun.* **1990**, 605. (c) Harwood, L. M.; Jackson, B.; Jones, G.; Prout, K.; Thomas, R. M.; Witt, F. T. *J. Chem. Soc., Chem. Commun.* **1990**, 608. (d) Brickwood, A. C.; Drew, M. G. B.; Harwood, L. M.; Ishikawa, T.; Marais, P.; Morisson, V. *J. Chem. Soc., Perkin Trans. 1* **1999**, 913.
23. Unbeknown to us at the time of submission of our paper (see ref. 16) on the high-pressure studies, Harwood had just published (see ref. 22a and 22d) a high-pressure study on IMDAF reactions involving systems containing 3, 4, and 5 carbon tethers.
24. We thank Prof. Carl Johnson (Wayne State University, Detroit, MI) for the use of his high-pressure apparatus.
25. Yates, P.; Eaton, P. *J. Am. Chem. Soc.* **1960**, *82*, 4436.
26. (a) Brion, F. *Tetrahedron Lett.* **1982**, *23*, 5299. (b) Laszlo, P.; Lucchetti, J. *Tetrahedron Lett.* **1984**, *25*, 4387. (c) Kotsuki, H.; Asao, K.; Ohnishi, H. *Bull. Chem. Soc. Spn.* **1984**, *57*, 3339. (d) McCulloch, A. W.; Smith, D. G.; McInnes, A. G. *Can. J. Chem.* **1974**, *52*, 1013.
27. Woo, S.; Keay, B. A. *Tetrahedron Lett.* **1992**, *33*, 2661.
28. As we became more experienced with the use of LA in the IMDAF reaction, it became apparent that when 1.1 equiv of MAC is used, equilibrium is usually reached within 1 h at –78 °C.
29. Laszlo, P.; Teston, M. J. *J. Am. Chem. Soc.* **1990**, *112*, 8750.
30. (a) Snider, B. B. *J. Am. Chem. Soc.* **1979**, *101*, 5283. (b) Snider, B. B. *Acc. Chem. Res.* **1980**, *13*, 426.
31. (a) Inukai, T.; Kasai, M. *J. Org. Chem.* **1965**, *30*, 3567. (b) Chapuis, C.; Jurczak, J. *Helv. Chim. Acta.* **1987**, *70*, 436. (c) Evans, D. A.; Chapman, K. T.; Bisaha, J. *J. Am. Chem. Soc.* **1988**, *110*, 1238.
32. (a) Roush, W. R. In *Advances in Cycloaddition*; Curran, D. P. Ed.; JAI, Greenwich CT, 1990, Vol. 2, pp. 91–146 and references therein. (b) Reich, H. J.; Eisenhart, E. K. *J. Org. Chem.* **1984**, *49*, 5282. (c) Shea, K. J.; Gilman, J. W. *Tetrahedron Lett.* **1983**, *24*, 657. (d) Sakan, K.; Craven, B. A. *J. Am. Chem. Soc.* **1983**, *105*, 3732.
33. (a) Marshall, J. A.; Audia, J. E. *J. Org. Chem.* **1984**, *49*, 5277. (b) Marshall, J. A.; Audia, J. E.; Grote, J.; Shearer, B. *Tetrahedron* **1986**, *42*, 2893. (c) Roush, W. R.; Gillis, H.R. *J. Org. Chem.* **1980**, *45*, 4267. (d) Roush, W. R.; Gillis, H. R.; Ko, A. I. *J. Am. Chem. Soc.* **1982**, *104*, 2269.
34. Houk, K. N. *J. Am. Chem. Soc.* **1973**, *95*, 4092.
35. (a) Houk, K. N.; Strozier, R. W. *J. Am. Chem. Soc.* **1973**, *95*, 4094. (b) Fleming, I. In *Frontier Orbitals and Organic Chemical Reactions*; Wiley: New York, 1976, p. 162.
36. Childs, R. F.; Mulholland, D. L.; Nixon, A. *Can. J. Chem.* **1982**, *60*, 801.
37. (a) Shambayati, S.; Schreiber, S. L. In *Comprehensive Organic Synthesis*; Trost, B. M., Fleming, I., Eds.; Pergamon: Oxford, 1991, Vol. 1, pp. 283–324. (b) Loncharich, R.J.; Schwartz, T. R.; Houk, K. N.; *J. Am. Chem. Soc.* **1987**, *109*, 14.
38. Reprinted with permission from Hunt, I. R.; Rogers, C.; Woo, S.; Rauk, A.; Keay, B. A. *J. Am. Chem. Soc.* **1995**, *117*, 1049. Copyright 1995 American Chemical Society.
39. This is the relative basicity of the substrate compared to Et_2O.
40. (a) Fratiello, A.; Schuster, R. E. *J. Org. Chem.* **1972**, *37*, 2237. (b) Fratiello, A.; Vidulich, G. A.; Chow, Y. *J. Org. Chem.* **1973**, *38*, 2309. (c) Schuster, R. E.; Bennett, R. D. *J. Org. Chem.* **1973**, *38*, 2904. (d) Frateillo, A.; Stover, C. S. *J. Org. Chem.* **1975**, *40*, 1244.
41. Maria, P. C.; Gal, J. F. *J. Phys. Chem.* **1985**, *89*, 1296.
42. Rauk, A.; Hunt, I. R.; Keay, B. A. *J. Org. Chem.* **1994**, *59*, 6808.
43. Mole, T.; Jeffery, E. A. In *Organoaluminum Compounds*; Elsevier: Amsterdam, 1972, Chap. 4, pp. 106–118.
44. (a) Lowry, T. H.; Richardson, K. S. In *Mechanism and Theory in Organic Chemistry*, 3rd ed.; Harper and Row: New York, 1987, p. 194. (b) Carey, F. A.; Sundberg, R. J. In *Advanced Organic Chemistry: Part A, Structure and Mechanism*, 2nd ed.; Plenum: New York, 1984, p. 176.

45. Calculations were performed using the AM1 semiempirical method within the SPARTAN[46] program package. ΔH_R values were evaluated from the calculated ΔH_f of the geometry optimized structures.
46. SPARTAN 2.0, Wavefunction, Inc, Irvine, CA. Carpenter, J. E.; Baker, J.; Hehre, W. J.; Kahn, S. D. SPARTAN User's Guide, 1991.
47. For example, the non-furan analogue of **13s→13p** with a 1,3-butadiene system has a $t_{1/2}$(0 °C) = 4 h: Oppolzer, W.; Snowden, R. L.; Simmons, D. P. *Helv. Chim. Acta* **1981**, *64*, 2002.
48. (a) Dolata, D. P.; Harwood, L. M. *J. Am. Chem. Soc.* **1992**, *114*, 10738. (b) Parrill, A.L.; Dolata, D. P. *Tetrahedron Lett.* **1994**, *35*, 7319.
49. Raimondi, L.; Brown, F. K.; Gonzalez, J.; Houk, K. N. *J. Am. Chem. Soc.* **1992**, *114*, 4796.
50. March, J. *Advanced Organic Chemistry*, 3rd. ed.; Wiley: New York, New York, 1985.
51. (a) Eliel, E. L.; Wilen, S. H. In *Stereochemistry of Organic Compounds*; Wiley: New York, 1994, pp. 682–684. (b) Jung, M. E.; Gervay, J. *J. Am. Chem. Soc.* **1991**, *113*, 22 and references therein.
52. March, J. *Advanced Organic Chemistry*, 3rd. ed.; Wiley: New York, 1985, p. 182–183.
53. Allinger, N. L.; Zalkow, V. *J. Org. Chem.* **1960**, *25*, 701.
54. Rogers, C. Ph.D. Dissertation, The University of Calgary, Calgary, AB, Canada, 1991.
55. Hunt, I. R.; Keay, B. A. Unpublished results.
56. Nieman, J. A.; Keay, B. A. Unpublished results.
57. Reprinted with permission from Hunt, I. R.; Rauk, A.; Keay, B. A. *J. Org. Chem.* **1996**, *61*, 751. Copyright 1996 American Chemical Society.
58. *Mathematica* is a trademark of Wolfram Research Inc., Champaign, IL.
59. Evaluation of K_{obs} was carried out using a simple Fortran program for the full version of Eq. 4 in terms of Eqs. 5 and 6.
60. Woo, S.; Keay, B. A. *Synlett* **1996**, 135.
61. (a) Fattorusso, E.; Magno, S.; Mayol, L. *Gazz. Chim. Ital.* **1979**, *109*, 589. (b) Bordoloi, M.; Shukla, V. S.; Nath, S. C.; Sharma, R. P. *Phytochemistry* **1989**, *28*, 2007. (c) Fraga, B. *Nat. Prod. Rep.* **1992**, *9*, 217.
62. This synthesis was finished prior to the discovery that catalytic quantities of Lewis acids provide better S:P ratios. Since 1.1 equiv of MAC provide no starting material and the products could be easily separated and the unwanted isomer recycled, we chose not to repeat the synthesis just to try the reaction with catalytic amounts of MAC.
63. Paquette, L.; Bulman-Page, P. C.; Pansegrau, P. D.; Wiedeman, P. E. *J. Org. Chem.* **1988**, *53*, 1450.
64. Mancuso, A. J.; Swern, D. *Synthesis* **1981**, 165.
65. Rhodes, A.; Fantes, K. H.; Boothroyd, B.; McGonagle, M. P.; Crosse, R. *Nature* **1961**, *192*, 952.
66. (a) Brufani, M.; Keller-Schierlein, W.; Loffler, W.; Mansperger, I.; Zahner, H. *Helv. Chim. Acta* **1968**, *51*, 1293. (b) Brufani, M.; Cerrini, S.; Fedeli, W.; Musu, C.; Cellai, L.; Keller-Schierlein, W. *Experientia* **1971**, *27*, 604.
67. Laatsch, H.; Kellner, M.; Lee, Y.-S.; Wolf, G. Z. *Naturforsch* **1994**, *49b*, 977.
68. (a) Linnett, P. E.; Beechey, R. B. *Methods Enzymol.* **1979**, *55*, 472. (b) Lardy, H.A.; Reed, R.; Lin, C.-H.C. *Fed. Proc., Fed. Am. Soc. Exp. Biol.* **1975**, *34*, 1707.
69. Akita, H.; Yamada, H.; Matsukura, H.; Nakata, T.; Oishi, T. *Tetrahedron Lett.* **1990**, *31*, 1731 and 1735.
70. Nakata, M.; Ohashi, J.; Ohsawa, K.; Nishimura, T.; Kinoshita, M.; Tatsuta, K. *Bull. Chem. Soc. Jpn.* **1993**, *66*, 3464.
71. Hoffmann, R. W.; Rolle, U. *Tetrahedron Lett.* **1994**, *27*, 4751.
72. Feng, F.; Murai, A. *Chem. Lett.* **1992**, 1587.
73. We postulated and showed, using D_3CLi, that some ethyllithium is formed due to a reaction between MeLi and DME at rt. For further information see ref. 60.
74. Chatgilialoglu, C. *Acc. Chem. Res.* **1992**, *25*, 188.
75. Barton, D. H. R.; McCombie, S. W. *J. Chem. Soc., Perkin Trans. 1* **1975**, 1574.
76. Woo, S.; Keay, B. A. *Tetrahedron: Asymm.* **1994**, *5*, 1411.

AN ALLENIC [2+2+1] CYCLOADDITION

Kay M. Brummond

I. INTRODUCTION

The Pauson–Khand (P–K) reaction (Eq. 1), first reported in the early 1970s,[1] is probably the most well-known of the [2+2+1] cycloadditions and has developed into a method of choice for the preparation of cyclopentenones. This formal

Advances in Cycloaddition
Volume 6, pages 211–237.

ISBN: 0-7623-0531-2

[2+2+1] cycloaddition, which was initially effected intermolecularly, has now been done intramolecularly,[2] catalytically,[3] asymmetrically,[4] at low temperatures,[5] and has been highlighted in the synthesis of several natural products,[6] attesting to its synthetic flexibility and usefulness.

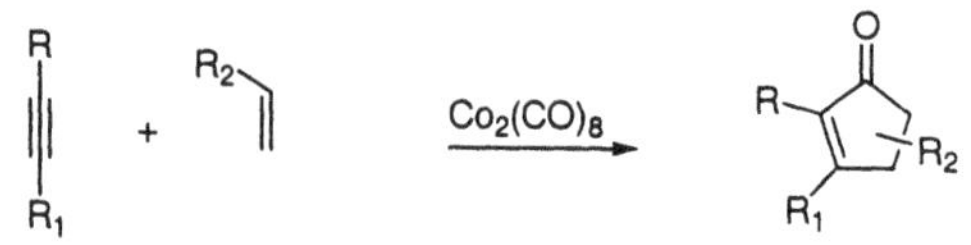

(1)

Conspicuously absent from the P–K reaction data base was the [2+2+1] cycloaddition of an alkyne and an allene, with an allene moiety functioning in the place of the olefin. Allenes are not newcomers to cycloaddition reactions; in fact, allenes have been used in place of olefins in nearly every type of cycloaddition process. A few of the more recent examples are: Cope ring expansion;[7] cobalt-mediated [2+2+2];[8] [2+2];[9] [4+2];[10] [4+1];[11] and ene reactions.[12] The absence of the use of the allene in the P–K cycloaddition can possibly be explained by the report that allenes polymerize in the presence of dicobaltoctacarbonyl.[13] Although it was reported in a review published by Pauson that a dicobalthexacarbonyl acetylene complex and 1,2-cyclononadiene react readily, the structure of the presumed cycloadduct was not firmly established.[14] With the advent of new metal carbonyl complexes to facilitate the [2+2+1] cycloaddition, the diversity in cycloaddition precursors, and consequently the cycloadducts, increased dramatically. In a single report by Aumann, the feasibility of using an allene moiety in place of an olefin in an *intermolecular* P–K type cycloaddition process was established (Eq. 2).[15] For more details, see Section III of this review.

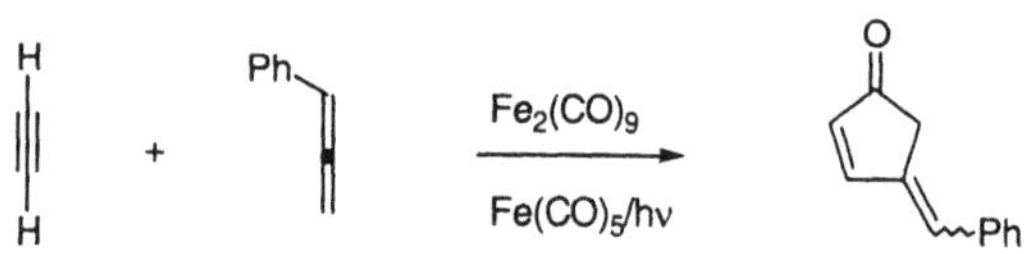

(2)

Alternatively, this allenic [2+2+1] cycloaddition, if effected *intramolecularly*, would give rise to interesting substructures that are present in a variety of natural products. Thus, we felt the intramolecular variant would be an interesting process to investigate, given the potential synthetic utility.

The π-system of the allene moiety possesses two sites at which the cycloaddition can potentially occur. If cycloaddition occurs with the internal π-bond (reaction pathway A, Scheme 1), the resulting cycloadduct will be an α-methylene cyclopentenone. Alternatively, if the cycloaddition occurs with the external π-bond of the allene (reaction pathway B, Scheme 1) the resulting adduct will be a 4-alkylidene cyclopentenone. At the outset of our work we felt that it would be both interesting and useful if one could control the cycloaddition pathway by variations in cycloaddition substrates or reaction conditions.

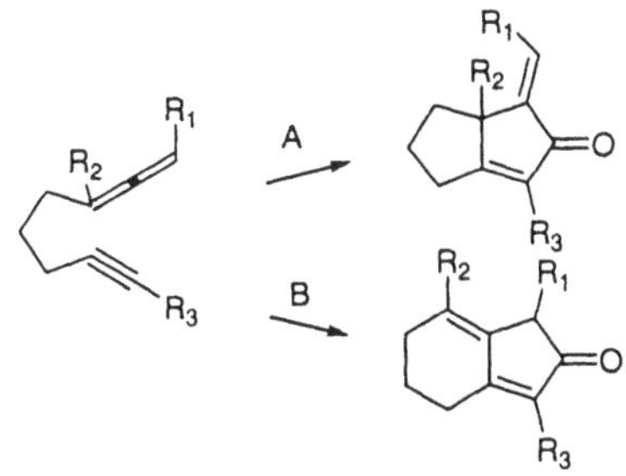

Scheme 1.

II. INTRAMOLECULAR ALLENIC [2+2+1] CYCLOADDITIONS

A. Monosubstituted Allenes

In order to test the feasibility of this intramolecular allenic [2+2+1] cycloaddition, a simple alkynyl allene was prepared and subjected to the standard P–K cycloaddition conditions. Alkynyl allene **1** was treated with dicobaltoctacarbonyl to give the corresponding dicobalthexacarbonyl alkyne complex. An unusual upfield shift of the allenyl protons was observed in the ^{1}H NMR spectrum. This upfield shift has tentatively been interpreted to result from the complexation of the allene to the dicobalthexacarbonyl complex. This type of intramolecular complexation between the cobalt–alkyne complex and a tethered group has been observed previously by Krafft.[16] Attempts to effect the cycloaddition using a variety of standard Pauson–Khand cycloaddition conditions (toluene, CO, 110 °C; Me_3NO, CH_2Cl_2; SiO_2, O_2, 50 °C; NMO, CH_2Cl_2, Ar) were unsuccessful.

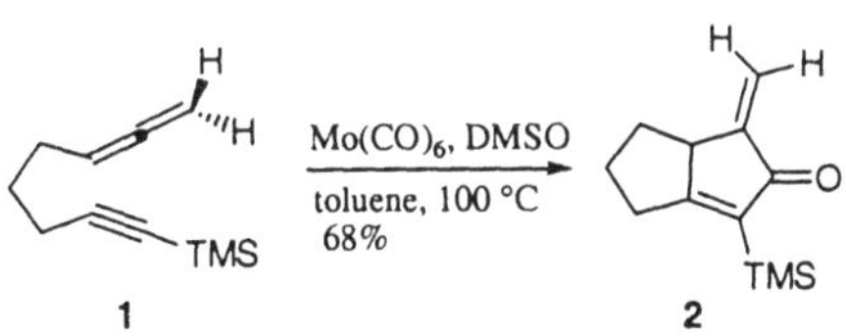

(3)

We quickly turned to alternative metals and reaction protocols to effect the cycloaddition. Conditions reported by Jeong and coworkers[17] proved to be quite successful in effecting the cycloaddition. Treatment of the alkynyl allene **1** with molybdenum hexacarbonyl [$Mo(CO)_6$] and dimethylsulfoxide at 100 °C gave the α-methylene cyclopentenone **2** in 68% yield (Eq. 3). The cycloaddition occurred exclusively with the internal π-bond of the allene, furnishing the α-methylene cyclopentenone as the only observed product.[18] Other monosubstituted allenes were also submitted to the cyclization conditions. Alkynyl allenes **3** and **5** were prepared possessing substituents on C-4 and C-5 of the tether (Eq. 4). When R = H (compound **3**), the cyclization occurred in 47% yield to give α-methylene cyclopentenone **4** as a 3:1 mixture of diastereomers, with the major isomer possessing a *trans* relationship between the proton at the ring fusion and the proton geminal to the hydroxyl moiety. Protection of the hydroxyl moiety as the MOM ether (compound **5**) did not produce significant changes in the yield (54%) but gave cycloadduct **6** as a 1:1 diastereomeric ratio of products. We next examined the effect of lengthening the tether. Alkynyl allene **7** cyclized to give the bicyclo[4.3.0]nonane **8** in 30% yield based upon recovered starting material (Eq. 5). The incorporation of a four-carbon tether in this P–K cycloaddition gives very low yields and is thought to be a result of the sensitivity of the cycloaddition to entropic effects and competing polymerization of the allene moiety. Similarly low conversions have also been observed in the standard alkenyl P–K cycloadditions. In all these examples using monosubstituted allenes, cycloaddition occurred exclusively with the internal π-bond of the allene (pathway A, Scheme 1).

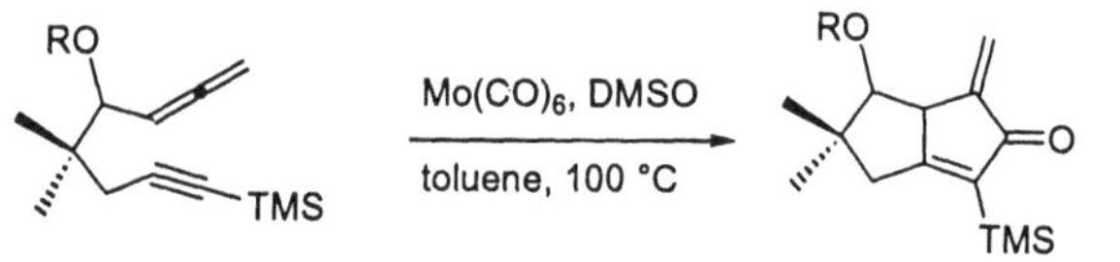

3 R = H
5 R = MOM

4 R = H, 47%
6 R = MOM, 54%

(4)

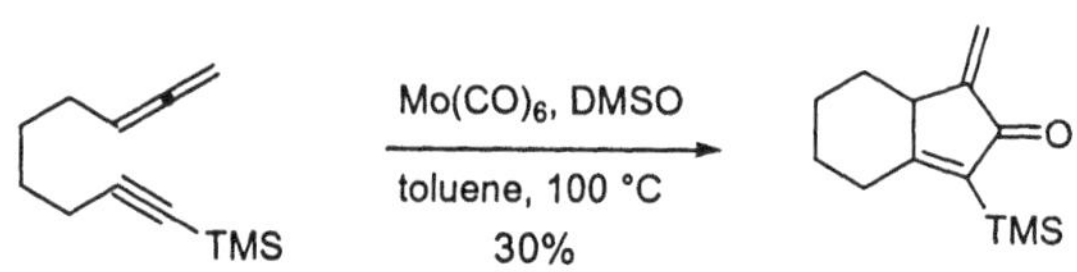

7 **8**

(5)

It is interesting to note that in subsequent studies carried out by Cazes et al., similar results with respect to the monosubstituted allenes were seen.[19] For example, treatment of the allene **9** to the standard P–K cycloaddition conditions [$Co_2(CO)_8$, NMO] afforded no defined cycloadducts (Eq. 6).

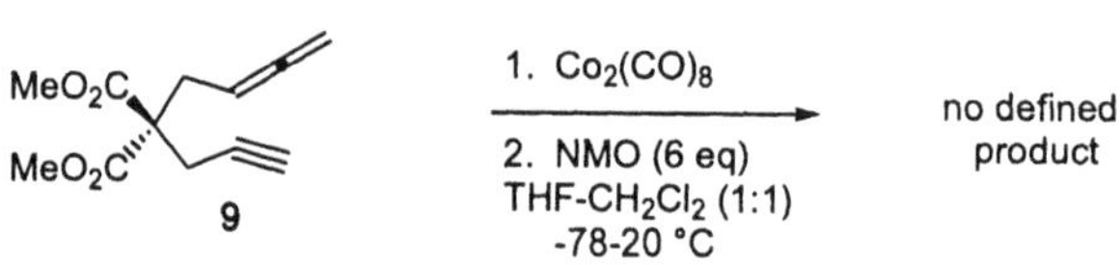

9

(6)

Narasaka has reported that when using iron carbonyl complexes to promote the cycloaddition reaction of a monosubstituted allene **10** only trace amounts of cycloadduct **11** were isolated (Eq. 7).[20]

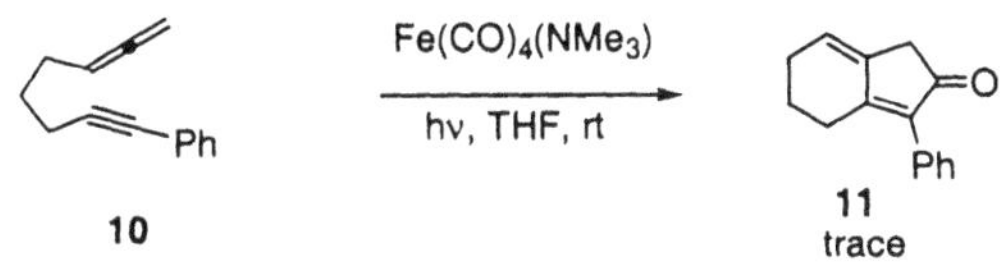

(7)

Buchwald has shown that a monosubstituted allene **12** undergoes cyclization using a titanocene-catalyzed cyclocarbonylation to afford the bicyclic dienone **13** in 86% yield (Eq. 8).[21] In both the Narasaka and Buchwald examples, cycloaddition occurs to afford the 4-alkylidene cyclopentenone (pathway B, Scheme 1).

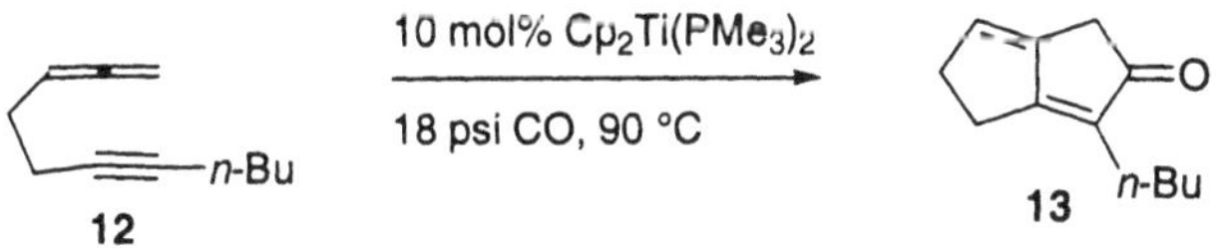

(8)

B. 1,3-Disubstituted Allenes

There are numerous natural products possessing substitution at the terminus of the exocyclic olefin moiety of the α-methylene cyclopentenone and the allenic [2+2+1] cycloaddition of a 1,3-disubstituted allene would provide a direct route to this structure. Consequently, our attention was turned to the cycloaddition of more substituted allenes (Eq. 9), where substitution patterns could be varied to determine the regio- and stereochemical outcome of the cycloaddition. When a 1,3-disubstituted alkynyl allene, 6,7-pentadecadien-1-yne (R = C_7H_{15}, Eq. 9) was treated with molybdenum hexacarbonyl/DMSO, cycloaddition occurred exclusively via pathway A to give the bicyclo[3.3.0]octane ring system as a mixture of *E*:*Z*-isomers (2:1 ratio) in a 75% yield (entry 1, Table 1).[22] It was subsequently shown that the *E*:*Z* ratio was not an artifact of the cyclization conditions.[22b] The phenyl-substituted allene (R = Ph, Eq. 9) cyclized with slightly higher stereoselectivity (5:1, *E*:*Z*) to form the α-methylene cyclopentenone (entry 2, Table 1). This modest increase in

Table 1. Intramolecular P–K Cycloadditions Using 1,3-Disubstituted Allenes

Entry	*R*	*E/Z Ratio*	*Yield (%)*
1	C_7H_{15}	2/1	75
2	Ph	5/1	70
3	$SiMe_3$	4/1	44
4	$SiMe_2Ph$	8/1	61
5	Si*t*-BuMe$_2$	6/1	49
6	*t*-Bu	1/1	56

stereoselectivity may be attributed to a stereoelectronic effect caused by hindered rotation around the allene phenyl bond.

Placement of a trimethylsilyl moiety on the terminus of the allene (R = $SiMe_3$, Eq. 9) gave only a 4:1, *E*:*Z* ratio of stereoisomers in a 44% yield (entry 3, Table 1). The silyl group offers some advantages since this moiety can potentially be removed after cyclization, so it was deemed important to look at alternative silyl substituents. The more bulky dimethylphenyl silyl moiety was placed on the terminus of the allene (R = $SiMe_2Ph$, Eq. 9). This cyclization afforded the α-methylene cyclopentenone in a 61% yield, with a higher *E*:*Z* selectivity (8:1) (entry 4, Table 1). Finally, placement of a *tert*-butyldimethylsilyl moiety at the terminus of the allene (R = Si(*t*-Bu)Me_2, Eq. 9) gives rise to the cycloadduct in a 49% yield and an 6:1, *E*:*Z* ratio (entry 5, Table 1). In an effort to see if we could get complete facial selectivity, the *tert*-butyl substituted alkynyl allene was synthesized. Surprisingly, this substrate cyclized to give a 1:1 ratio of *E*:*Z*-isomers (entry 6, Table 1). Additional work is required to determine if this result is a product of a total lack of facial selectivity or isomerization of the cyclization intermediate.

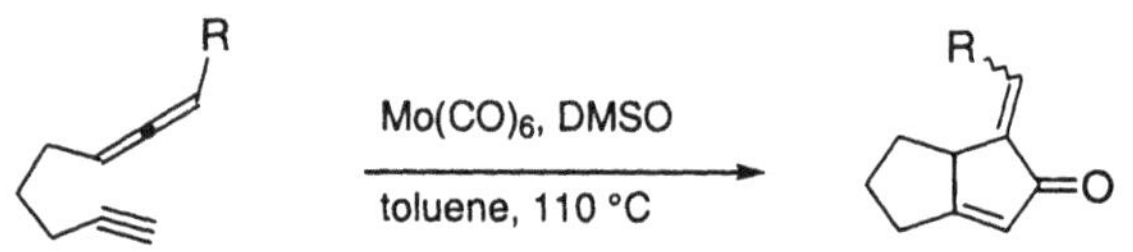

(9)

In the examples listed in Table 1, it is likely that the *Z*-isomer results from the addition of the metal–alkyne complex to the same face of the allene as the R group designated by structure **B**. The *E*-isomer results from the addition of the alkyne–metal complex to the face opposite of the R group designated by structure **A** (Scheme 2). In an effort to increase the facial selectivity, we examined other metals to promote the cycloaddition. We first turned to dicobaltoctacarbonyl [$Co_2(CO)_8$] as a metal promoter since metal carbonyl species has been used successfully in the selective formation of diastereomers.[23] Complexation of 6,7-(1-trimethylsilyl)pentadecadien-1-yne with $Co_2(CO)_8$ under standard conditions affords the desired metal complex **14** (Eq. 10). Subjection of this cobalt–alkyne complex to a variety of cycloaddition conditions indeed resulted in reactions that proceeded with much

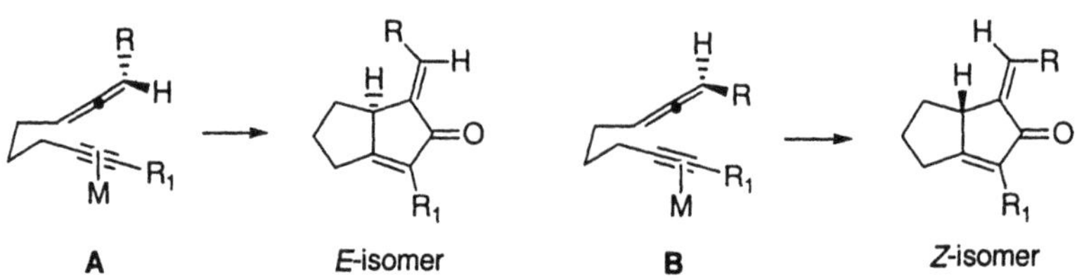

Scheme 2.

higher facial selectivity. However, the π-bond selectivity eroded, resulting in nearly 1:1 mixtures of bicyclo[4.3.0]nonane **15** and the bicyclo[3.3.0]octane ring systems **16** (Table 2). In the cobalt-mediated cyclizations, the highest yields were obtained when the alkynes were precomplexed with dicobaltoctacarbonyl and purified before being subjected to cyclization conditions (compare entries 1–4 to 5 and 6, Table 2) since the isolated cobalt complexes were obtained in nearly quantitative yields. The reaction proceeded much more rapidly when trimethylamine-*N*-oxide was used as a promoter instead of DMSO, but the observed facial selectivity was lower (compare entries 1, 2, and 3, Table 2). The use of cyclohexylamine as a promoter provided a low yield of the cycloadducts as a 1:1 mixture of the 5/5:6/5 ring systems (entry 4, Table 2). However, the facial selectivity was excellent in this case affording the predominantly *E*-isomer in a 19/1:*E*:*Z* ratio. In entries 5 and 6 (Table 2) the cobalt complexes were not isolated but submitted to cycloaddition conditions directly. When the reaction was carried out at higher temperatures (80 °C vs. 40 °C) the facial selectivity appeared to be higher (compare entries 5 and 6). However, since the yields and regioselectivities were consistently low, the use of dicobaltoctacarbonyl as a metal mediator was ultimately abandoned.

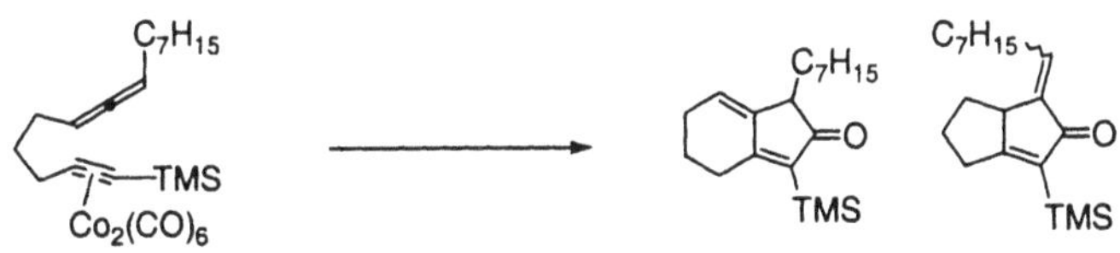

14 **15** **16** (10)

Similar results were reported by Cazes, where mixtures of bicyclic ring systems were obtained in low to moderate yields when the cyclization of the alkynyl allenes was conducted with dicobaltoctacarbonyl in the presence of *N*-methylmorpholine *N*-oxide (Eqs. 11 and 12).[19]

Table 2. Intramolecular [2+2+1] Cycloadditions Using 1,3-Disubstituted Allenes and Dicobaltoctacarbonyl as the Metal-Mediator

Entry	*Conditions*	*E/Z Ratio*	*5/5:6/5 Ratio*	*Yield (%)*
1	DMSO, air, CH_2Cl_2 40 °C, 22h	10:1	1.2:1	60
2	Me_3NO, Ar, CH_2Cl_2 25 °C, 1.3h	5:1	3:2	62
3	Me_3NO, Ar, CH_2Cl_2 –78–25 °C, 2h	6:1	2:1	52
4	cyclohexylamine CH_2ClCH_2Cl, 83 °C 15 min	19:1	1:1	39
5	1. $Co_2(CO)_8$, benzene 2. DMSO, air, benzene 80 °C, 20h	19:1	2:1	44
6	1. $Co_2(CO)_8$, benzene 2. DMSO, air, benzene 40 °C, 28h	3.5:1	1:1	33

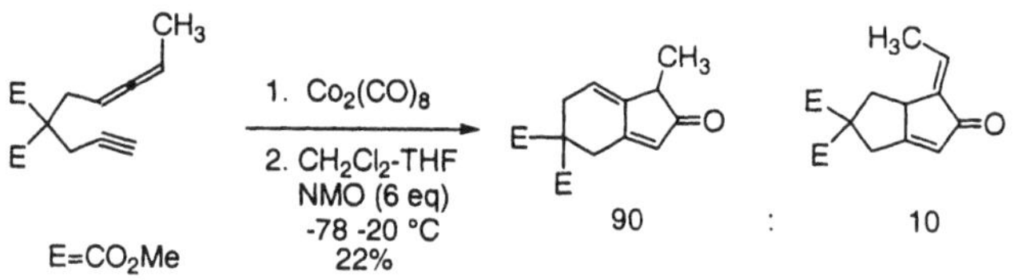

(11)

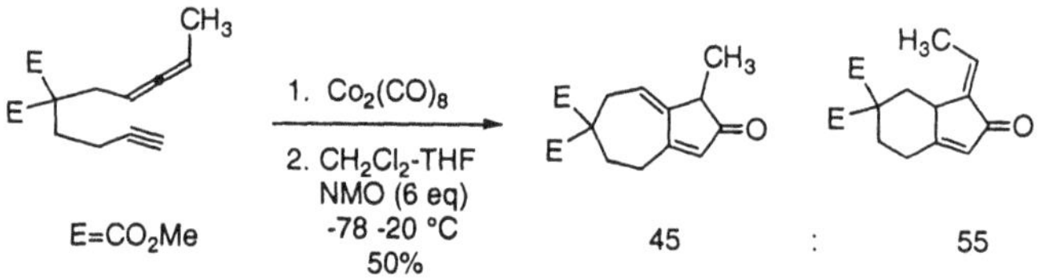

(12)

Finally, we turned to "Cp_2Zr" which is postulated to operate via a mechanism different from that of either molybdenum hexacarbonyl or dicobaltoctacarbonyl. When the cyclization of alleneyne **17** was effected in the presence of Cp_2Zr,[24] the bicyclo[3.3.0]octane **18** was isolated as the major product with only minor amounts of the bicyclo[4.3.0]nonane **19** in moderate yields (48%, 19 : 1) (Eq. 13). Under the reaction conditions, only the *E*-isomer of **18** was isolated but this stereoselectivity was shown to be an artifact of the workup conditions. The conditions used to work up the Negishi zirconium reaction (3M HCl) result in isomerization of the (*Z*)-bicyclo[3.3.0]octane to the *E*-isomer. This was shown by treatment of the (*Z*)-bicyclo[3.3.0]octane (independently synthesized) to these workup conditions which resulted in the complete isomerization of the *Z*- to *E*-isomer in less than 15 minutes. In an effort to determine the true stereochemical outcome of the zirconium-mediated cyclization, we isolated and hydrolyzed zirconacycle intermediate **20** and obtained *cis,trans*-**21** and the *trans,trans*-**22** dienes in a 10:1 ratio (Eq. 14).[22b] This corresponds to an *E*:*Z* ratio of 10:1 for cycloadduct **18**. The good facial and chemoselectivity make this an appealing synthetic method.

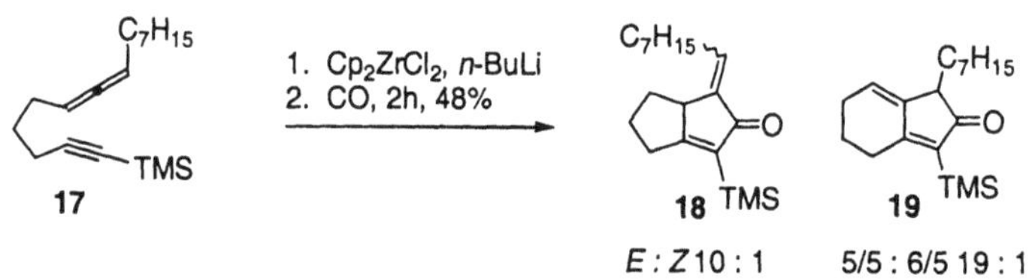

(13)

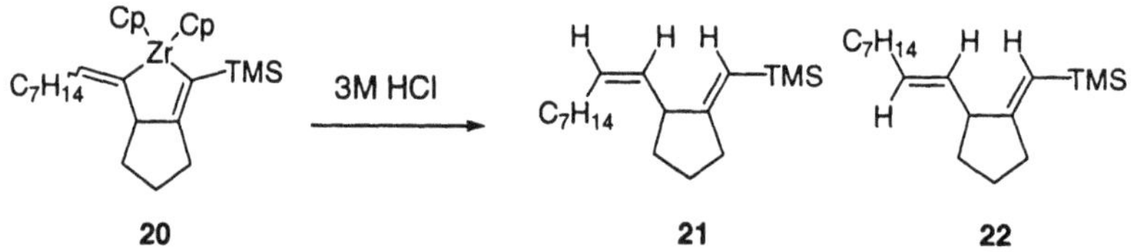

(14)

C. 3,3-Disubstituted Allenes

Cyclizations of 3,3-disubstituted allenes were investigated.[22] The inefficient processing of 2,2-disubstituted olefins appears to be a weakness in many standard P–K systems and the analogous 3,3-disubstituted allenes were predicted to be affected similarly. However, unlike the olefinic P–K reaction, in the allenic variant cycloaddition can occur with an alternate π-bond. Treatment of 3-*n*-butyl-1,2-octadien-7-yne under molybdenum cyclizations resulted in the formation of the bicyclo[4.3.0]nonane ring system as the only observed product in a 60% yield (entry 1, Table 3). This product arises from cyclization with the external π-bond of the allene (pathway B, Scheme 1). *This result demonstrates a dependence of the π-bond selectivity in the allenic* [2+2+1] *reaction upon the substrate structure.* To date, there are only a few examples of this type of substrate dependence in the [2+2+1] reaction.[25] In an attempt to sterically direct the cyclization reaction toward the internal double bond of the allene, a trimethylsilyl moiety was placed at the terminus of the allene of entry 1 in Table 3. Treatment of this trisubstituted alkynyl allene under molybdenum cyclization conditions gave the desilylated bicyclo[4.3.0]nonane ring system in 59% yield (product in entry 1). Based upon GC-MS data the desilylation occurred prior to cyclization. It is likely that a more robust silyl moiety will give cycloaddition without desilylation but those cycloaddition precursors have not yet been prepared. Cyclization of a more functionalized precursor occurred to give the [5.6.5] ring system in 42% yield (entry 2, Table 3).

The [2+2+1] cycloaddition of 3,3-disubstituted allenes has also been used to prepare some interesting carbocyclic skeletons possessing functionality that can easily be manipulated to other substrates (entries 3–6, Table 3). These alkynyl allenes were prepared using a method developed in our laboratories for the direct conversion of ketones to allenes (Scheme 3).[26] The conjugate addition of organometallic reagent **24** to 1-acetyl-1-cyclopentene or 1-acetyl-1-cyclohexene **23** was effected using catalytic manganese (II) chloride (30%) and copper (I) chloride (3%) followed by an in situ trap of the enolate with diethyl chlorophosphate to afford the desired enol phosphates **25** (Scheme 3). Elimination of the phosphate moiety to give the allene was then carried out by the addition of LDA at low temperature. The trimethylsilyl moiety could be removed from the alkyne terminus with tetra-*n*-butylammonium fluoride to afford compound **26**. Exposure of the alkynyl allenes to molybdenum hexacarbonyl/DMSO affords the respective cycloadducts where cycloaddition occurs predominantly with the less-substituted π-bond of the allene in moderate to good yields (entries 4–6, Table 3). The subjection of the alkynyl allene (entry 6, Table 3) to molybdenum [2+2+1] cycloaddition conditions gave only the linear [5.5.5] ring system in a 66% yield with no evidence of the angular [5.4.5] ring system. The rapid assembly of these ring systems demonstrates the applicability of these two methods and provides skeletons visible in naturally occurring compounds.

Table 3. Allenic [2+2+1] Cycloadditions Using 3,3-Disubstituted Allenes

Entry	*Substrate*	*Conditions*	*Product*	*Ratio A/B*	*Yield (%)*
1	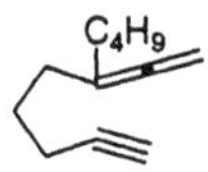	$Mo(CO)_6$, DMSO toluene, Ar, 100 °C 24h	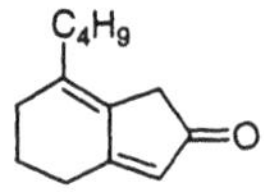		60
2	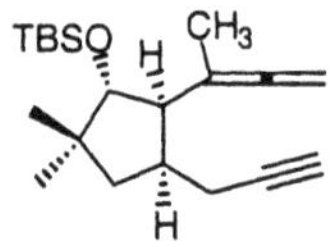	$Mo(CO)_6$, DMSO toluene, Ar, 100 °C 19h	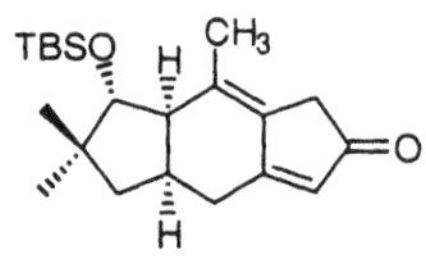		42
	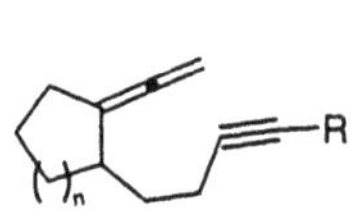		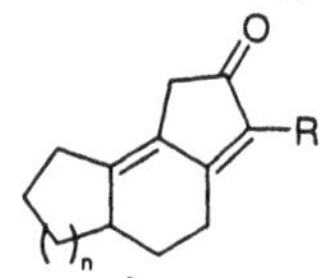 A 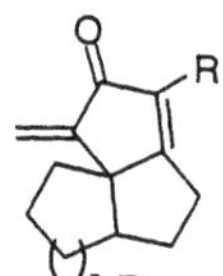 B		
3	n = 1, R = H	15h		7/1	73
4	n = 1, R = TMS	12h		1/0	50
5	n = 2, R = H	15h		4/1	62
6	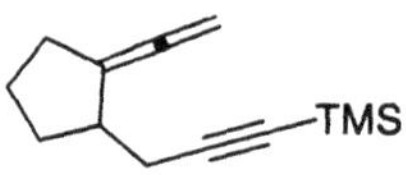	5h	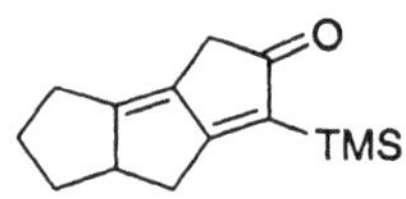		66
7	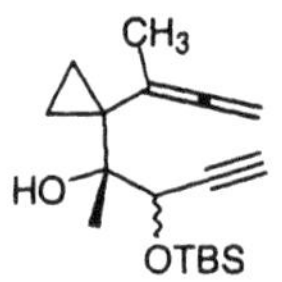	30 min	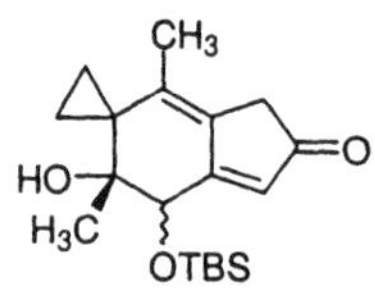		69
8	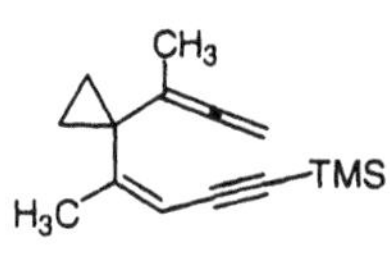	10 min	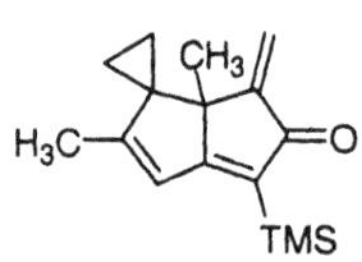		66
9	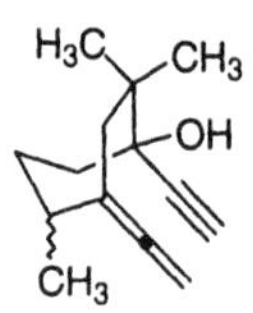	3h	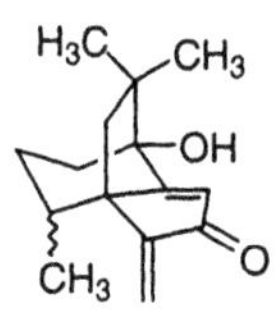		60

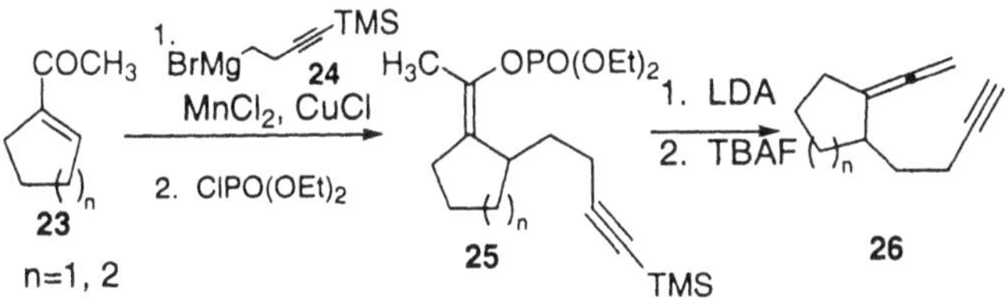

Scheme 3.

Entries 7, 8, and 9 of Table 3 involve intermediates in the synthesis of natural products and will be discussed in more detail in Section V of this review. As predicted in entry 7 the cycloaddition of the alkynyl allene occurs exclusively with the less-substituted double bond of the allene. Surprisingly, but not totally unexpectedly, the cycloaddition of alkynyl allenes in entries 8 and 9 occurred with the more-substituted π-bond of the allene. It was reasoned that, due to a more favorable orbital overlap in these conformationally constrained substrates, cycloaddition could only occur with the more-substituted olefin of the allene. Cazes has also shown that the intramolecular allenic P–K reaction of 3,3-disubstituted allenes occurs predominantly with the less-substituted olefin of the allene moiety (Eqs. 15 and 16).[19]

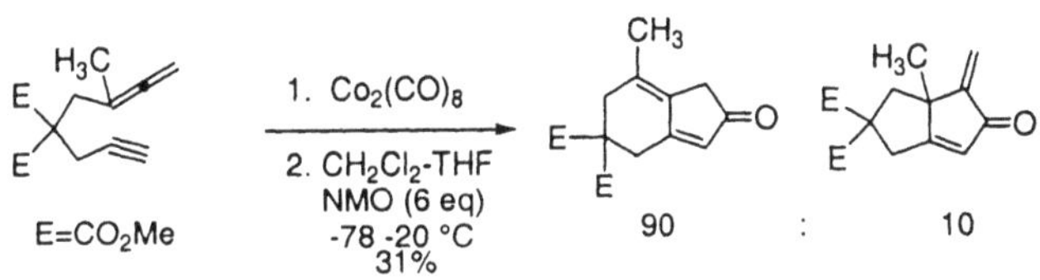

(15)

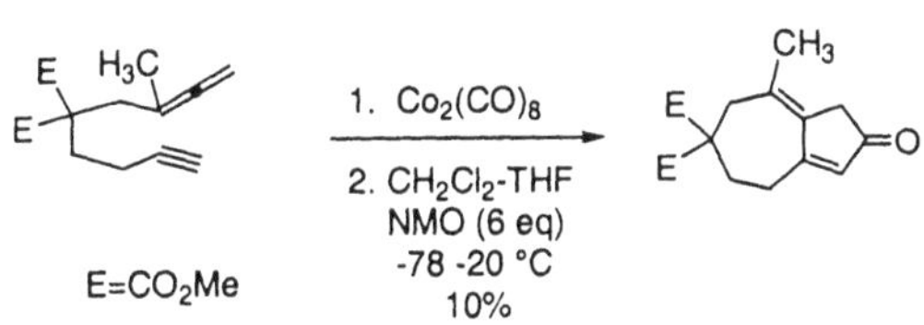

(16)

Narasaka has demonstrated that the iron-mediated reaction of 3,3-disubstituted allenes afford cycloadducts, whereby the reaction occurs with the less-substituted double bond of the allene.[20] In some cases a mixture of products was obtained where complexed cyclopentadienes were formed. Various 1-(ω-alkynyl) propadienylsulfides **27** (R_1 = SMe, Eq. 17, entries 1–4, Table 4) were converted into bicyclo [*n*.3.0] dienones ($n = 0 - 2$) **28** and **29** using the reaction conditions described in Eq. 17. In an effort to determine the effect of the alkylthio group in this iron carbonyl-me-

Table 4. Intramolecular Allenic [2+2+1] Cycloadditions Using 3,3-Disubstituted Allenes

Entry	R_1	R_2	*n*	**28/29**	*Yield (%)*
1	SMe	H	0	1/0	60
2	SMe	Ph	1	1.6/1	49
3	SMe	H	2	1/0	15
4	SMe	H	1	1/0	50
5	Si*t*-BuMe$_2$	Ph	1	14/1	30
6	CH$_3$	Ph	1	1/0	22

diated cycloaddition, other alkynyl allenes possessing a *tert*-butyldimethylsilyl (R_1 = Si(*t*-Bu)Me$_2$, Eq. 17, entry 5, Table 4) and methyl groups (R_1 = Me, Eq. 17, entry 6, Table 4) on the allene were exposed to the cycloaddition conditions. The yields of the non-sulfur containing cycloadducts (entries 5 and 6, Table 4) are lower than that of the thioether (entry 2, Table 4), but the product ratio is much higher. The results indicate that the electron-releasing alkylthio group is beneficial to the cycloaddition.

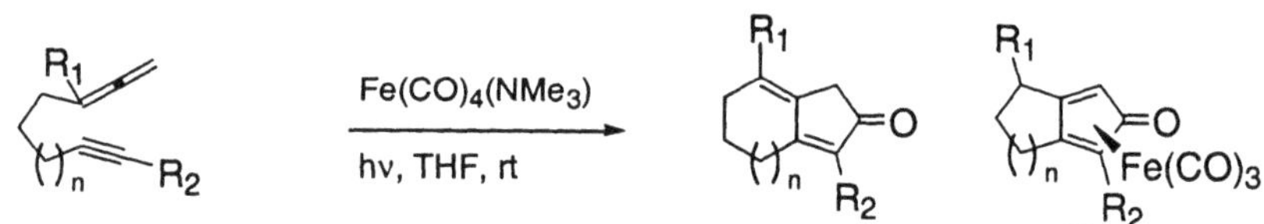

27 28 **29**

(17)

Table 5. Allenic [2+2+1] Cycloadditions Using Trisubstituted Allenes

Entry	*Allenyne*	*Cyclopentanone*	*Conditions*	*Yield (%)*
1	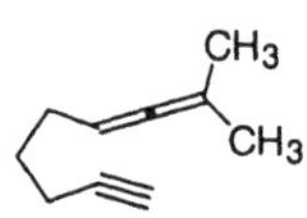	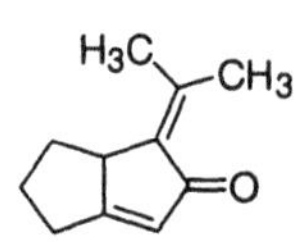	Mo(CO)$_6$, DMSO tol, Ar, 100 °C, 10h	59
2	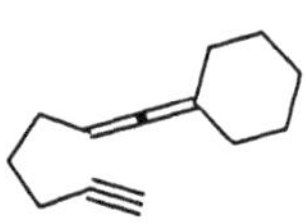	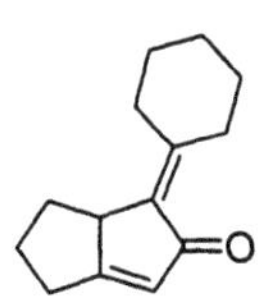	Mo(CO)$_6$, DMSO tol, Ar, 100 °C, 10h	58

D. 1,1,3-Trisubstituted Allenes

We have shown that trisubstituted allenes undergo cyclization exclusively with the less-substituted π-bond of the allene (entries 1 and 2, Table 5).[22]

Livinghouse has demonstrated that cycloadditions of trisubstituted allenes can be done catalytically using dicobaltoctacarbonyl and a carbon monoxide atmosphere (Eq. 18).[27] Cyclization occurs in excellent yields with the less-substituted olefin of the allene.

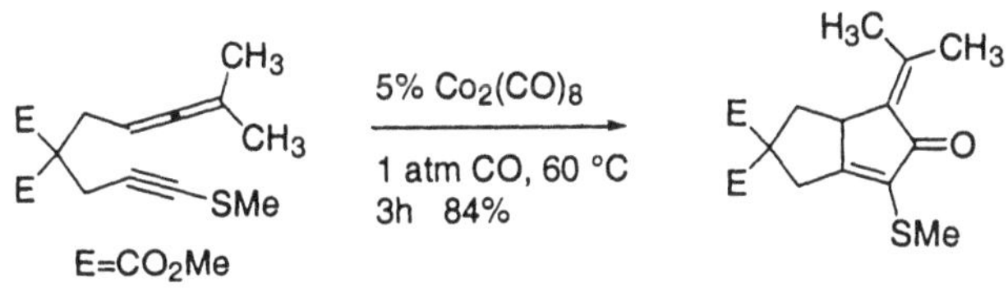

(18)

On the other hand, Cazes has observed that the cobalt-mediated cycloaddition using NMO as a promoter results in cycloaddition with the more-substituted double bond to afford predominantly the 4-alkylidenecyclopentenone (Eq. 19).[19]

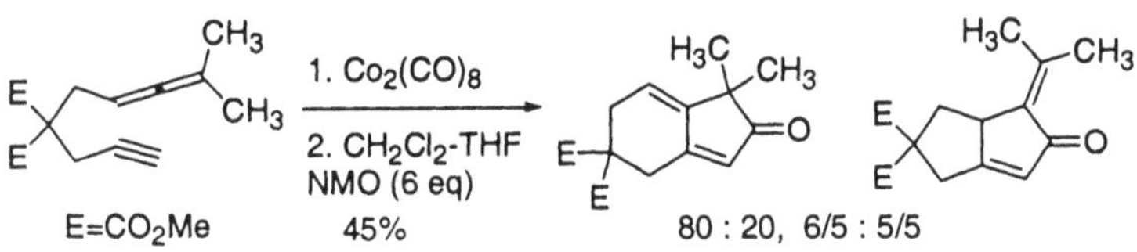

(19)

III. INTERMOLECULAR ALLENIC [2+2+1] CYCLOADDITIONS

As previously mentioned (Section I), Aumann was the first to establish the feasibility of the intermolecular [2+2+1] cycloaddition.[15] Treatment of a variety of alkynes and allenes with carbonyl iron complexes affords 4-alkylidene cyclopentenones **30**, **31**, **32**, and **33** in a regioselective fashion in yields ranging from 15–56% (Eq. 20 and Table 6).

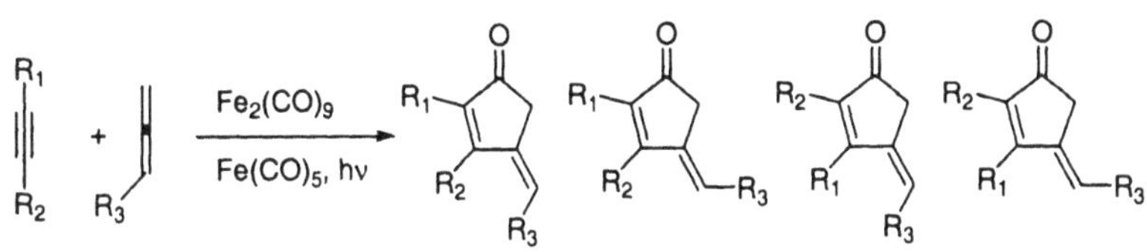

30 **31** **32** **33**

(20)

A substantial amount of effort has been devoted to the intermolecular allenic [2+2+1] cycloaddition by Cazes and coworkers. They were able to effect an

Table 6. Intermolecular [2+2+1] Cycloadditions

Entry	R_1	R_2	R_3	*Products*	*Yield (%)*
1	H	H	Ph	**30, 31**	56
2	Ph	H	Ph	**31**	18
3	Ph	Ph	Ph	**30, 31**[a]	47
4	Ph	H	H	**30, 32**	15
5	Ph	Ph	H	**30**[a]	25
6	Ph	H	CH_2Ph	**31**	25
7	Ph	Ph	CH_2Ph	**31**[a]	25

Note: [a]Other products were obtained.

intermolecular allenic P–K cycloaddition using dicobaltoctacarbonyl and NMO.[28] In this report, the cobalt complex of alkynes were prepared and reacted with a variety of allenes (Eq. 21). In all cases, cycloadducts were isolated corresponding to the reaction taking place with the less substituted π-system of the allene moiety producing the 4-alkylidene cyclopentenones in 33–69% yields (Table 7).

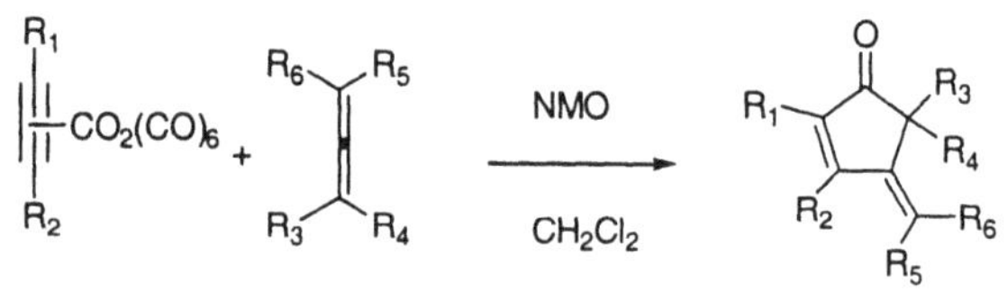

(21)

Entries 1–5 (Table 7) involve the use of a symmetrical alkyne, affording only the (*E*)-4-alkylidene cyclopentenones in the cases where stereochemistry is an issue. Entries 6 and 7 (Table 7) involve the use of an unsymmetrical alkyne and, as is the case in most intermolecular P–K cycloadditions, the reactions are regioselective with respect to the alkyne affording the product with the larger substituent α to the carbonyl moiety. Also when asymmetrical alkynes are used the products are obtained as a mixture of [*E*,*Z*]-stereoisomers. Finally, a simple allene, 1,2-propadiene, is cyclized with 2-butyne to afford a single 4-methylene cyclopentenone in a 59% yield (entry 8, Table 7).

In a later report, Cazes demonstrated an intermolecular allenic P–K cycloaddition with allenes possessing a variety of heteroatom substituents (Eq. 22).[29] In these studies the cobalt complexes were prepared and subjected to NMO and the allene to afford the corresponding cycloadducts in 30–75% yields (Table 8). As is evidenced by Table 8, an interesting array of substituted allenes prove to be suitable in the preparation of these highly unsaturated cyclopentanes. It would appear that in the case of allenes possessing electron-rich substituents, the reaction occurs predominantly with the less sterically congested double bond of the allene (entries 1–4, Table 8). But for the allenes possessing electron-withdrawing substituents, the

Table 7. Intermolecular Cobalt-Mediated [2+2+1] Cycloadditions

Entry	R_1	R_2	R_3	R_4	R_5	R_6	*Product*	*E/Z*	*Yield (%)*
1	C_3H_7	C_3H_7	H	H	H	C_6H_{13}	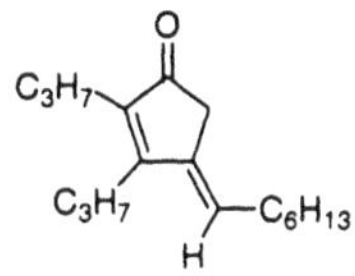	100/0	69
2	C_3H_7	C_3H_7	H	H	H	Ph	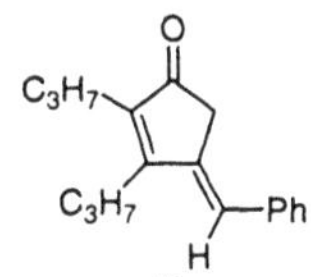	100/0	33
3	C_3H_7	C_3H_7	H	H	H	TMS	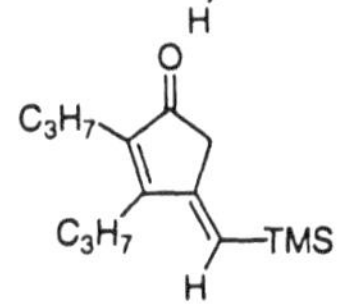	100/0	40
4	C_3H_7	C_3H_7	H	H	CH_3	CH_3	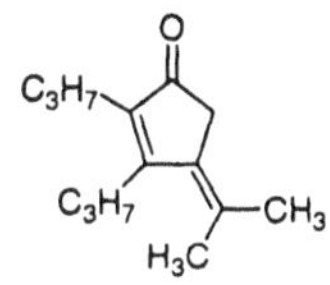		61
5	C_3H_7	C_3H_7	C_5H_{11}	H	C_5H_{11}	H	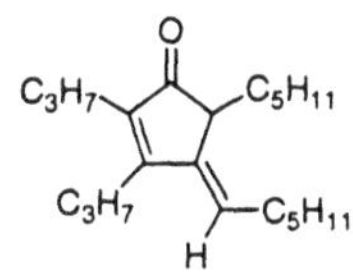	100/0	66
6	C_5H_{11}	H	H	H	H	C_6H_{13}	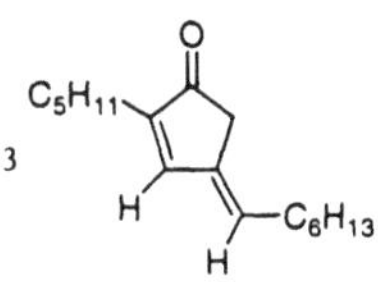	79/21	54
7	Ph	H	H	H	H	C_6H_{13}	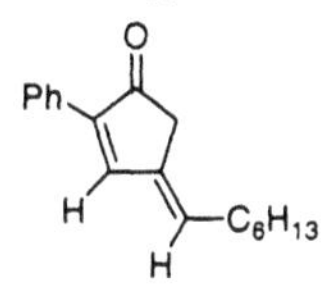	75/25	52
8	CH_3	CH_3	H	H	H	H	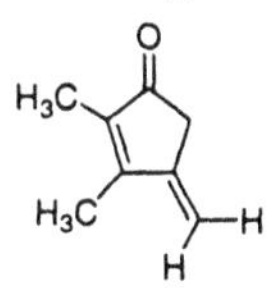		59

Table 8. Intermolecular Cobalt-Mediated Allenic [2+2+1] Cycloadditions with Functionalized Allenes

Entry	R_1	R_2	R_3	R_4	R_5	R_6	*Product*	*Yield (%)*
1	CH_3	CH_3	H	H	H	O-tBu	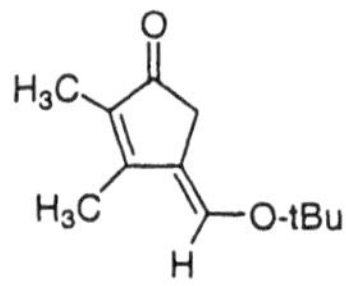	30
2	CH_3	CH_3	H	H	H	$SiPhMe_2$	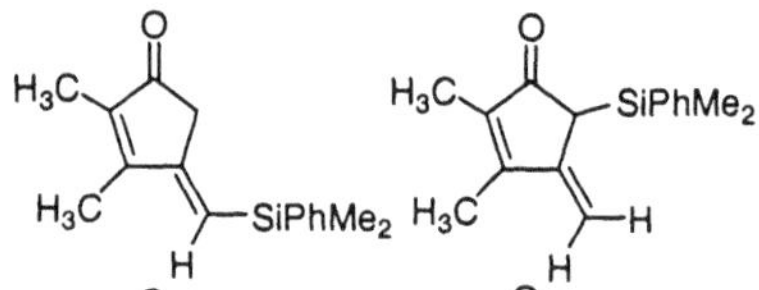	74 95/5
3	CH_3	CH_3	H	CH_3	H	$SiPhMe_2$	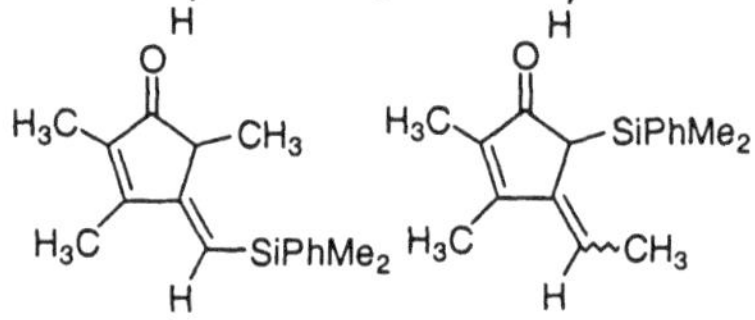	75 70/30
4	CH_3	CH_3	H	H	H	$SnBu_3$	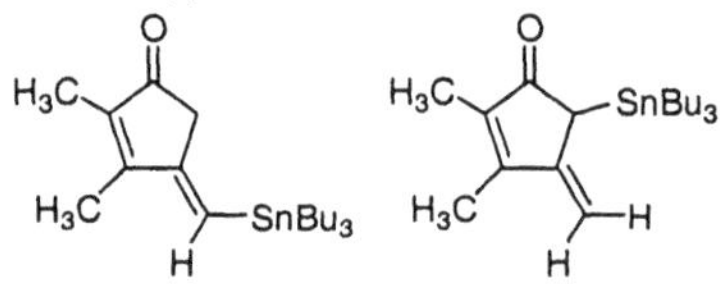	26 72/28
5	C_3H_7	C_3H_7	H	H	H	CO_2Et	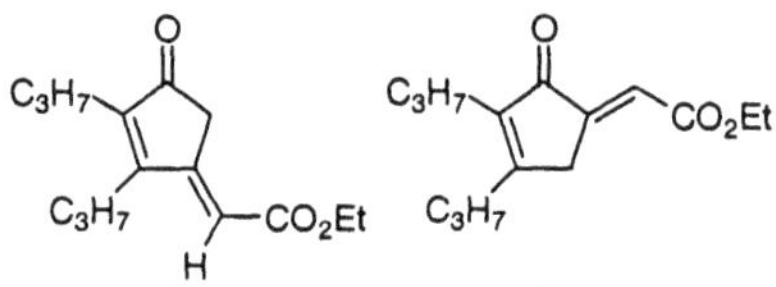	47 70/30
6	C_3H_7	C_3H_7	H	H	H	SO_2Ph	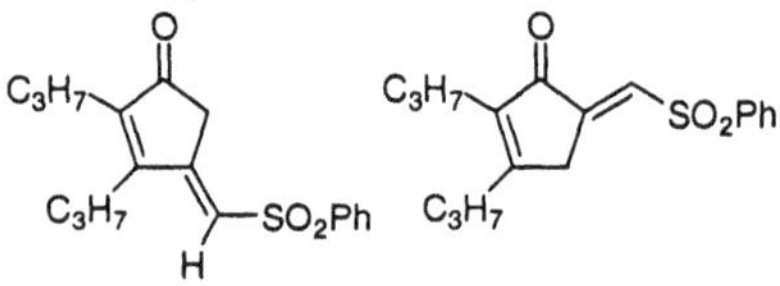	35 70/30
7	CH_3	CH_3	H	C_4H_9	H	CO_2Et	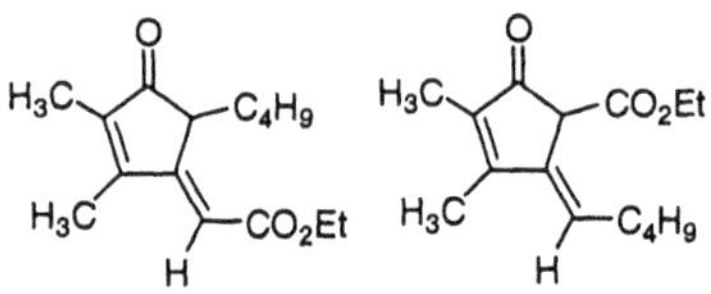	37 70/30
8	C_3H_7	C_3H_7	CH_3	CH_3	H	CO_2Et	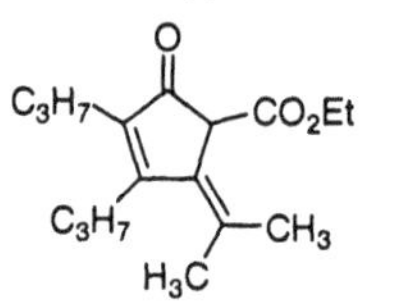	30

reaction occurs predominantly with the more electron-rich olefin (entries 5 and 6, Table 8). In an effort to determine whether sterics or electronics was a more controlling force, two more experiments were performed. Ethyl octa-2,3-dienoate gave a 70/30 mixture of regioisomeric 4-alkylidene cyclopentenones (entry 7, Table 8). Ethyl (3-methyl)-penta-2,3-dienoate gave a single cyclopentenone (entry 8, Table 8), where the cyclization occurred with the more electron-poor double bond. Based upon these last two examples, it was reasoned that sterics is a more influential force than electronics in the intermolecular allenic [2+2+1] cycloaddition.

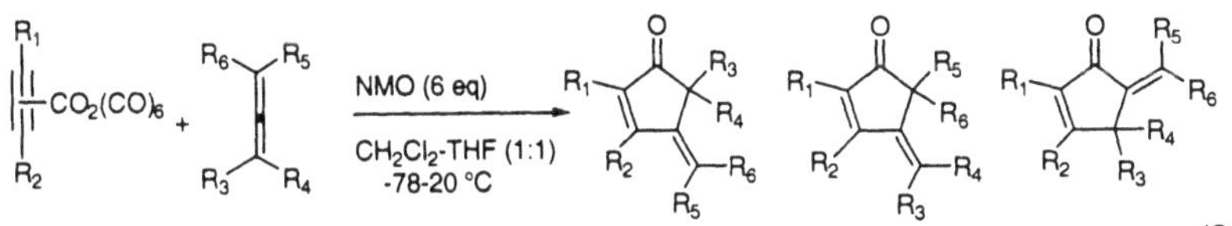

(22)

Cazes offers an mechanistic rationale for these observed regioselectivities (Scheme 4). For electron-releasing groups (Y = alkyl, OR, SiR_3, SnR_3), the allene is inserted into one of the C–Co bonds at the *sp*-hybridized carbon of the allene, leading to a π-allyl organocobalt intermediate which can undergo a CO insertion via either pathway **a** or **b** to afford the respective cylclopentenones. CO insertion in the less sterically hindered side of the π-allyl system is favored (path **a**), leading to the observed product ratios. Alternatively, if the allene possesses an electron-withdrawing group (Y = CO_2R, SO_2Ph), the oxidative addition of the allene to the C–Co bond occurs with the more electron-rich double bond, leading to an α-methylene cyclopentenone via path **c** (Scheme 4).

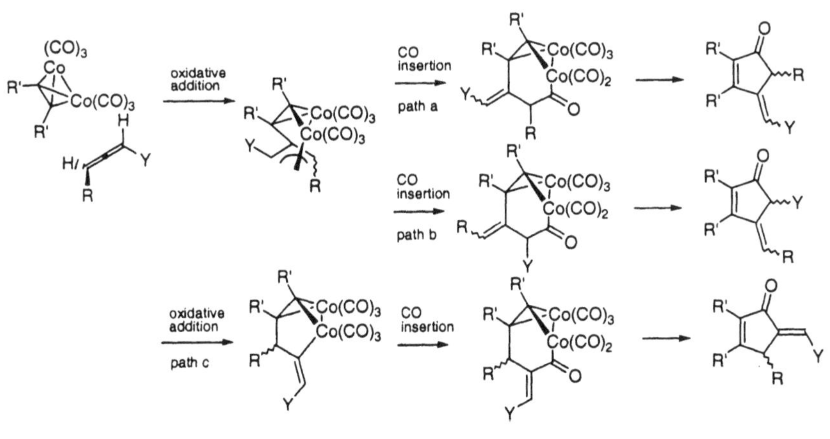

Scheme 4.

Table 9. Intermolecular Iron-Mediated Allenic [2+2+1] Cycloadditions

Entry	R_1	R_2	R_3	R_4	R_5	R_6	*Product*	*Yield (%)*
1	Ph	H	H	H	$(CH_2)_3Ph$	TMS	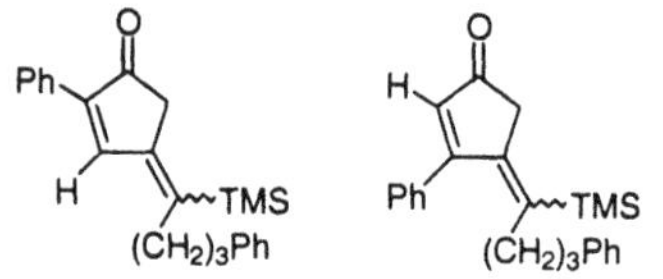	73 3.9/1
2	Ph	H	H	H	H	Ph	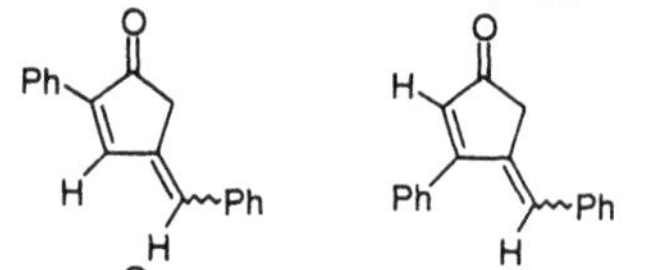	66 4.5/1
3	Ph	Ph	H	H	$(CH_2)_3Ph$	TMS	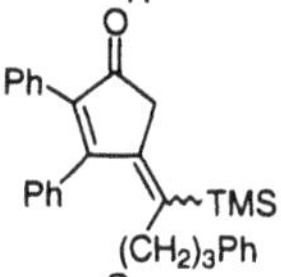	60
4	Bu	H	H	H	$(CH_2)_3Ph$	TMS	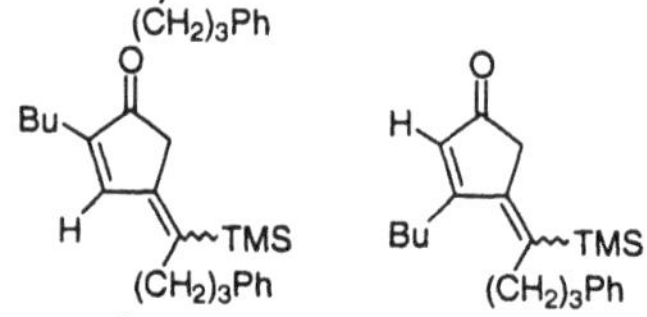	61 9/1
5	*n*-Pr	*n*-Pr	H	H	$(CH_2)_3Ph$	TMS	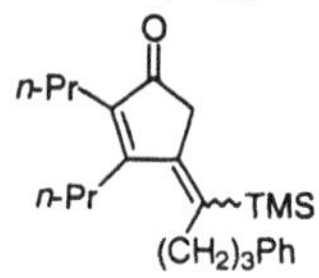	57

In 1995, Narasaka reported a successful intermolecular carbonylative coupling reaction between allenes and alkynes (Eq. 23).[20] In this report, a variety of acetylenes and propadienyl silanes were reacted to afford 4-alkylidene cyclopentenones (Table 9). This reaction was mediated by iron(0) tetracarbonyltrimethylamine with photoirradiation. In all cases where an asymmetrical alkyne was used (entries 1, 2 and 4, Table 9) the predominant product possessed the larger substituent of the precursor alkyne in the position α to the carbonyl but mixtures of regioisomers were obtained. The regiochemistry of the cycloaddition with respect to the allene occurs with the less-substituted double bond of the allene in all examples.

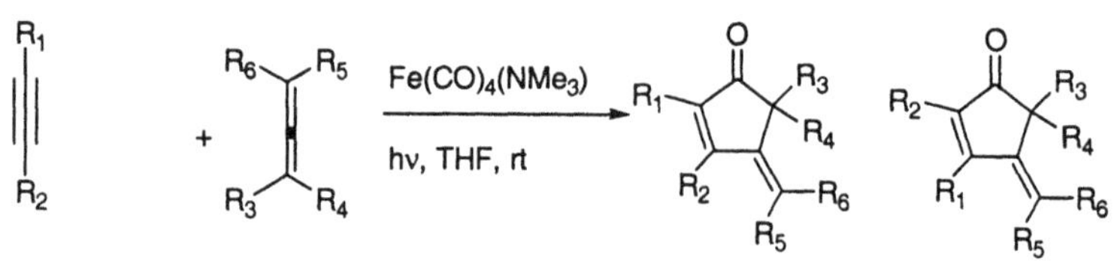

(23)

IV. TRANSFER OF CHIRALITY IN THE ALLENIC [2+2+1] CYCLOADDITION

In conjunction with the development of the intramolecular allenic [2+2+1] cycloaddition, we reasoned that this reaction would be ideally suited to produce chiral α-methylene cyclopentenones via a stereoselective cycloaddition onto one face of a chiral allene. In particular, for 1;3-disubstituted alleneynes, good facial and regioselectivities were obtained when the cycloaddition reaction was mediated by "Cp_2Zr" (Eq. 13). Therefore, cycloadditions incorporating a chiral allene should undergo a transfer of chirality to the resulting cycloadduct. This transfer of chirality is expected to arise from a bias for the addition of the metal acetylene complex to the less sterically encumbered face of the chiral allene (Scheme 5).

This type of chirality transfer has been previously demonstrated in other cycloaditions reactions. For instance, inter-[30] and intramolecular[10c] Diels–Alder reactions, [2+2] cycloadditions[9c] and ene reactions[12] have all afforded products containing stereogenic centers derived from the axial chirality element of an allene precursor. Most relevant to the chemistry being carried out in our group is the cycloaddition of an alleneyne reported by Sato (Eq. 24).[31] Treatment of the chiral alleneyne **34** (≤86% ee) with (η^2-propane)Ti(O-i-Pr)$_2$ and carbon monoxide gave optically active bicyclic cyclopentenone **35** in 68% yield and 86% ee after hydrolysis of the oxa(titana)cyclopropane. This single carbonylative cyclization example demonstrates the feasibility of using chiral allenes to prepare chiral α-methylene cyclopentenones.

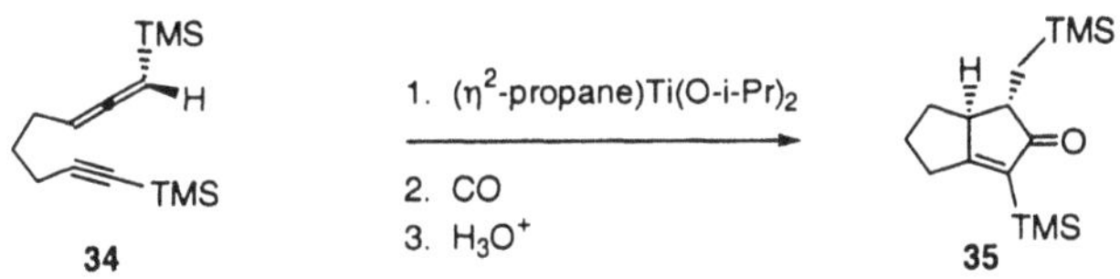

(24)

The objective of this study centered on the choice of a substrate to demonstrate an efficient transfer of chirality. In order to test this objective a variety of chiral allenes were prepared and subjected to cyclization conditions (Scheme 6). Using a method reported by Vermeer,[32] 3-(methylsulfonyloxy)-1-undecyne (**36**) was treated with the organocopper reagent prepared from 5-(trimethylsilyl)-4-pentynylmagnesium chloride, copper bromide and lithium bromide to afford (R)-1-trimethylsilyl-

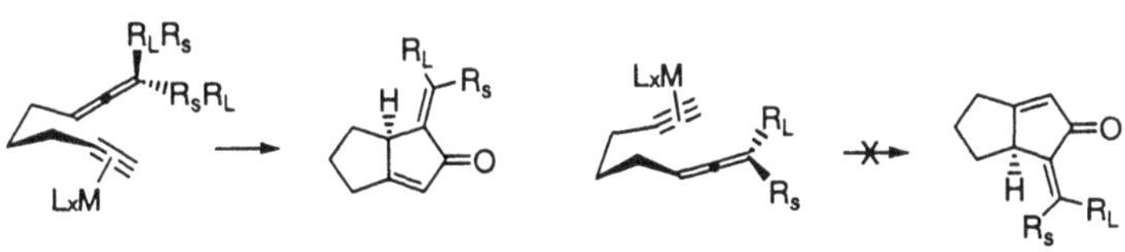

Scheme 5.

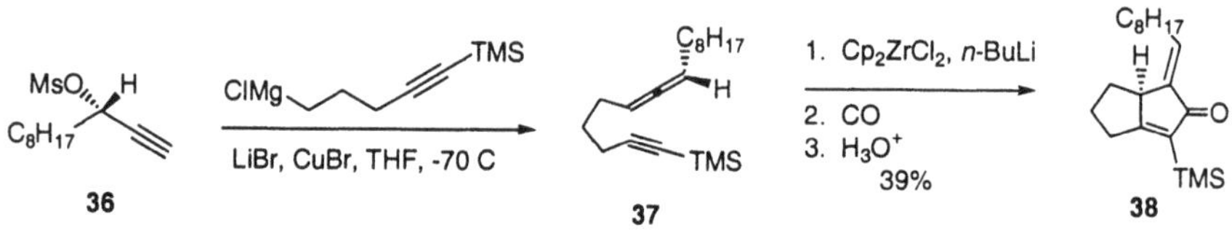

Scheme 6.

6,7-pentadecadien-1-yne (**37**) in 90% yield and 95% ee. Exposure of the chiral allene to the Negishi cycloaddition protocol resulted in α-methylene cyclopentenone **38** with a transfer of chirality (85% ee as determined by chiral shift reagent) in accordance with the facial selectivity that had been previously observed (Eq. 13). While this transfer of chirality is synthetically appealing, the chemical yield is low (39%) and all attempts to optimize proved fruitless.

Despite the higher degree of facial selectivity obtained with the zirconium protocol, the yields from this protocol were low and these conditions were not particularly user-friendly. For instance, rigorous measures must be taken to ensure the absence of oxygen or moisture in the reaction and an external source of carbon monoxide, sometimes requiring the use of high pressures. Since molybdenum hexacarbonyl has been a very effective metal in facilitating the allenic P–K cycloaddition, our attention turned to improving the facial selectivity of the molybdenum-mediated cycloaddition. In addition, chiral allenes possessing silyl substituents were investigated since the silyl group can potentially function as a traceless controlling group and the diversification of the groups attached to the silicon moiety can easily be accomplished.

In the initial test case, the requisite chiral allene was prepared from the (*S*)-propargyl alcohol **39** using the method reported by Myers[33] followed by removal of the TMS moiety from the terminus of the alkyne to afford **40** (Scheme 7). The

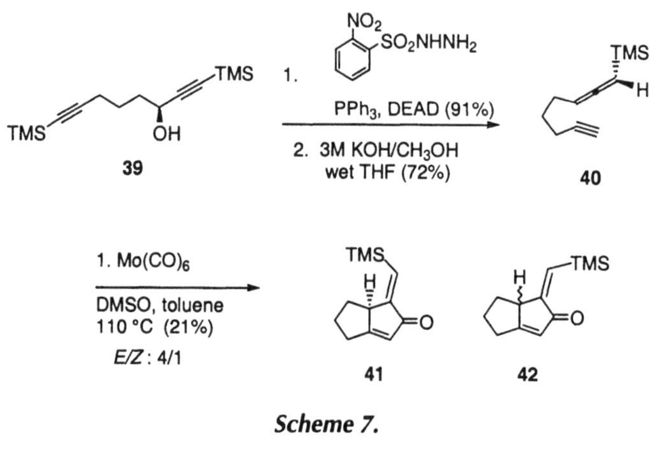

Scheme 7.

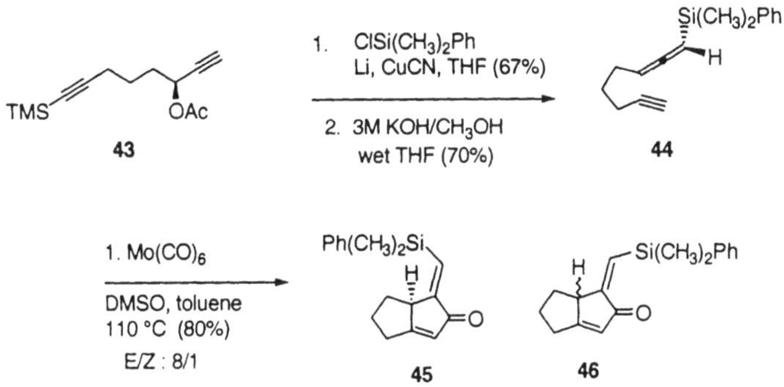

Scheme 8.

cyclization precursor (*R*)-1,2-(1-trimethylsilyl)-octadien-7-yne (**40**) (obtained in 99% ee as determined by GC) was subjected to the standard molybdenum conditions to give compounds **41** and **42** as a mixture of (*E*,Z)-isomers (4/1) in low yield (44%). The (*E*,Z)-isomers were separated using column chromatography and it was subsequently determined that the *E*-α-methylene cyclopentenone **41** was obtained in 99% ee but the *Z*-α-methylene cyclopentenone **42** in only 79% ee. Given the low yield and the lack of facial selectivity for the trimethylsilyl-substituted allene, it was reasoned that a more robust and bulkier silyl moiety would be more appropriate.

An alleneyne possessing a dimethylphenylsilyl moiety on the terminus of the allene was prepared via an S_N2' addition of the organocopper silyl compound to the chiral propargylic acetate **43** followed by removal of the TMS moiety from the terminus of the alkyne to afford chiral allene **44** (Scheme 8). Subjection of the alleneyne **44** to the standard molybdenum conditions gave a good yield (80%) of α-methylene cyclopentenones **45** and **46** in an *E*:*Z* ratio of 8/1. The (*E*)-isomer was obtained in greater than 95% ee as determined by chiral shift reagent ($Eu(hfc)_3$), resulting from a complete chirality transfer and the (Z)-isomer was obtained in only 63% ee.

At the present time we are investigating a second objective; alternative conditions for effecting a complete transfer of chirality in this intramolecular [2+2+1] cycloaddition.

V. APPLICATION OF THE ALLENIC [2+2+1] CYCLOADDITION TO THE SYNTHESIS OF BIOLOGICALLY RELEVANT MOLECULES

A. Hydroxymethylacylfulvene (HMAF)

The allenic [2+2+1] cycloaddition provides access to substructures that are not easily accessible using other synthetic methods. Of particular interest to our group is a substructure embedded within the illudin family of natural products, which should be readily accessible via the allenic [2+2+1] cycloaddition. The naturally occurring sesquiterpenes illudin M (**47**) and S (**48**), have been shown to possess potent antitumor activity, but when tested *in vivo* were found to have a poor therapeutic index.[35] Subsequently, illudin analogues have been prepared that show greatly improved efficacy when compared to the parent compounds.[36] One analogue in particular, hydroxymethylacylfulvene (HMAF) (**49**; Scheme 9), has generated a tremendous amount of excitement since it has proven effective against breast, lung, and colon tumors in animal models while exhibiting dramatically reduced toxicity.[37]

Based upon our preliminary investigations, the substrate regioselectivity observed in the [2+2+1] cycloadditions of 3,3-disubstituted allenes was ideally suited for the synthesis of the illudane carbon skeleton. The general features of the synthesis of the illudin analogue, HMAF, are outlined retrosynthetically below (Scheme 10). The conversion of **50** to HMAF (**49**) has been previously reported.[38] It was reasoned that fulvene **50** could be obtained from the 4-alkylidene cyclopentenone **51** via the addition of a methyl anion to the ketone followed by dehydration of the newly formed tertiary alcohol. The 4-alkylidene cyclopentenone **51** could in turn be obtained from a regioselective allenic P–K type cycloaddition of the very functionalized alkynyl allene **52**.

Our first approach to the synthesis of HMAF involved the preparation of alleneyne **53** (Scheme 11). Installation of the hydroxyl moieties via an asymmetric dihydroxylation would be performed subsequent to the cycloaddition due to problems encountered in trying to effect this oxidation prior to cyclization. Examination of models indicated that the overlap of the π-bond of the internal double bond of the allene with the alkyne is more favorable than that of the external double bond of the allene. However, all preliminary investigations suggested that the cycloaddition would occur with the less-substituted double bond to provide the desired alkylidene cyclopentenone **54**. When compound **53** was subjected to the standard allenic [2+2+1] reaction conditions, none of the desired [3.6.5] ring system **54** was observed. Instead cyclization proceeded in less than 30 min to give the [3.5.5] tricyclic compound **55** in 66% yield. While the desired cycloadduct was not obtained from this experiment, this result was not entirely unexpected and provided

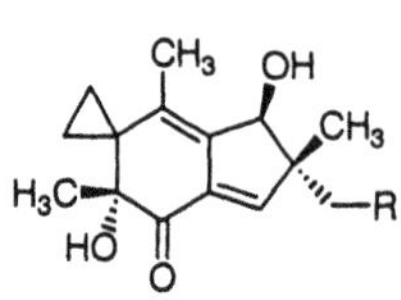

(47) illudin M R = H
(48) illudin S R = OH

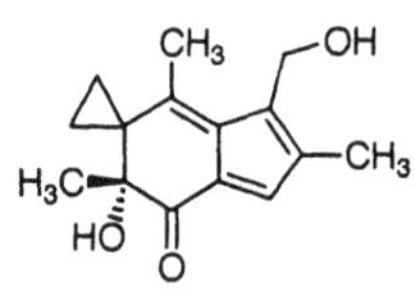

(49) hydroxymethylacylfulvene (HMAF)

Scheme 9.

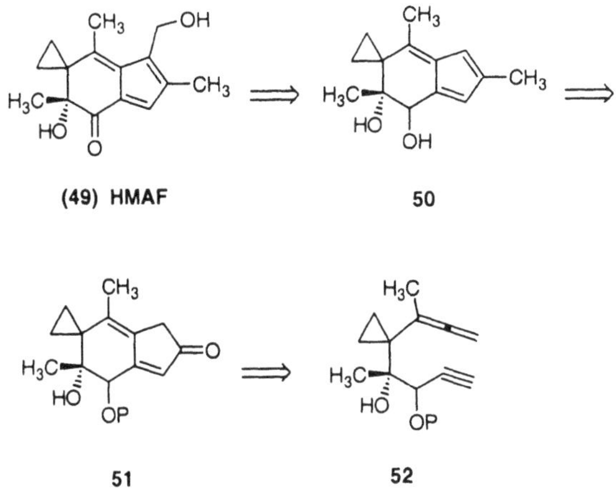

Scheme 10.

us with a better understanding of the stereoelectronic requirements of this reaction. Based upon this result it was determined that a more flexible P–K precursor was required.

To that end, the readily available 1,1-diacetylcyclopropane (**56**) was treated with the lithio derivative of the *tert*-butyldimethylsilyl ether of 3-trimethylsilylpropyn-1-ol (**57**) (Scheme 12) to afford ketone **58** as a 1.3:1 mixture of diastereomers in 57% yield. Next, addition of ethynylmagnesium bromide to ketone **58** in the presence of cerium trichloride gave the desired propargyl alcohol **59a** in 97% yield. Selective formation of the propargylic acetate of the less-hindered tertiary alcohol gave diyne **59b** in 98% yield. Treatment of propargylic acetate **59b** with $[CuH(PPh_3)]_6$ gave the allene **60a** in 54% yield. Finally, the trimethylsilyl moiety was removed from the alkyne terminus using a standard protocol to afford the desired cyclization precursor **60b** in 95% yield. We were pleased to discover that

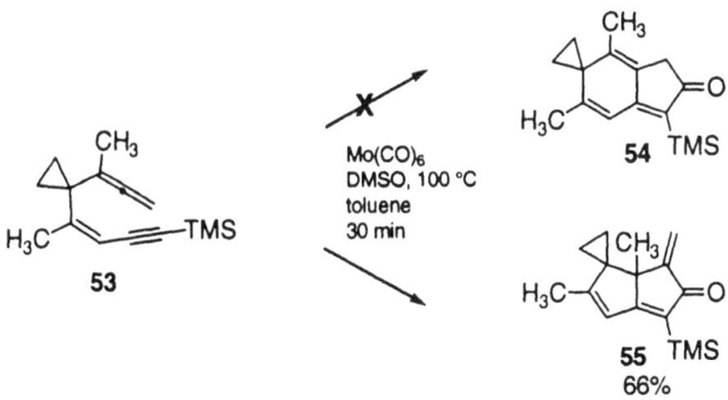

Scheme 11.

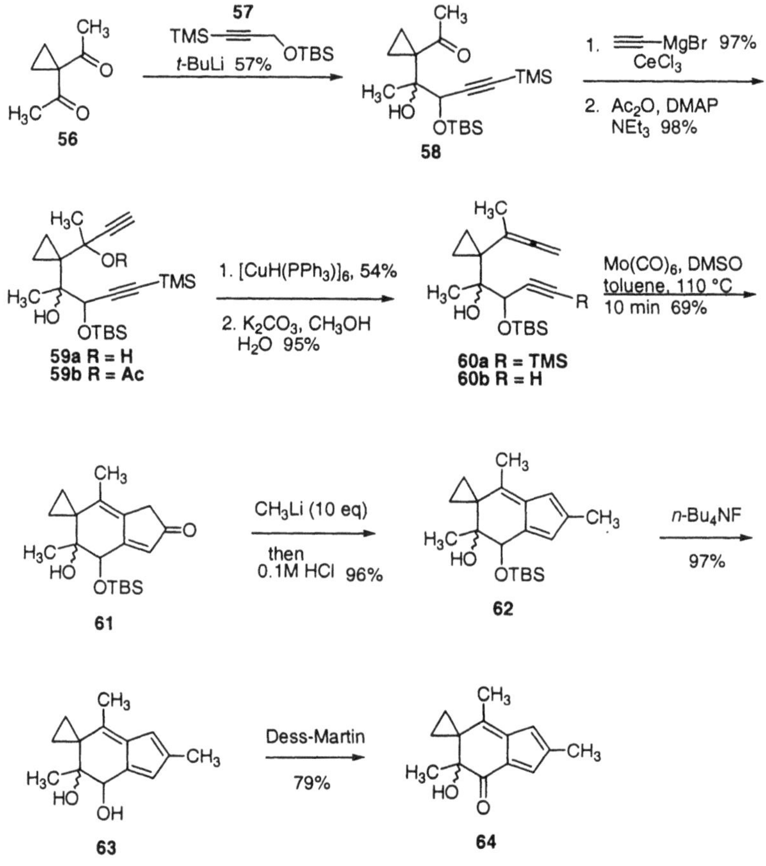

Scheme 12.

alkynyl allene **60b** undergoes a rapid cycloaddition (10 min) under the standard allenic P–K conditions [$Mo(CO)_6$, DMSO, toluene, 110 °C] to produce the desired 4-alkylidene cyclopentenone **61** as the only observed cycloadduct in 69% yield. Treatment of the ketone moiety of alkylidene cyclopentenone **61** with excess methyllithium in the presence of cerium trichloride gave the desired tertiary alcohol which underwent dehydration upon acidic workup to afford fulvene **62** in 96% yield. Removal of the TBS protecting group of the silyl ether was effected using tetra-*n*-butylammonium fluoride which provided diol **63** in 97% yield. In order to compare our synthetic material to an authentic sample, the secondary alcohol of compound **63** was oxidized to the ketone to give the acylfulvene **64** in 79% yield. The synthesis of HMAF was completed using the previously reported procedure, whereby the hydroxymethyl moiety is introduced by treatment of acylfulvene **64** with paraformaldehyde and sulfuric acid in acetone/water to afford HMAF in 75% yield.[39]

B. Suberosenone

The conversion of alleneyne **53** to the α-methylene cyclopentenone **55** provided the necessary precedent for the launching of our next natural product synthesis (Scheme 13). Suberosenone (**65**) was isolated in 1996 from an organic extract of the gorgonian *Subergorgia suberosa* (Scheme 13).[40] The structure was determined by spectral methods and was found to be a sesquiterpene with a quadrone (**66**)-like carbocyclic skeleton. Suberosenone represents the first member of the quadrone class of sesquiterpenes to be isolated from a marine organism or nonfungal source and it has demonstrated potent, differential cytotoxicity in the human tumor-based primary screen against ovarian, renal, and melanoma cell lines when tested by the National Cancer Institute.

Molecular models indicate that optimal orbital overlap occurs with the internal olefin of the allene. Thus cycloaddition was predicted to occur at the more substituted π-system. Gratifyingly, we have successfully effected the [2+2+1] cycloaddition of the alleneyne **67** to afford the very strained tricyclic α-methylene cyclopentenone **68** in 60% yield (Eq. 25). The cyclization was extremely facile and represents the first example of using a P–K cycloaddition to generate a bicyclo[3.2.1] carbocycle.[41] Completion of the synthesis of suberosenone requires a selective reduction of the endocyclic double bond from the least hindered face of the ring system and removal of the tertiary hydroxyl moiety.

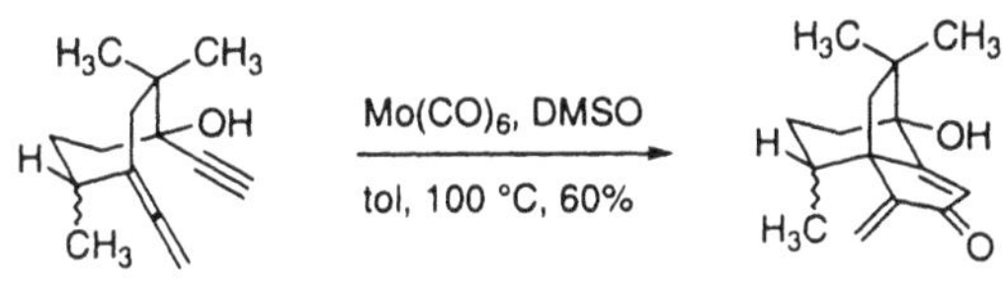

67 **68** (25)

In summary, the inter- and intramolecular and allenic [2+2+1] cycloaddition is extremely useful in the preparation of α-methylene and 4-alkylidene cyclopentenones. This reaction has been performed with a variety of allene substrates to

Suberosenone (**65**) Quadrone (**66**)

Scheme 13.

afford a diverse pool of cycloadducts. Based upon the examples provided within this review, one should be able to predict with a high degree of confidence the regio- and stereoselectivity of the cycloaddition process.

ACKNOWLEDGMENTS

The author thanks the group members who have made this work possible: Angela D. Kerekes, Dr. Jianliang Lu, Honghe Wan, and Dr. Junquan Wang. Special thanks goes to Professor Joseph L Kent. Finally, this work would not have been possible without the generous support of the NSF-EPSCoR and the NIH (GM54161).

REFERENCES

1. (a) Khand, I. U.; Knox, G. R.; Pauson, P. L.; Watts, W. E. *J. Chem. Soc., Perkin Trans. 1* **1973**, 975. (b) Khand, I. U.; Knox, G. R.; Pauson, P. L.; Watts, W. E.; Foreman, M. I. *J. Chem. Soc., Perkin Trans. 1* **1973**, 977.
2. For the first intramolecular P–K cycloaddition, see: Schore, N. E.; Croudace, M. C. *J. Org. Chem.* **1981**, *46*, 5436.
3. Sugihara, T.; Yamaguchi, M. *J. Am. Chem. Soc.* **1998**, *120*, 10782. Jeong, N. *Organometallics* **1998**, *17*, 3642. Mitsudo, T. *Yuki Gosei Kagaku Kenkyusho Koenshu* **1998**, *12*, 12. Koga, Y.; Kobayashi, T.; Narasaka, K. *Chem. Lett.* **1998**, *3*, 249. Kim, J. W.; Chung, Y. K. *Synthesis* **1998**, 2, 142.
4. For a recent review on asymmetric P–K reactions, see: *Org. Prep. Proced. Int.* **1998**, *30*, 121.
5. Sugihara, T.; Yamada, M.; Ban, Hitoshi; Yamaguchi, M.; Kaneko, C. *Angew. Chem., Int. Ed. Engl.* **1998**, *36*, 2801. Rajesh, T.; Periasamy, M. *Tetrahedron Lett.* **1998**, *39*, 117.
6. (a) Pauson, P. L. *Tetrahedron* **1985**, *41*, 5855. (b) Schore, N. E. *Chem. Rev.* **1988**, *88*, 1081. (c) Schore, N. E. *Org. React.* (NY) **1991**, *41*, 1. (d) Schore, N. E. In *Comprehensive Organic Synthesis*; Trost, B. M.; Fleming, I., Eds; Pergamon: Oxford 1991, Vol. 5, pp. 1037. (e) Schore, N. E. In *Comprehensive Organometallic Chemistry II*; Wilkinson, G.; Stone, F. G. A.; Abel, E. W., Eds; Pergamon: Oxford 1995, Vol. 12, p. 703.
7. Vedejs, E.; Cammers-Goodwin, A. *J. Org. Chem.* **1994**, *59*, 7561.
8. Aubert, C.; Llerena, D.; Malacria, M. *Tetrahedron Lett.* **1994**, *35*, 2341.
9. (a) Pasto, D. J.; Sugi, K. D. *J. Org. Chem.* **1991**, *56*, 3795. (b) Carriera, E. M.; Hastings, C. A.; Shepard, M. S.; Yerkey, L. A.; Millward, D. B. *J. Am. Chem. Soc.* **1994**, *116*, 6622. (c) Shepard, M. S.; Carreira, E. M. *J. Am. Chem. Soc.* **1997**, *119*, 2597.
10. (a) Jung, M. E.; Zimmerman, C. N.; Lowen, G. T.; Khan, S. I. *Tetrahedron Lett.* **1993**, *34*, 4453. (b) Ikeda, I.; Honda, K.; Osawa, E.; Shiro, M.; Aso, M.; Kanematsu, K. *J. Org. Chem.* **1996**, *61*, 2031. (c) Gibbs, R.; Bartels, K.; Lee, R. W. K.; Okamura, W. H. *J. Am. Chem. Soc.* **1989**, *111*, 3717.
11. Sigman, M. S.; Kerr, C. E.; Eaton, B. E. *J. Am. Chem. Soc.* **1993**, *115*, 911.
12. Borzilleri, R. M.; Weinreb, S. M.; Parvez, M. *J. Am. Chem. Soc.* **1995**, *117*, 10905. Jin, J.; Weinreb, S. M. *J. Am. Chem. Soc.* **1997**, *119*, 5773.
13. Landor, S. R. *The Chemistry of Allenes*; Academic: 1982, Vol. 2, pp. 321–322.
14. See ref. 6a, note 24.
15. Aumann, R.; Weidenhaupt, H.-J. *Chem. Ber.* **1987**, *120*, 23.
16. Krafft, M. E.; Scott, I. L.; Romero, R. H.; Feibelmann, S.; Van Pelt, C. E. *J. Am. Chem. Soc.* **1993**, *115*, 7199.
17. Jeong, N.; Lee, S. J.; Lee, B. Y.; Chung, K. Y. *Tetrahedron Lett.* **1993**, *34*, 4027.
18. Kent, J. L.; Wan, H.; Brummond, K. M. *Tetrahedron Lett.* **1995**, *36*, 2407.

19. Ahmar, M.; Locatelli, C.; Colombier, D.; Cazes, B. *Tetrahedron Lett.* **1997**, *38*, 5281–5284.
20. Shibata, T.; Koga, Y.; Narasaka, K. *Bull. Chem. Soc. Jpn.* **1995**, *68*, 911–919.
21. Hicks, F. A.; Kablaoui, N. M.; Buchwald, S. L. *J. Am. Chem. Soc.* **1996**, *118*, 9450.
22. (a) Brummond, K. M.; Wan, H. *Tetrahedron Lett.* **1998**, *39*, 931. (b) Brummond, K. M.; Wan, H.; Kent, J. L. *J. Org. Chem.* **1998**, *63*, 6535.
23. Magnus, P.; Principe, M. *Tetrahedron Lett.* **1985**, *26*, 4851.
24. Negishi, E.-I.; Holmes, S. J.; Tour, J. M.; Miller, J. A. *J. Am. Chem. Soc.* **1985**, *107*, 2568.
25. Borodkin, V. S.; Shapiro, N. A.; Azov, V. A.; Kochetkov, N. K. *Tetrahedron Lett.* **1996**, *37*, 1489.
26. Brummond, K. M.; Dingess, E. A.; Kent, J. L. *J. Org. Chem.* **1996**, *61*, 6096.
27. Belanger, D. B.; Livinghouse, T. Newsletter and Abstracts from the 216th ACS National Meeting, Boston, MA, August 23–27, 1998, abstract #505.
28. Ahmar, M.; Antras, F.; Cazes, B. *Tetrahedron Lett.* **1995**, *36*, 4417.
29. Ahmar, M.; Chabanis, O.; Gauthier, J.; Cazes, B. *Tetrahedron Lett.* **1997**, *38*, 5277.
30. Ikeda, I.; Honda, K.; Osawa, E.; Shiro, M.; Aso, M.; Kanematsu, K. *J. Org. Chem.* **1996**, *61*, 2031.
31. Urabe, H.; Takedo, T.; Hideura, Sato, F. *J. Am. Chem. Soc.* **1997**, *119*, 11295.
32. Kleijn, H.; Elsevier, C. J.; Westmijze, H.; Meijer, J.; Vermeer, P. *Tetrahedron Lett.* **1979**, *33*, 3101.
33. Myers, A. G.; Zheng, B. *J. Am. Chem. Soc.* **1996**, *118*, 4492.
34. Nesmeyanov, A. N.; Krivykh, V. V.; Kaganovich, V. S.; Rybinskaya, M. I. *J. Organomet. Chem.* **1975**, *102*, 185.
35. (a) McMorris, T. C.; Kelner, M. J.; Wang, W.; Diaz, M. A.; Estes, L. A.; Taetle, R. *Experientia* **1996**, *52*, 75. (b) McMorris, T. C.; Kelner, M. J.; Wang, W.; Yu, J.; Estes, L. A.; Taetle, R. *J. Nat. Prod.* **1996**, *59*, 896.
36. (a) McMorris, T. C.; Yu, J.; Hu, Y.; Estes, L. A.; Kelner, M. J. *J. Org. Chem.* **1997**, *62*, 3015. (b) McMorris, T. C.; Kelner, M. J.; Wang, W.; Estes, L. A.; Montoya, M. A.; Taetle, R. *J. Org. Chem.* **1992**, *57*, 6876.
37. (a) MacDonald, J. R.; Muscoplat, C. C.; Dexter, D. L.; Mangold, G. L.; Chen, S.-F.; Kelner, M. J.; McMorris, T. C.; Von Hoff, D. D. *Cancer Res.* **1997**, *57*, 279. (b) Kelner, M. J.; McMorris, T. C.; Estes, L. A.; Wang, W.; Samson, K. M.; Taetle, R. *Investigational New Drugs* **1996**, *14*, 161.
38. McMorris, T. C.; Hu, Y.; Yu, J.; Kelner, M. J. *Chem. Commun.* **1997**, 315.
39. Brummond, K. M.; Lu, J. *J. Am. Chem. Soc.* **1999**, 121, 5087.
40. Bokesch, H. R.; McKee, T. C.; Cardellina, J. H.; Boyd, M. R. *Tetrahedron Lett.* **1996**, *37*, 3259.
41. Filippini, M.-H.; Rodriguez, J. *Chem. Rev.* **1999**, *99*, 27.

INDEX

www.ingramcontent.com/pod-product-compliance
Ingram Content Group UK Ltd.
Pitfield, Milton Keynes, MK11 3LW, UK
UKHW020804290926
15470UKWH00003B/949

9 780762 305315